The Anthropology of Climate Change

Wiley Blackwell Anthologies in Social And Cultural Anthropology

Series Editor: Parker Shipton, Boston University

Drawing from some of the most significant scholarly work of the nineteenth and twentieth centuries, and sometimes earlier, the *Wiley Blackwell Anthologies in Social and Cultural Anthropology* series offers a comprehensive and unique perspective on the ever-changing field of anthropology. It represents both a collection of classic readers and an exciting challenge to the norms that have shaped this discipline over the past century.

Each edited volume is devoted to a traditional subdiscipline of the field such as the anthropology of religion, linguistic anthropology, or medical anthropology; and provides a foundation in the canonical readings of the selected area. Aware that such subdisciplinary definitions are still widely recognized and useful – but increasingly problematic – these volumes are crafted to include a rare and invaluable perspective on social and cultural anthropology at the onset of the twenty-first century. Each text provides a selection of classic readings together with contemporary works that underscore the artificiality of subdisciplinary definitions and point students, researchers, and general readers in the new directions in which anthropology is moving.

1. *Linguistic Anthropology: A Reader, 2nd Edition*
   Edited by Alessandro Duranti
2. *A Reader in the Anthropology of Religion, 2nd Edition*
   Edited by Michael Lambek
3. *The Anthropology of Politics: A Reader in Ethnography, Theory, and Critique*
   Edited by Joan Vincent
4. *Kinship and Family: An Anthropological Reader*
   Edited by Robert Parkin and Linda Stone
5. *Law and Anthropology: A Reader*
   Edited by Sally Falk Moore
6. *The Anthropology of Development and Globalization: From Classical Political Economy to Contemporary Neoliberalism*
   Edited by Marc Edelman and Angelique Haugerud
7. *The Anthropology of Art: A Reader*
   Edited by Howard Morphy and Morgan Perkins
8. *Feminist Anthropology: A Reader*
   Edited by Ellen Lewin
9. *Ethnographic Fieldwork: An Anthropological Reader, 2nd Edition*
   Edited by Antonius C. G. M. Robben and Jeffrey A. Sluka
10. *Environmental Anthropology*
    Edited by Michael R. Dove and Carol Carpenter
11. *Anthropology and Child Development: A Cross-Cultural Reader*
    Edited by Robert A. LeVine and Rebecca S. New
12. *Foundations of Anthropological Theory: From Classical Antiquity to Early Modern Europe*
    Edited by Robert Launay
13. *Psychological Anthropology: A Reader on Self in Culture*
    Edited by Robert A. LeVine
14. *A Reader in Medical Anthropology: Theoretical Trajectories, Emergent Realities*
    Edited by Byron J. Good, Michael M. J. Fischer, Sarah S. Willen, and Mary-Jo DelVecchio Good
15. *Sexualities in Anthropology*
    Edited by Andrew Lyons and Harriet Lyons
16. *The Anthropology of Performance: A Reader*
    Edited by Frank J. Korom
17. *The Anthropology of Citizenship: A Reader*
    Edited by Sian Lazar
18. *The Anthropology of Climate Change: An Historical Reader*
    Edited by Michael R. Dove

The Anthropology of Climate Change

An Historical Reader

Edited by

Michael R. Dove

WILEY Blackwell

This edition first published 2014

Registered Office
John Wiley & Sons, Ltd, The Atrium, Southern Gate, Chichester, West Sussex, PO19 8SQ, UK

Editorial Offices
350 Main Street, Malden, MA 02148-5020, USA
9600 Garsington Road, Oxford, OX4 2DQ, UK
The Atrium, Southern Gate, Chichester, West Sussex, PO19 8SQ, UK

For details of our global editorial offices, for customer services, and for information about how to apply for permission to reuse the copyright material in this book please see our website at www.wiley.com/wiley-blackwell.

Library of Congress Cataloging-in-Publication Data applied for

Hardback ISBN: 978-1-118-38355-1
Paperback ISBN: 978-1-118-38300-1

A catalogue record for this book is available from the British Library.

Cover image: Central Javanese farmer assessing the prospects for essential rain from distant clouds.
Photo by © Michael R. Dove.
Cover design by Nicki Averill Design.

Set in 9/11pt Sabon by SPi Publisher Services, Pondicherry, India

1 2014

Contents

Acknowledgments to Sources

1 Broad, Kenneth and Ben Orlove. 2007. Channeling Globality: The 1997–98 El Niño Climate Event in Peru. *American Ethnologist* 34(2): 285–302.

2 Burger, Richard L. 2003. El Niño, Early Peruvian Civilization, and Human Agency: Some Thoughts from the Lurin Valley. In *El Niño in Peru: Biology and Culture Over 10,000 Years*. J. Haas and M. O. Dillon, eds. *Fieldiana Botany* 43: 90–107.

3 Carey, Mark. 2008. The Politics of Place: Inhabiting and Defending Glacier Hazard Zones in Peru's Cordillera Blanca. In *Darkening Peaks: Glacier Retreat, Science, and Society*. B. Orlove, E. Wiegandt, and B.H. Luckman, eds. pp. 229–240. Berkeley, CA: University of California Press.

4 Colson, Elizabeth. 1957. Rain-Shrines of the Plateau Tonga of Northern Rhodesia. In *The Plateau Tonga of Northern Rhodesia (Zambia): Social and Religious Studies*. pp. 84–101. Manchester, UK: Manchester University Press, for the Institute for Social Research, University of Zambia.

5 Cruikshank, Julie. 2007. Melting Glaciers and Emerging Histories in the Saint Elias Mountains. In *Indigenous Experience Today*. Marisol de la Cadena and Orin Starn, eds. pp. 355–378. Oxford: Berg.

6 Hippocrates. 1923. *Airs, Waters, Places*. W. H. S. Jones, trans. In *Hippocrates*, Volume I, pp. 71–73, 105–137. Cambridge, MA: Harvard University Press.

7 Ibn Khaldûn. 1958. *The Muqaddimah: An Introduction to History*. Franz Rosenthal, trans. Frontispiece and pp. 103–109 and 167–183. New York: Bollingen Foundation.

8 Janković, Vladimir. 2007. Gruff Boreas, Deadly Calms: A Medical Perspective on Winds and the Victorians. *Journal of the Royal Anthropological Institute* (N.S.): S147–S164.

9 Lahsen, Myanna. 2004. Transnational Locals: Brazilian Experiences of the Climate Regime. In *Earthly Politics: Local and Global in Environmental Governance*. Sheila Jasanoff and Marybeth L. Martello, eds. pp. 151–172. Cambridge, MA: MIT Press.

10 McGovern, Thomas H. 1994. Management for Extinction in Norse Greenland. In *Historical Ecology: Cultural Knowledge and Changing Landscapes*. Carole L. Crumley, ed. pp. 127–154. Santa Fe, NM: School of American Research Press.

11 Meggers, Betty J. 1957. Environment and Culture in the Amazon Basin: An Appraisal of the Theory of Environmental Determinism. In *Studies in Human Ecology*. pp. 71–89. Washington, DC: The Anthropological Society of Washington and the General Secretariat of the Organization of American States.

12 Montesquieu, Charles de Secondat, Baron de. 1989. Book 14: On the Laws in Their Relation to the Nature of the Climate. In *The Spirit of the Laws*. Anne M. Cohler, Basia Carolyn Miller, Harold Samuel Stone, trans./eds. pp. 231–237, 242–245. Cambridge: Cambridge University Press. (Original: *Esprit des loix*. Geneva.)

13 Ratzel, Friedrich. 1896–1898. Nature, Rise, and Spread of Civilization. In *The History of Mankind*. A. J. Butler, trans. pp. 20–30. London: Macmillan.

14 Sanders, Todd. 2008. The Making and Unmaking of Rain and Reigns. In *Beyond Bodies: Rainmaking and Sense Making in Tanzania*. pp. 69–102. Toronto: University of Toronto Press.

15 Scheper-Hughes, Nancy. 2005. The Disaster and its Doubles. *Anthropology Today* 21 (6): 2–4.

16 Schneider, David M. 1957. Typhoons on Yap. *Human Organization* 16: 10–15.

17 Shaw, Rosalind. 1992. "Nature", "Culture" and Disasters: Floods and Gender in Bangladesh. In *Bush Base, Forest Farm: Culture, Environment and Development*. Elisabeth Croll and David Parkin, eds. pp. 200–217. London: Routledge.

18 Solway, Jacqueline S. 1994. Drought as "Revelatory Crisis": An Exploration of Shifting Entitlements and Hierarchies in the Kalahari, Botswana. *Development and Change* 25(3): 471–498.

19 Spillius, James. 1957. Natural Disaster and Political Crisis in a Polynesian Society: An Exploration of Operational Research II. *Human Relations* X: 113–125.

20 Theophrastus. 1926. Concerning Weather Signs. In *Enquiry into Plants, and Minor Works on Odours and Weather Signs*, Volume II, pp. 391–407. Sir Arthur Hort, trans. Cambridge, MA: Harvard University Press.

21 Weiss, Harvey and Raymond Bradley. 2001. What Drives Societal Collapse? *Science* 291(5504): 609–610.

22 Zimmermann, Francis. 1988. The Jungle and the Aroma of Meats: An Ecological Theme in Hindu Medicine. *Social Science & Medicine* 27(3): 197–206.

About the Editor

Michael R. Dove is the Margaret K. Musser Professor of Social Ecology in the School of Forestry and Environmental Studies, Professor in the Department of Anthropology, Director of the Tropical resources Institute, and Curator of Anthropology at the Peabody Museum, Yale University. His interests include the cultural and political aspects of natural hazards and disasters; political dimensions of resource degradation; indigenous environmental knowledge; and the study of developmental and environmental institutions, discourses, and movements. He is author of *The Banana Tree at the Gate: The History of Marginal Peoples and Global Markets in Borneo* (2011), co-editor of *Complicating Conservation: Beyond the Sacred Forest* (2011) and *Environmental Anthropology: A Historical Reader* (Wiley Blackwell, 2008), and editor of *Southeast Asian Grasslands: Understanding a Folk Landscape* (2008).

Preface

The purpose of this volume is to illustrate the contributions that anthropology can make to contemporary research and policy regarding climate change through reprinting, discussing, and putting into conversation with one another a number of key, canonical works in the history of the anthropological study of climate and society. I have evenly divided my selections among early anthropological works, recent ones, and those in between. I have selected papers that are, or will become, classics, by prominent scholars, which make important contributions to academic and policy discussions concerning climate change and, often, to wider theoretical and policy debates as well. I have selected works that are still not only readable but interesting and relevant. I have tried to select "memorable" works, which deliver an argument in such a way that a reader will still recall it five or ten years hence. I have selected works that are neither strictly theoretical essays nor derivative critiques of the works of others, in favor of original, ethnographic, case studies. I have selected works that have a clear, central theme, which relates to one of the four major sections of the book. This approach stems in part from my decision to organize this volume not around historic eras or schools of climate research, but around a number of persistent, cross-cutting, and inter-linked themes, which span eras. I have selected papers that can be thematically linked to multiple other papers in the volume, thereby constituting a sort of intra-volume "dialogue" that reflects the larger one that has characterized the development of the field of climate studies itself. To further this dialogue, I have organized the volume into a series of paired papers, each one of which speaks to the other in a way that is hopefully stimulating for the reader. In some cases, this "conversation" extends across decades, centuries, or millennia, which makes it all the more powerful. I have selected works with balanced, global coverage. I have restricted my selection of papers to those written by anthropologists, defined as scholars either trained as anthropologists or whose work came to focus to such a degree on anthropological topics as to give them a professional identity as anthropologists, with the exception of a number of pre-twentieth-century scholars whose work marks them as the intellectual ancestors of modern anthropologists. Inevitably, there are gaps in the coverage afforded by the papers selected. I have sought to remedy this with a comprehensive Introduction, which reviews the wider literature on the topics taken up in each reading and on the four wider themes of the book.

I selected papers that could be reprinted in their entirety, without abbreviation or other amendation, so that they can serve as authoritative sources for students and scholars, without the need for recourse to the original publications. For reasons of space, however, I had to violate this rule in a minority of cases, as follows:

Chapter 1 Hippocrates. 5th century B.C. Airs, Waters, Places
This work comprises two distinct parts: following an Introductory Chapter I, Chapters II–XI deal with the effects of local climate upon health, and Chapters XII–XXIV deal with the effects of regional climate upon character. For reasons of space, I reprinted here only Chapters I and XII–XXIV, which focus most directly on Hippocrates' comparative analysis of climate and society, although Chapters II–XI also are relevant to this volume. Also, I deleted notes from the translator concerned solely with questions of translation from Greek to English.

Chapter 2 Charles de Secondat Montesquieu. 1748. On the Laws in Their Relation to the Nature of the Climate
Montesquieu's "The Spirit of the Laws" is a large and wide-ranging work on law and society, comprising six "Parts" and thirty-one "Books." Montesquieu's thoughts on climate and society extend through Books 14–17 in Part 3, but the material of greatest theoretical interest to this volume's study of climate and society is in Book 14, titled as above, which contains 15 chapters, of which I have reprinted 1–6 and 13–15 as being of most direct relevance.

Chapter 3 Ibn Khaldûn. 1370. The Muqaddimah: An Introduction to History
This is a sweeping study of history, geography, ethnography, and political science. The material on climate and society is concentrated in one of its six chapters: Chapter I: Human Civilization in General, which is in turn divided into six "Prefatory Discussions." The Second, Third, Fourth, and Fifth Prefatory Discussions are most relevant to this volume and are reprinted here in their entirety, except for the Second, of which only the "Supplementary Note to the Second Prefatory Discussion" is included, the remainder being largely a detailed exegisis of the map reprinted as Figure 3.1.

Chapter 5 Theophrastus. 4th century B.C. Concerning Weather Signs
The text used here is part of a two-volume edition of Theophrastus, "Enquiry Into Plants," the most extensive botanical treatise of the classical era. "Concerning Weather Signs," and another work published alongside it, "Concerning Odours," are not properly part of "Enquiry Into Plants," but are separate "minor works" dealing largely with non-botanical topics. "Concerning Weather Signs" comprises five sections: "Introductory: General Principles," "The Signs of Rain," "The Signs of Wind," "The Signs of Fair Weather," and "Miscellaneous Signs." For reasons of space, only the first two sections are reprinted, although all are relevant to the subject of this volume.

Chapter 20 Todd Sanders. 2008. The Making and Unmaking of Rains and Reigns
This is Chapter 2 of Sanders' book *Beyond Bodies: Rainmaking and Sense Making in Tanzania*. The remainder of the book is an ethnography of an African society, focusing on issues of gender and religion. For reasons of space, some of the extensive notes to Chapter 2, many of them dealing with historical matters, were either deleted or abbreviated, retaining just the references to works cited.

The following chapters were not abridged in any way but are part of larger works.

Chapter 4 Francis Zimmermann. 1988. The Jungle and the Aroma of Meats: An Ecological Theme in Hindu Medicine
This is a synopsis of Zimmermann's 1987 book of the same title, much of which – dealing with the ecological/climatic dimensions of the ancient Vedic teachings – is relevant to the themes of this volume.

Chapter 7 Ratzel, Friedrich. 1896–1898. Nature, Rise, and Spread of Civilization
Ratzel's three-volume 1885–1888 *Völkerkunde*, a sweeping study of humankind and civilization, was translated and published in English as the six-volume *The History of Mankind*. "Nature, Rise, and the Spread of Civilization" is Chapter 4 in Book I, "Principles of Ethnography," of Division/Volume I of this work. This chapter contains Ratzel's clearest statements regarding environmental/climatic determinism, but relevant material is also found elsewhere in the six volumes.

Chapter 11 James Spillius. 1957. Natural Disaster and Political Crisis in a Polynesian Society: An Exploration of Operational Research II
This is the second of a two-part article published on this topic by Spillius. The first part is a detailed ethnographic account of the involvement of him and Raymond Firth in disaster relief efforts. This too is relevant to the subject of this volume, but the second part was chosen for reprinting because it succinctly pulls out of the ethnography the ethical issues of scholarly engagement with climate-related disasters.

Chapter 13 Elizabeth Colson. 1957. Rain-Shrines of the Plateau Tonga of Northern Rhodesia
This is Chapter 3 of Colson's monograph, *The Plateau Tonga of Northern Rhodesia (Zambia)*. Its subject is the means – one of which is the rain-shrines – by which this "stateless" society is held together.

Michael R. Dove
Killingworth, Connecticut
August 2013

Acknowledgments

Earlier versions of the text for this volume were presented and discussed in my advanced seminar in Yale's School of Forestry and Environmental Studies, "Disaster, Degradation, Dystopia: Social Science Approaches to Environmental Perturbation and Change" (Spring 2010 and Spring 2011), and in an undergraduate class in Yale College's Environmental Studies major, "Anthropology of Climate Change" (Fall 2012). The students in these classes, and especially my Associate in Teaching Catherine (Annie) Claus in the last-mentioned class, were wonderful interlocutors for my efforts to develop the themes in this book. I have also been ably assisted in my library research for this volume by several research interns, Katie Hawkes, Julia Fogerite, and Emily Schosid. With administrative and financial matters, I have relied upon the industry of two administrative assistants, Laurie Bozzuto and Julie Cohen.

None of the aforementioned people or organizations necessarily agrees with anything said in this volume, however, for which I am alone responsible.

Introduction: The Anthropology of Climate Change

Six Millennia of Study of the Relationship between Climate and Society

Michael R. Dove

Background

Clarence J. Glacken writes, in his magisterial 1967 (p. vii) *Traces on the Rhodian Shore: Nature and Culture in Western Thought from Ancient Times to the End of the Eighteenth Century*, that Western thinking about humans and the earth has been dominated by three persistent questions:

> Is the earth, which is obviously a fit environment for man and other organic life, a purposefully made creation? Have its climates, its relief, the configuration of its continents influenced the moral and social nature of individuals, and have they had an influence in molding the character and nature of human culture? In his long tenure of the earth, in what manner has man changed it from its hypothetical pristine condition?

Glacken further asserts that these questions have been central not just to thinking about the environment, but also to the development of critical thought itself: "In exploring the history of these ideas from the fifth century B.C. to the end of the eighteenth century, it is a striking fact that virtually every great thinker who lived within this 2,300-year period had something to say about one of the ideas, and many had something to say about all of them" (Glacken 1967: 711). That is to say, pondering on the relationship between nature and culture was a key project in the development of civilization in the West (and indeed, throughout the world). However unique modern anthropogenic climate change may be, therefore, a discourse of climate and culture has

The Anthropology of Climate Change: An Historical Reader, First Edition. Edited by Michael R. Dove.

been prominent within human society for millennia. Indeed, it might be said to have been an integral part of *the* discourse of civilization itself.

Anthropology has played a central role in this discourse. Thinking of the intellectual forebears of the discipline from the classical era to modern times, as well as anthropology proper over the past two centuries, theorizing regarding the relationship between nature and culture, between environment and society, has been central to the development of anthropology as a field. Consider as an example what is known as "climate theory," referring to the idea that climate determines human character, culture, and the rise and fall of civilizations. One of its earliest known developments was in the Hippocratic school 2,400 years ago. After it had been periodically reiterated over the succeeding two millennia, a remarkably similar theory was promulgated by two modern scholars who are often claimed as belonging to contemporary anthropology: the French enlightenment political thinker Montesquieu in the eighteenth century and the German geographer and ethnographer Ratzel in the nineteenth. A reaction against simplistic environmental determinism then set in, leading to what Rayner (2003: 286) has called an eighty-year gap in social science studies of climate. By the mid-twentieth century, explicit anthropological studies of climate were limited to very modest analyses of correlations between climate and human biology (Mills 1942; Gladwin 1947; Whiting 1964).

This perceived move by anthropology away from climate was more apparent than real, however. Throughout the twentieth century, anthropologists were very much concerned with climate through their studies of subsistence practices of hunting and gathering, fishing, herding, and agriculture (e.g., Evans-Pritchard 1940; Richards 1948). Classic studies in environmental anthropology by the likes of Steward (1955), Mauss (1979 [1950]), and Conklin (1957) delved deeply into emic or native views of climate. Anthropologists built on this experience when, later in the twentieth century, more explicitly climatic topics emerged, like degradation and desertification (Spooner and Mann 1982; Little and Horowitz 1987). The questions being debated in these studies are as theoretically robust as any that have ever concerned anthropology. More recent, and with more immediate relevance to contemporary concerns about climate change research and policy, has been the contribution of anthropology to a new generation of disaster studies (Vayda and McCay 1975; Oliver-Smith 1996). Rejecting an earlier focus on individual ability or inability to cope with disaster, and the view of disaster as a "break" in the normal (Wallace 1956), the new studies ask how coping ability is affected by the dynamics of the wider society and, further, the role that society plays in determining who does or does not become a disaster victim in the first place (Hewitt 1983; Wisner 1993). As the social dimension of disasters became clear, anthropologists realized that there is a politics of knowledge associated with them (Harwell 2000; Mathews 2005), which historical studies show to have roots in the colonial era (Grove 1995; Davis 2001; Endfield and Nash 2002).

Margaret Mead (1977) is reputed to have been the first anthropologist to talk about climate change. For the past two decades, anthropologists have been involved in a significant way with research on climate change (Crate 2011), whether the involvement is measured by meetings and conferences, or grants and publications, including some noteworthy edited collections (Strauss and Orlove 2003; Casimir 2008; Crate and Nuttall 2009). Initially, this involvement built on traditional anthropological expertise with small, local communities, for example studying issues of risk and vulnerability (Ribot, Magalhães, and Panagides 1995) and the reality or prospects for adaptation (Berkes and Jolly 2001; Finan and Nelson 2001; Eakin 2006). From there anthropologists moved to related topics such as REDD (Reduced Emissions from Deforestation and Degradation), drawing on the field's expertise on indigenous, forest-dwelling peoples in the tropics (Schwartzman and Moutinho 2008).

A separate and important subgenre of the anthropological study of climate is the emerging field of the history and especially prehistory of human society and climate change. Some anthropologists have drawn on novel oral historical materials to contribute to this study (McIntosh 2000; Cruikshank 2001); but most work has come from archaeology. A long-established interest in the impact of climate change on ancient societies has been greatly reinvigorated by contemporary climate change debates (Bawden and Reycraft 2000), with special interest in the El Niño–Southern Oscillation (ENSO) phenomenon as a proxy for climate change (see Chapter 14, this volume).

From these beginnings in familiar ground, anthropologists have moved to such non-traditional topics as the international institutions involved in climate change research and policy, for example the IPCC (O'Reilly 2012), the meetings at which the global climate change community attempts to hammer out policy (Doolittle 2010), and thorny issues of communication and public skepticism (Diemberger et al. 2012). Beyond anthropology, there is a voluminous literature on climate change. Of special interest is apposite scholarship in the humanities on climate beliefs embedded in literature and the arts (Mentz 2010), and collections on global governance and climate change (Jasanoff and Martello 2004; Roberts and Parks 2007; Hulme 2009).

These new directions notwithstanding, anthropologists insist that their work on climate change – which some have called "climate anthropology" (Nelson and Finan 2000) or "climate ethnography" (Crate 2011) – takes advantage of the traditional strengths of the field, which Roncoli, Crane, and Orlove (2009) refer to as "being there" and the capacity to provide insight into perceptions, knowledge, valuation, and response. There are a number of dimensions to contemporary climate change that require these sorts of insights: (i) climate change has a reality at the local level; (ii) global debates about climate change policy are affected by North–South post-colonial histories; (iii) climate change has likely been imbricated in the evolution of human society; and (iv) the knowledge, science, and understanding of climate change is itself a social phenomenon, which affects the prospects for mitigation and adaptation. No other discipline matches the capacity to illuminate such issues of anthropology, which thus has something unique to offer to contemporary debates about climate change research and policy (Magistro and Roncoli 2001).

The aim of the current volume is to illustrate the scope and relevance of anthropological work on climate change, and in particular its intellectual roots and historic development. In none of the contemporary work has there been any effort to examine the history of anthropological work on climate and society, much less earlier apposite traditions of scholarly work on this topic. This is a serious gap in the anthropology of climate change. Scholars with an anthropological bent, and indeed human society in general, have been thinking about climate and society for millennia; and this history is a valuable resource for coping with twenty-first century climate change. To assess this resource, this volume presents twenty-two different examples of anthropological work on climate and society, organized into four principal sections. The first, "Continuities," presents papers that illustrate intellectual continuities from the classical era, through the Enlightenment, and up to the present, focusing on "climate theory." The second section, "Societal and Environmental Change," is dedicated to papers dealing with an important corollary question for climate theory: When climate changes, does society follow suit? The third section of the book, "Vulnerability and Control," contains papers that ask how societies attempt to cope with the impact of extreme climatic events, and how social differentiation affects this impact. The fourth and final section, "Knowledge and its Circulation," looks at epistemological issues, in particular the factors that determine how climatic perturbation is interpreted.

In the remainder of this Introduction, I will review in detail these four principal parts of the volume and their contents.

Part I: Continuities

The volume begins with an examination of deep historical continuities in thinking about climate and society, beginning with the "climate theory" of the classical era, then looking beyond the Greco-Roman tradition to other civilizations, and then examining some historical currents in that most anthropological of methods, the ethno-scientific study of other conceptual systems.

Climate Theory

One of the most enduring ways of thinking about the relationship between climate and human society is the so-called "climate theory," which derives the character of society from climate.

1 Hippocrates. Fifth century B.C. *Airs, Waters, Places*

Important commentaries on environmental matters have been noted in the writings of many scholars of the classical era, including Aristotle, Herodotus, Thucydides, and Pliny. But in terms of an extended, in-depth analysis of the relationship between culture and nature, the work of Hippocrates (born 460 B.C.) is perhaps unsurpassed. His *Airs, Waters, Places* is a seminal work on the linkages between climate, landscape, physique, and temperament: "Some physiques resemble wooded, well-watered mountains, others light, dry land, others marshy meadows, others a plain of bare, parched earth" (p. 42). Glacken (1967: 81–2) calls this "the earliest systematic treatise concerned with environmental influences on human culture . . . ." It was a radical work in explicitly replacing the gods with nature as a causal agent.

The Hippocratic tradition presents two distinct bodies of theory regarding disease and, more generally, the relationship between society and environment. As Glacken (1967: 80) writes, "From early times there have been two types of environmental theory, one based on physiology (such as the theory of the humours) and one on geographical position; both are in the Hippocratic corpus." The two types of theory are related but still separate; *Airs, Waters, Places* has different chapters for biological man and cultural man, for medicine and ethnography. The independent variable in the case of biology is seasonally driven variation in weather; the independent variable in the case of culture is latitudinally and topographically driven variation in climate. Health and culture are problematized and explained in this formulation, not nature. Nature is the independent variable, and health and culture are the dependent ones. Explanation of difference in the world was sought by examining the impact of nature on people, therefore, not that of people on nature.[1]

Jones (1923: 66) observes that the second portion of *Airs, Waters, Places*, which is reprinted here, is "scarcely medical at all, but rather ethnographical." This refers to Hippocrates' use of the comparative method, which was to enjoy such a long and productive history within environmental anthropology (Steward 1955). In the first, "medical" part of *Airs, Waters, Places*, not reprinted here, Hippocrates relates human health to the characteristics of the site or locale, whereas in the second, "ethnographic" part, he generalizes from this causal relationship to explain the correlation between the character of entire societies and the regions within which they live: "Now I intend to compare Asia and Europe, and to show how they differ in every respect, and how the nations of the one differ entirely in physique from those of the other."

Hippocrates uses comparative analysis to explain human difference, not similarity. Early in the ethnographic part of *Airs, Waters, Places*, Hippocrates (p. 42) informs his readers that "The races that differ but little from one another I will omit, and describe the condition only of those which differ greatly, whether it be through nature or through custom." His attempt to explain different peoples in terms of different climates was a search for an answer to the age-old question: Why is the "other" different? The earth itself, the geomorphology of which produces not sameness but infinite diversity, has throughout human history provided ready at hand one answer to this question.

Hippocrates' question assumes an underlying common humanity. The "other" is different but still human. There was neither felt need nor actual effort to explain in terms of climate the difference from humans of the fabled races of antiquity, like the cyclops for example. It was only the existence of "others" like us, yet unlike us in *some* respects, that posed a question. In contrast, the existence of "others" like us, in all respects, posed no question at all. Hippocrates could have seen like people in unlike environments as posing an equally logical question, but he did not. The historic ramifications of his choice were enormous; as Glacken (1967: 85) notes, "If Hippocrates had shown an interest in accounting for similarities rather than differences, the history of environmental theories would have been entirely different."

The intent of *Airs, Waters, Places* is prognostic not programmatic (Glacken 1967: 81): the aim in the first, medical part is to predict the effects of the seasons of the year on human health; whereas the intent of the second, ethnographic part is to predict the effect of the climatic regions of the earth on human character. Hippocrates' (p. 41) intent is "to show" these differences; it is not to present an agenda for action. By offering an explanation of human differences, including problematic differences, this rationalization of the status quo has proved to be politically powerful down to the present day. As Glacken (1967: 258) writes, this explanation is "serviceable in accounting for cultural, and especially for racial, differences"; and, thereby, it helps to justify privilege. It is all the more powerful because it de-privileges others on the basis not of their own character but rather of that of their environment. The continued power of such explanations can be seen in the great current popularity of works of global geographic determinism (e.g., Diamond 1997, 2005). It perhaps also can be seen in the self-privileging stances being taken by the industrialized nations toward the late-industrializing nations with less historic responsibility for, but greater current vulnerability to, climate change, as in distinctions being made between countries with high versus low "adaptive capacity" (Moore 2010). Today, as 2,500 years ago, therefore, climate is an instrument in segmentary politics.

2 *Charles de Secondat Montesquieu. 1748. On the Laws in Their Relation to the Nature of the Climate*

The Hippocratic work on climate and society was, as Glacken writes (1967: 502), "dramatically revived" during the Enlightenment by Montesquieu in his *The Spirit of the Laws*. Montesquieu (1689–1755) published this in 1748, and considered it to be his life's work (Cohler 1989: xi). The writing is clearly reminiscent of *Airs, Waters, Places* (Montesquieu: p. 48): "You will find in the northern climates people who have few vices, enough virtues, and much sincerity and frankness. As you move toward the countries of the south, you will believe you have moved away from morality itself . . . ."

Montesquieu read widely and although he does not specifically refer to Hippocrates in *The Spirit of the Laws*, he had a version of *Airs, Waters, Places* in his library and its impact on his work is generally acknowledged (Levin 1936: 26–39; Cohler 1989: xx). Also suggestive is the fact that one of the key intellectual constructs in Montesquieu's climate theory is based on his freezing and thawing of a sheep's tongue, and his observation of the attendant contracting and lengthening of its "papillae," thereby explaining the supposed greater sensitivity of people in warm versus cold climates – which parallels Hippocrates' own experiment in freezing and thawing water.[2]

Anthropologists claim Montesquieu as an intellectual forebear because of the marked element in *The Spirit of the Laws* of cultural relativism with respect to non-Christian religions and practices like polygamy (Nugent 1752: 6–7; Launay 2010). As Neuman (1949: xxxii) writes, Montesquieu believed that "The reconciliation of might and right must be achieved differently in different cultures." More generally, Montesquieu developed what seems today to often be a social scientific approach to his subject. Referring to his effort to interrelate all of the elements – morals, customs, principles of government, and the spirit of the nation – that shape the character of a country, Neuman (1949: xlvi) remarks that "It is, as one would say today, an attempt to develop the principles of a cultural anthropology." The founder of French sociology/anthropology, Emile Durkheim, devoted his 1892 dissertation to an assertion of the foundational contributions of Montesquieu to this field (Neuman 1949: xxxiii, n.4).

In addition to its political thought, *The Spirit of the Laws* is famed for its theorizing regarding the relationship between nature and culture, in particular between climate and law. Glacken (1967: 653) asserts that "By his advocacy of climatic influences, Montesquieu in the *Esprit des Pois* had provoked some of the most searching thought on social and environmental questions that had yet appeared in Western civilization . . . ." Montesquieu's thesis, based on a tradition of thinking that can be traced back to Aristotle and Plato, is that law-making should be suited to the character of the society and that this is influenced by the character of the environment. He writes "If it is true that the character of the spirit and the passions of the heart are extremely different in the various climates, laws should be relative to the differences in these passions and to the differences in these characters" (p. 47). As Neuman (1949: xliv) writes, "He tries to establish a direct, causal relationship between climate, the physiological condition of man, his character, and the structure of political society."

The ideal relationship between climate and political control, or law, according to Montesquieu, is for the latter to temper the ill-effects of the former. The role of government, thus, is to negate the un-civilizing influences of environment. Montesquieu (p. 49) writes, "That bad legislators are those who have favored the vices of the climate and good ones are those who have opposed them." Montesquieu wrote in a tradition of thought that has continued nearly unchecked down to the present day, which postulates that one of the achievements of the advance of civilization has been to lessen the vulnerability of human society to the climate, the environment. Accordingly, the tempering effect of law on the effects of climate was thought to be most needed in the less-developed parts of the world: "As a good education is more necessary to children than to those of mature spirit, so the people of these climates [the sub-continent] have greater need of a wise legislator than the peoples of our own" (p. 49).

Since Montesquieu, like those who went before him in developing and applying "climate theory," compared nature and culture across space, not time, the resulting studies do not easily accommodate cultural change. Hence Voltaire's challenge to Montesquieu: since the climate has not changed, how does he explain the difference between modern Greece and Athens of the Periclean Age (Glacken 1967: 582)?[3] Some modern scholars claim that this charge of determinism is based on a faulty reading of Montesquieu. For example, Neumann (1949: xlv) writes, "That he did not attempt to derive political conditions exclusively or even primarily from climatic conditions is clear to everyone who takes the trouble of reading what he wrote. He was not a geopolitician." Kriesel (1968: 574) categorizes Montequieu as an early "possibilist" (like Wissler and Kroeber), not an early "determinist" or Ratzelian (see Chapter 7, this volume) .

Montesquieu put the comparative study of politics and environment on an empirical, historical basis (Neumann 1949: x). In several instances he attempts to isolate and examine the influence of what he perceives to be explanatory or independent variables, which distinguishes him from nearly all of his predecessors. For example, he looks at what happens when people living in one environment move to a different one: when the Visigoths migrated from the region of Germany to the Spanish Peninsula; when northern Europeans fought as soldiers in southern Europe in the war of the Spanish succession; and when European colonists reared their children in India. In all cases, Montesquieu maintained, the migrant group took on the character of their new environment.

Beyond the Greco-Roman Tradition

Although little known to audiences in Europe and North America, there are hoary intellectual traditions regarding climate and society outside of the Western tradition, which have had and continue to have an important influence in other parts of the world. McIntosh (2000) represents an unusual effort to tease out from the archaeological record the Mande vision of long-term climatic patterns in West Africa, and Freidel and Shaw (2000) have attempted something similar with the Maya of Central America. The papers reprinted in this section concern equally unusual records from North Africa and the subcontinent.

3 *Ibn Khaldûn. 1958. The Muqaddimah: An Introduction to History*

Ibn Khaldûn, like Montesquieu, was a scholar-politician trained in law, but in a non-Western tradition. Born in 1332 in Andalusia in southern Spain, to a Moslem family that had migrated there from Yemen in the eighth century, he died in 1406 in Cairo. Formally trained as a *faqîh* jurist,

Ibn Khaldûn was also an *adîb* or man of letters, and it is for his scholarship that he is still known to us today, in particular his *The Muqaddimah: An Introduction to History*. First published in Arabic in 1370 in Cairo, Toynbee (1935: III, 322) has extravagantly called it "undoubtedly the greatest work of its kind that has ever yet been created by any mind in any time or place." Written at the end of the intellectual development of medieval Islam, it captured the historical depth and conceptual heights of this development (Rosenthal, 1958: cxiii).

The medieval Islamic renaissance was distinct from but not unconnected to Western intellectual traditions – and the links to classical Greek scholarship were explicit. As Lawrence (2005: xi) writes of Ibn Khaldûn, "He engaged the full spectrum of sciences that were known in Arabic translations from Greek sources by the ninth century." Ibn Khaldûn himself writes that "The sciences of only one nation, the Greeks, have come down to us . . ." (III: 78). Ibn Khaldûn refers in multiple places to translations of Greek works (II: 203, III: 130, 151, 250), and he explicitly states that "Muslim scientists assiduously studied the (Greek sciences)" (III: 116). Part of the Greek corpus that was passed on involved climate theory. Ibn Khaldûn is seen as a forerunner of Montesquieu and Bodin among others (Rosenthal 1958: lxvii, 86n) and even as a bridge between Hippocrates and these enlightenment thinkers (Gates 1967).

Ibn Khaldûn is one of the earliest historic figures claimed as a direct ancestor of modern anthropology. Of special interest to anthropologists is his theorizing regarding the dynastic cycles of the Islamic states of North Africa, the Maghreb, which Ibn Khaldûn claimed characteristically run their full course – from ascent to decline – in just three generations (Rosenthal 1958: lxxxii; cf. Launay 2010). The driver of this dynastic cycle is the dynamic relationship between the two fundamentally contrasting socio-ecologies of the region – the urban and the hinterland. Going beyond Hippocrates, Ibn Khaldûn does not just distinguish the two socio-ecologies, he examines the relationship between them, their integration into a single system in effect.

As Lawrence writes (2005: x), Ibn Khaldûn's thesis is that "civilization is always and everywhere marked by the fundamental difference between urban and primitive, producing a tension that is also an interplay between nomad and merchant, desert and city, orality and literacy." Ibn Khaldûn argues that state formations arise from desert roots, spawn an urban society, forget and then become vulnerable to the values of the desert from whence they sprang, and so collapse. Ibn Khaldûn's formulation of this dichotomy is one of the most important antecedents of modern political geography.

Ibn Khaldûn was interested in the causes as well as the consequences of the difference between the ways of the desert tribesman and the ways of the urban citizen. He cites some socio-historical determinants of human difference, like descent and custom, but mostly he cites environmental, climatic variables. His thesis is that temperate versus intemperate climates produce, respectively, temperate versus intemperate societies, encompassing "the sciences, the crafts, the buildings, the clothing, the foodstuffs, the fruits, even the animals" and "the bodies, colour, character qualities, and general conditions" of the human inhabitants (p. 59). The temperate zone encompasses the Maghreb, western India, China, Spain, Galicia, and Iraq and Syria.

Just as Hippocrates sees people affected by climatic differentiation at local as well as global scales, so does Ibn Khaldûn see the global, zonal dichotomy reproduced at the local level of desert versus hills. Comparing the peoples of the resource-poor desert and the resource-rich hills of the Maghreb, he says of the former, "Their complexions are clearer, their bodies cleaner, their figures more perfect and better, their characters less intemperate, and their minds keener as far as knowledge and perception are concerned" (p. 63). The principal determinant of these distinctions is material abundance versus dearth, which is partly explained by more localized environmental differences like fertile versus infertile soils. At the global level, this distinction is climatic and based on the forces of hot and cold.

Ibn Khaldûn employs a comparative method to discern and illuminate the dichotomy between temperate and intemperate regions and peoples. As he writes, "If one pays attention to this sort of thing in the various zones and countries, the influence of the varying quality of the climate upon the character of the inhabitants will become apparent" (p. 62). As others have done in applying climate theory to the explanation of human difference (e.g., Montesquieu), Ibn Khaldûn tests the validity of his thesis by looking at people who move between zones. He asserts that skin color, which reflects temperateness, lightens among the descendants of people who move from South to North and darkens among the descendants of people who move in the opposite direction.

Throughout, the *Muqaddimah* statements are buttressed with the prefatory comment, "Based on observation and continuing tradition . . . ." This seemingly innocuous phrase is nothing less than Ibn Khaldûn's effort to balance the respective authorities of science and religion. He is balancing *inshâ* norm or Tradition, which cannot be qualified; and *khabar* Event, which must be confirmed, qualified, or refuted (Lawrence 2005: xxi). Ibn Khaldûn was not merely developing science, therefore; he was also developing the conceptual space for it within Moslem societies.

4 *Francis Zimmermann. 1988. The Jungle and the Aroma of Meats: An Ecological Theme in Hindu Medicine*

Another important, non-Western tradition, with its own unique development of "climate theory," is that of ancient India. The Vedic texts on which this tradition is based were composed 2,000–4,000 years ago; but they are living texts, still cited today in Ayurvedic and related teachings. Zimmermann's paper reprinted here is a synopsis of his 1987 book of the same title, which will also be referred to here. It is based on an exacting interpretation of the ancient Sanskrit texts and their ecological significance.[4] Francis Olivier Zimmermann, born in France in 1942, is an anthropologist and currently Directeur d'études à l'Ecole des Hautes Etudes en Sciences Sociales in Paris.

Zimmermann argues that the ancient Vedic texts describe a cosmological divide between the semi-arid savanna (*jāṅgala*) of western India and the perennially wet forests (*ānūpa*) of eastern India, which is in turn based on a fundamental underlying polarity between *agni* (fire) and *soma* (water). As in Ibn Khaldûn's Maghreb, the principal axis of comparison here is wet versus dry. Like the other climatic divides discussed earlier, whether Greek or Moslem, there is a normative, even political dimension to this one – one zone is the abode of the civilized Aryan, whereas the other is that of the uncivilized barbarians. As Zimmermann writes (1987a: 18), "The *jāṅgala* incorporated land that was cultivated, healthy, and open to Aryan colonization, while the barbarians were pushed back into the *ānūpa*, the insalubrious, impenetrable lands." The polarity between the savanna and the rainforest "is a matter not of physical facts but of brahminic norms," and thus is not just descriptive but prescriptive (Zimmermann 1987a: 29).

Unlike the Greek and Moslem cases, this socio-climatic divide was partly human-made: the *jāṅgala* was at least partly anthropogenic in character. It both preceded and followed the spread of the Aryan civilization in India; it was both reflection and consequence of Aryan land-use practices. Zimmermann (1987a: 44) suggests that degradation of forests and abandonment of overused lands led unintentionally to the creation, or at least spread, of the savanna. There is evidence to suggest, however, that the savanna-ization of western India was more purposive than this, that it was the product of continual, active land-management, especially burning and grazing by the livestock-oriented Aryans (Dove 1992: 235–7).

In both the Indian and Greek traditions, environmental difference maps onto the body as well as the society, but in the latter case body and society are still distinct. The Vedic sages in India, in contrast, treat the two topics in a much more integrated fashion. As Zimmermann (1987a: 20) writes, "The texts invite a double reading, or, to put it another way, one text is enmeshed in the other: a discourse on the world (natural history) is contained within a discourse on man (medicine)."

Vedic climatic or environmental theory can also be distinguished from the Greek and Islamic traditions based on its empirical character. As Zimmermann (1987a: 198) writes, "The idea of a 'science' of Nature is altogether alien to India or, to be more precise, in India it is formulated in a radically different fashion." Explicit, direct empirical inquiry was deemed unnecessary. Zimmermann (p. 76) writes, "There was no experimental method, or methodic reasoning other than on the basis of traditional teachings." Whereas the Greeks developed a method and theory of natural history, therefore, what was developed in India was essentially a lexicon. An example of the unusual empirical character of this lexicon is its inclusion side by side with both real and mythical creatures (e.g., the *karāla* or musk deer on the one hand and on the other the *makara* or sea dragon) (1987: 103).

A normative loading was associated with the semi-empirical character of the Vedic lexicons. As Zimmermann (p. 75) writes of the Sankrit treatises, "They were normative texts that transcribed authentic, orthodox knowledge." Whereas Ibn Khaldûn constantly cites both "observation and continuing tradition," for example, the Vedic teachings cite tradition alone.

Appadurai (1988: 207) asks, if Ayurveda really has so little empirical content, how could it have persisted for four millennia? He asks if we should see Ayurveda not as non-scientific, therefore, but rather as an alternative discourse? Zimmermann (p. 79) grants that the Ayurvedic physician "was first of all a man of the soil." There clearly was an empirical human-ecological basis to the *jāṅgala/ānūpa* polarity (Dove 1992). Perhaps it would be most useful to say that, as Zimmermann himself suggests, the manner of producing and articulating this reality was a radically different one.

Ethno-climatology

One of anthropology's core methodologies is the close study of local, native, indigenous systems of knowledge – ethno-botany, ethno-ecology, and ethno-zoology all being examples of this. The latest addition to this tradition of work is ethno-climatology, which is a product of the surge of interest in climate studies over the past two decades or so. An early effort is Bharara's (1982) analysis of the recollection and prediction of drought in Rajasthan. A "stellar" example is the analysis by Orlove, Chiang, and Crane (2002) of the native Andean system of basing forecasts of seasonal patterns of precipitation on changes in the visibility of the Pleiades.[5] There are also broader studies of the entire spectrum of climate-related knowledge and practice in other cultures, such as Sillitoe's (1994) study in Papua New Guinea. One of the "thickest" (Geertz 1973) ethno-climatological studies in existence is Zimmermann's (1987b) analysis of the many dimensions of the monsoon in traditional Indian culture.

5 Theophrastus. Fourth century B.C. *Concerning Weather Signs*

There are many classic Greco-Roman texts on climate and weather, which are authoritatively surveyed by Sider and Brunschön (2007: 5–29). Major works on weather include the familiar ones – Virgil's *Georgics*, Pliny's *Natural History*, and Aristotle's *Meteorologica*. Specifically having to do with the signs of weather, the major classical works are Hesiod's *Works and Days* (1914) and the study by Theophrastus reprinted here. Whereas Hesiod focused on the meaning of regular occurrences for the annual weather cycle, Theophrastus focused on the meaning of irregular occurrences for immediate weather conditions (Sider and Brunschön 2007: 3–4). As was typical of these classical works, neither addressed long-term changes in climate. Nor were any of these ancient works self-conscious ethno-climatological studies, although all drew and reported on what was essentially local, folk knowledge.

Theophrastus, a student of Plato and then Aristotle, was born about 370 B.C. in Eresos in Lesbos and died about 285 B.C. "Concerning Weather Signs" is a listing of all of the then-known signs – primarily from either astronomical phenomena or animal behavior – from which weather could be forecast in the short term. The work consists of, first, a prologue, followed by signs of rain, wind, storms, and fair weather. Only the prologue and signs of rain are reprinted here. The forecasts are strictly meteorological in character: that is, they predicted changes in weather, not, at least not directly, changes in the fortunes of humans (Sider and Brunschön 2007: 36).

"Concerning Weather Signs" is rich in "thick description" like the following: "It is a sign of storm or rain when the ox licks his fore-hoof; if he puts his head up towards the sky and sniffs the air, it is a sign of rain" (pp. 85–86). This is based on a celebration of fine-grained knowledge of the sort that environmental anthropologists and human ecologists have treated as major discoveries in recent decades: "It is a sign of rain if ants in a hollow place carry their eggs up from the ant-hill to the high ground, a sign of fair weather if they carry them down" (p. 86). As is the case in most good studies in environmental anthropology and human ecology, this knowledge is locally grounded and place-specific. Theophrastus makes multiple references to observable conditions on the flanks or summits of particular, named mountains, and he explicitly underscores the importance of this locale-specific orientation: "Wherefore good heed must be taken to the local conditions of the region in which one is placed" (pp. 84, 86). He goes on to emphasize the concomitant need for local expertise: "The signs of rain, wind, storm and fair weather we have described so far as was attainable, partly from our own observation, partly from the information of persons of credit" (p. 84). Theophrastus (p. 84) then lists a half-dozen named individuals known to him as "good astronomers." He also offers a remarkable, early statement regarding the value of such informants with locally situated knowledge: "It is indeed always possible to find such an observer, and the signs learnt from such persons are the most trustworthy" (p. 84).

"Concerning Weather Signs" lacks any meteorological explanation as to why the signs *work*, in contrast to Orlove, Chiang, and Crane (2002), for example. Sider and Brunschön (2007: 4) say that this absence may be due to mischance and the likely abridgement of the text: "[S]ince what we have is largely the signs stripped of any philosophical underpinning or scientific framework that Aristotle or Theophrastus would surely have supplied . . . ."

6 *Vladimir Janković. 2007. Gruff Boreas, Deadly Calms: A Medical Perspective on Winds and the Victorians*

Traditionally a subject of little interest except as part of the most detailed ethno-ecological studies (Conklin 1957; Gladwin 1970), wind is drawing increasing attention due to the surge of interest in things climatic (Low and Hsu 2007). These studies of wind fall within the burgeoning literature on climate and culture (Golinski 2007; Hulme 2009). Developing along parallel and often intersecting lines there is also a literature on the history of meteorology (Fleming, Janković, and Coen 2006; Fleming and Janković 2011).

Vladimir Janković, the Wellcome Research Lecturer at the Centre for the History of Science, Technology and Medicine at the University of Manchester in England, studies the cultural history of weather, climate, and meteorology in Britain from the seventeenth through nineteenth centuries (e.g., Janković 2000). The focus of his paper reprinted here is the cultural history of "wind," which in the nineteenth century was popularly perceived as a threatening, boundary-crossing, mysterious, and heatedly debated phenomenon, much like today's greenhouse gases. Janković (p. 89) says that his topic is "what might be termed a *meteorological pathogenesis* … ." As he writes, "In this paper I propose to reflect on the medical meanings of the nineteenth-century winds" (p. 89).[6]

Although Janković is writing about an era separated from that of Theophrastus by two millennia, he employs a similar ethno-climatological method. Like Theophrastus, the data that Janković gathers and analyzes are in part folk beliefs, specifically beliefs concerning the medical properties of winds. This harkens back to the classical period, and to Hippocrates' (p. 41) interest in ". . . the hot winds and the cold, especially those that are universal, but also those that are peculiar to each particular region." During the period under study here, Victorian Britain, the Hippocratic tradition of "medical topography" was still very much alive.

Janković draws his data not only from scholarly but also from popular sources, including the novels of Sir Walter Scott, Charlotte Brontë, and Jane Austen. This was a time of great debate over the role of wind in the causality of illness – something that was missing from the Hippocratic corpus, which posited but did not analyze the influence of winds. This debate was part of a wider intellectual development that ultimately brought an end to a vernacular meteorology in which winds had geographic footprints and peculiar ways of blowing.[7] Vernacular and commonly accessible public knowledge gave way to the scientific knowledge and obscurity of elites.

Whereas most contemporary ethno-climatological studies compare distant, non-Western beliefs with the more familiar scientific meteorological knowledge of modern, industrialized Western societies, Janković is studying the recent and decidedly non-modern meteorological beliefs of a historic West European population.[8] Focusing as it does on Victorian-era views of wind, this comes closer to problematizing the cultural reality of the researcher himself. Janković thereby historicizes a modern Western system of climate knowledge that has evinced little interest in studying its own past. His study denaturalizes contemporary, Western views of weather by showing their relative recency.

Janković talks about the nineteenth century as a period of epistemic competition over climate that ended with the decisive rise of the scientific paradigm and marginalization of folk theories. Unexpectedly, this may prove to have been not the end of the epistemic competition but merely a pause in it. The early twenty-first century has seen the unanticipated re-emergence of such a competition with respect to global climate change (Demeritt 2001; Smith and Leiserowitz 2012). Any scientific statement on climate change today that enters a public forum will be passionately debated by critics, who do not hesitate to put forth not only their critiques but also their own alternative theories and interpretations of the data.

Part II: Societal and Environmental Change

Historic change is the subject of the papers in this next section of the book, especially concurrent change in human society and natural environment. Its analysis raises questions about the role of environmental difference and change in the rise and demise of civilizations, and the relevance of the past to the present and future.

Environmental Determinism

The first set of papers have to do with the closest equivalent to climate theory in the modern era, environmental determinism, a more self-consciously academic articulation of the associations noted by Hippocrates, Ibn Khaldûn, and others. This is the theory that a major determinant of the character of human society is the character – the potential and resources – of the bio-physical environment.

7 Friedrich Ratzel. 1896–1898. Nature, Rise, and Spread of Civilization

The apogee of environmental determinism in the late nineteenth and early twentieth centuries was reached in the development of the field of "anthropogeography," a domain largely founded by the German geographer and ethnographer, Friedrich Ratzel (1844–1904), who long taught at the University of Leipzig. His most important contributions were *Anthropogeographie* (1882–1891) and *Völkerkunde* (1885–1890), an English translation of the latter being published in 1896 as *The History of Mankind*.

The History of Mankind was a broadly comparative work, in method and global scope much like *Airs, Waters, Places* and *The Muqaddimah*. It differed from *Airs*, in particular, in that it combined its spatial analysis with an historic one: "Ethnography must acquaint us not only with what man is, but with the means by which he has become what he is, so far as the process has left any traces of its manifold inner workings. . . . The geographical conception of their surroundings, and the historical consideration of their development, will thus go hand in hand" (1896: 3). In short, Ratzel was approaching cultural differences, mapped across the regions of the globe, as also historical differences.

Ratzel was primarily interested in the difference between what he calls the "civilized" versus uncivilized societies, and the historic process that led from the latter to the former – his premise being that civilized societies were once uncivilized and sometimes could become so again, and savage societies could someday become civilized.[9] Ratzel (p. 110) glosses the uncivilized societies as "natural races," which he defines by a human developmental metaphor, reminiscent of Montesquieu: "There is a distinction between the quickly ripening immaturity of the child and the limited maturity of the adult who has come to a stop in many respects. What we mean by 'natural' races is something much more like the latter than the former." The most important factor in differentiating the natural races from the civilized ones is environmental. As he writes (1896: 14), "We speak of natural races, not because they stand in the most intimate relations with Nature, but because they are in bondage to Nature." Ratzel is not suggesting that the civilized races are less dependent upon nature, only that their dependence is different.

As with earlier climate theorists, there is a broad, latitudinal dimension to Ratzel's analysis of the differing influence of nature upon different societies; and it is similarly based on a temperate/intemperate distinction. As he notes, "The real zone of civilization, according to all the experience which history up to the present day puts at the disposal of mankind, is the temperate" (p. 114). Ratzel suggests that the evolution of mankind took place in the "soft cradle" (p. 113) of the resource-rich tropics, but civilization developed in the less favorable conditions of the temperate zones. The temperate zone is more stimulating of the development of civilization than the tropical zone because, in part, and echoing the earlier work of Hippocrates, Ibn Khaldûn, and the Vedic sages, it is more austere. As Ratzel says when writing about the development of agriculture, ". . .[W]ant is more favourable than abundance" (1896: 88). A second aspect of the temperate latitudes that stimulates the development of civilization is its higher density of population. This is desirable because, as Ratzel writes, striking an unexpectedly anti-Malthusian note, "In density of population lies not only steadiness of and security for vigorous growth, but also the immediate means of promoting civilization" (p. 113).

Ratzel is conscious of the fact that such generalizations are challenged by exceptional cases. Of agricultural development, for example, he says, "But it is unsafe to say with Buckle that there is no example in history of a country that has become civilized by its own exertions without possessing some one of those [natural] conditions in a highly favourable form" (1896: p. 27). More pointedly, he suggests that the force of humans may or may not overwhelm the influence of natural endowments and conditions. As proof of this, he maintains that there is no exact correlation between the natural races and the environment, even if there is between civilized races and the environment: "Nothing gives a more striking lesson of the way in which the utilisation of Nature depends upon the will of man than the likeness of the conditions in which all savage races live in all parts of the earth, in all climates, in all altitudes" (1896: 14).

Early in the twentieth century, the academic tide turned decisively against Ratzel, forever dimming his legacy. One interlocutor of Ratzel's, Marcel Mauss, writes of the anthropogeographers, "They have, however, attributed to this factor [land] a kind of perfect efficacy, as if it were capable of producing effects on its own without interacting with other factors that might reinforce or neutralize its effects either partially or entirely" (1979 [1950]: 21). Some scholars attribute the reaction against Ratzel at least in part not to his own scholarship but to the subsequent reworking of it into a more extreme form of environmental determinism by his students and followers, notably Ellen Churchill Semple. As Kroeber, an admirer of Ratzel's, wrote in later years of him: "But he did conceive of culture as more than an incidental phenomenon, and was far from being the crass environmentalist which Semple's misrepresentatively selected adaptation makes him out to be" (1947: 7).

The eclipse of Ratzel's determinism a century ago casts in curious relief the modern popularity of the work of the eco-physiologist Jared Diamond (1997, 2005). Like Ratzel's, Diamond's environmental determinism is focused on explaining the historic socio-economic ascendance and continued political-economic dominance of the northern latitudes, but in direct contrast to Ratzel as well as Ibn Khaldûn and the Vedic sages, Diamond attributes this not to the resource dearth but rather the resource abundance of these climes.

8 *Betty J. Meggers. 1957. Environment and Culture in the Amazon Basin: An Appraisal of the Theory of Environmental Determinism*

Although environmental determinism fell out of fashion in academia early in the twentieth century, a fascination with the tropics and their potential or lack thereof for human development did not, the Amazon often being a central test case. One of the most important contributors to the mid-twentieth century anthropological literature on the Amazon was Betty J. Meggers (1921–2012), a Columbia-trained archaeologist, who was long affiliated with the National Museum of Natural History, in the Smithsonian Institution.

Like Kroeber and others in twentieth-century anthropology, Meggers sought to rethink the discipline's post-Ratzel rejection of environmental determinism: "Ridicule of this overembellishment [of environmental determinism] brought about the disgrace of the theory, with little serious effort to determine whether or not the core was sound" (p. 116). Her efforts to determine this included an influential 1971 book and the article reprinted here. The research question that drove Meggers was much the same as in Ratzel's case, namely ". . . the understanding of how and why culture develops when, where and as it does" (p. 125). Equally important in the mid-twentieth century, dominated as it was by the conception of the "Third World" and the developed/underdeveloped dichotomy, was the question when and where culture does *not* develop.

Meggers believes that the environments of the world offer unequal potential for human exploitation, with the tropical forest offering one of the lowest potentials. Whereas Ratzel saw the tropics as being too rich, therefore, Meggers sees it as being too poor. This is due to a high, even temperature that favors bacteria and thus not the accumulation of humus; abundant annual rainfall and thus leaching of nutrients from the soil; intensity of rainfall and thus soil erosion; and variability in rainfall which can stress crops. Any human activity that entails total clearing of the tropical forest exposes the land to all of the ills of these characteristics. This contrasts with what Meggers calls "slash and burn" agriculture: its lack of tillage keeps erosion to a minimum; the brief period between clearing the forest and planting keeps humus destruction to a minimum; leaving the burned and unburned vegetation on the field promotes the return to the soil of nutrients from the cleared vegetation; and the brief period of cropping followed by natural afforestation promotes the recovery of the original fertility of the land.

Although slash and burn agriculture has a beneficial impact on the land, Meggers suggests that it does not have a beneficial impact on the culture: "This type of food production has a conservational effect on soil and soil fertility, which is desirable, but also exercises a conservative influence on the culture, keeping it in a relatively simple level of development" (p. 123). In slash and burn agriculture, the period of cultivation is brief but the period of fallow is long, which means "1) that a relatively large amount of land per capita must be available for agricultural use, and 2) that the settlement cannot remain permanently in one place" (p. 123). She sees this as placing critical limits on settlement size: "[G]enerally speaking 1000 individuals is a large population for villages in the South American tropical forest, and settlements with less than 300 people are typical" (p. 124). This means no differentiation in production and consequently no technological development. The validity of this line of reasoning, Meggers avers, is attested to by the historic lack of development of advanced cultures in the tropical forest based on slash and burn agriculture. As a result, "[I]t seems acceptable to conclude that the Tropical Forest Type of culture characteristic of the Amazon Basin shows the effects of environmental determinism" (p. 125).

Seminal works like those by Freeman (1955 [1970]) and Conklin (1957) all supported Meggers' thesis that slash and burn agriculture was well adapted to the forest and climate of the tropics. However, some of the most influential work by anthropologists in the Amazon took issue with Meggers' findings that slash and burn agriculture inhibited social development. Carneiro's (2008 [1960]) quantitative analysis of carrying capacity under slash and burn agriculture showed that communities up to 500 persons in size could be permanently supported at the same site by this system of cultivation. Dumond (1961) more directly assailed Meggers' thesis by arguing that there is historic evidence of slash and burn agriculture supporting state development in the Amazon.

Meggers' biggest oversight pertains to the political dimension of slash and burn agriculture. She writes, "The disrepute in which it [slash and burn agriculture] is held does not stem from a conservational effect on the landscape, but rather from the conservative influence it exerts over the local culture" (p. 123). In fact, this "disrepute" stems from the illegibility of slash and burn agriculture vis-à-vis centralized states and its consequent ability to frustrate, and thrive beyond the reach of, state control (Scott 2009; Dove 2011). Meggers elides these issues: "If we accept the premise that the standard climate determines that agricultural exploitation must have certain features, then man's problem is to find a solution that fulfills these requirements and in addition meets the demands of modern civilization" (p. 127). The premise that a single new "solution" can be devised that will reward both slash and burn agriculturalists and state elites was typical of Meggers' era.

Climate Change and Societal Collapse

The studies in this section look at the question not of the rise but of the demise of civilizations, and they do so within the histories of specific times and places, which complicate simple deterministic explanations.

9 Thomas H. McGovern. 1994. Management for Extinction in Norse Greenland

The first study in this section is McGovern's history of Norse settlement in Greenland, focusing on the question of adaptation to the so-called "Little Ice Age." Adaptation to climate change is a subject of increasing interest (Roncoli 2006; Ayers and Forsyth 2009; Moore 2010). The Little Ice Age, referring to the cooling of global temperatures by 2–3 degrees centigrade from the mid-fourteenth through the end of the seventeenth centuries, has attracted increasing academic attention both for its intrinsic historic interest and for use as a proxy measure of the impact of the modern climatic perturbation (Nunn et al. 2007; Bulliet 2009; White 2011). With respect to this and related questions, the climatic relations of sub-Arctic aboriginal societies have emerged as a topic of considerable interest (cf. Chapter 19, this volume). Thomas H. McGovern is an archaeologist at Hunter College, City University of New York, who specializes in the study of Norse and Inuit societies in the North Atlantic and Arctic (1981, 1988, 1991). The Norse of Greenland have become one of the most widely cited cases of climatic determinism (Diamond 2005; McAnany and Yoffee 2010).

During the period A.D. 800–1000, Norse seafarers colonized an area stretching from western Norway to eastern North America, and including Greenland around A.D. 985. In that latter island they developed an economy based on the raising of livestock on pastures in the inner fjords and hunting migratory seals in the outer fjords. They also hunted walrus in the north of Greenland to trade their ivory and skins to elites in Europe. In the last quarter of the fifteenth century, however, one half-millennium after their founding, the Norse settlements in Greenland disappeared. Danish expeditions sent in the seventeenth century to re-establish contact were surprised to find only long-abandoned settlements. Many scholars have attributed their disappearance to the Little Ice Age. As McGovern writes, "Many climate impact theories have been proposed, but most may be reduced to the simple statement 'it got cold and they died'" (p. 141). McGovern faults this explanation on multiple grounds: historic records show that this collapse did not happen abruptly; it did not happen in a resource-poor environment; and it did not affect the aboriginal Inuit who were sharing occupation of Greenland at the time.

When the Little Ice Age made their already marginal economy of localized herding and hunting less productive, McGovern argues, the Norsemen did not adapt, change, diversify. They did not shift their subsistence base to greater dependence on marine mammals; they did not shift their trade-oriented activities from walrus hunting to commercial fishing; and they did not adopt any Inuit technologies for exploiting the island's natural resources. The Inuit not only utilized a wider range of Greenland's resources, but they also adapted as the climate shifted, in particular by moving up and down the coastline as resource zones shifted. In contrast, "The Norse were far less mobile" (p. 144). At the end of the tenth century the Norse developed a successful adaptation to Greenland's environment, but with the onset of the Little Ice Age in the fourteenth century this adaptation became less successful, and the Norse did not adapt.

McGovern attributes the Norse failure to adapt to cultural factors. One problem was their unwavering commitment to their subsistence herding base, which was the source of authority for political and religious elites and thus not easily relinquished. A second problem was their hostile, Christian view of Greenland nature and natural resources, which inhibited their exploitation of the environment to its fullest extent, in contrast to the view taken by the Inuit. A third problem was their cultural distance from the Inuit. As McGovern says, the Norsemen were "culturally preprogrammed to reject all innovations from the Inuit, fatally ignoring tainted technology and alien expertise and keeping closer and closer to home, hearth, and church" (p. 147).

McGovern sees the exercise of human agency here. As he writes, "It is hardly an accident that these life-saving skills were so systematically rejected – it took a great deal of effort on somebody's part" (p. 148). The decision to stick to the traditional but ever less productive subsistence system was a *choice*, in short. The outcome could have been different. McGovern notes that Norse and aboriginal social and economic systems were successfully integrated among the Danes and Inuit in eighteenth-century Greenland and in more recent times among the Norwegians and Sami (cf. Ratzel pp. 113–114, this volume)

McGovern concludes by suggesting that "The case of Norse Greenland may have some disquieting parallels in the modern world" (p. 149). One parallel is the arrogance of certain world views: "Like the Norse elites, we are today very certain of the complete adequacy of a particular world view"(p. 149). Another is the problematic confluence between the self-interest of political–economic elites and dysfunctional environmental relations. But McGovern also says that, like the Norsemen, we have agency. As he writes, "Like the Norse Greenlanders, however, we are not inevitably the prisoners of history and culture. Like them, we have many potential options" (p. 149).

10 Harvey Weiss and Raymond Bradley. 2001. What Drives Societal Collapse?

Whereas the disappearance of the Norse settlements looks at first glimpse like a "collapse," other cases complicate both the concept of collapse and its causes. The lead author of this selection, Harvey Weiss, an archaeologist and Professor of Anthropology and Near Eastern Studies at Yale University, has been a pioneer in critiquing orthodox thinking on climatic perturbation and the collapse of civilizations. Like Meggers, he is an anti, anti-determinist. His fieldwork focuses on climate and society in third millennium B.C. Mesopotamia (Weiss 1993, 1996, 2000). His co-author, Raymond Bradley, is a geoscientist at the University of Massachusetts, at Amherst.

There are numerous cases in the archaeological and historical records of seeming societal collapse, which Weiss and Bradley describe as having "frequently involved regional abandonment, replacement of one subsistence base by another (such as agriculture by pastoralism), or conversion to a lower energy sociopolitical organization (such as local state from interregional empire)" (p. 152). Orthodox opinion long held that these cases were owing to "combinations of social, political, and economic factors" (p. 152). The failure to cite climate as one determinant was associated with the belief that global climate during the past 11,000 years was "uneventful" (p. 154).

Weiss and Bradley directly challenge this view, drawing on new sources of "high-resolution paleoclimatic data" (p. 152), which paint a picture not of stability but instability. They identify a number of episodes of extreme climatic fluctuation, especially abrupt and extended droughts, that had sweeping social consequences. The most notable Old World examples include the 10,800–9,500 B.C. cooling and drying of the Younger Dryas, which impacted Natufian hunters and gatherers in the Levant and Northern Mesopotamia; the 6,400 B.C. drought in the same region; the 3,200–3,000 B.C. drought at Uruk in Southern Mesopotamia; and the 2,200 B.C. drought and cooling from the Aegean to the Indus. Examples of climate-related collapse from the New World include the sixth century A.D. Moche in Peru, the ninth century A.D. Classic Maya, the tenth century A.D. Tiwanaku in the Andes, and the Anasazi of the thirteenth century A.D. in the U.S.

One reason prehistoric and early historic societies were vulnerable to such climatic events is that they occurred with such infrequency that no cultural memory of them was retained during the intervening centuries or millennia. As far as the affected societies were concerned, these were completely unfamiliar events, without precedent. Thus, there were no institutions at the ready to deal with them. Entirely new ways of life had to be developed, and relatively quickly. Under such circumstances, the erasure of existing institutions, and the shifting to radically different life-ways, was the only option available. Collapse was, as Weiss and Bradley put it, "an adaptive response to otherwise insurmountable stresses" (p. 152).

Weiss (2000) elsewhere presents a detailed study of one such case based on his fieldwork in northern Mesopotamia. He describes the late third millennium B.C. climatic perturbation in this region as including an "abrupt onset, ca. 300-year duration, radical increase in airborne dust, major aridification, cooling forest removal, *Sanguisorba minor* 'land degradation,' and possible alterations in seasonality" (2000: 83–4). These environmental changes provoked sweeping changes in social-ecological organization, consisting of "collapse to less extractive political organization, directed habitat-tracking to regions where agriculture was sustainable, and the abandonment of reduced-production cultivation for pastoralism" (2000: 88). Weiss (ibid.) sees collapse of the social order and flight as adaptive under these circumstances, which moves the debate away from an unrealistic and unproductive dichotomy between societal "collapse" and its logical opposite of unchanging, timeless societal "persistence."

The collapse of the current world order would doubtless reduce greenhouse gas emissions, but is this "adaptive"? The lesson from history is that traumatic social change does provide adaptation to climate change, but at a huge social cost. The object of enlightened modern society, then, is perhaps to bring about adaptive social changes of the same order of magnitude but in a planned and therefore less costly way.

Climatic Events as Social Crucibles

The papers in this last set on climate change and social collapse examine the changes that take place in the immediate aftermath and as a consequence of climatic events. Several related theses underlie their analyses: climatic perturbations take their greatest toll on economically and politically marginal peoples, at the same time as they offer opportunities for resource capture by political and economic elites; such perturbations throw into relief the stresses and contradictions of society; and climatic perturbations can accelerate ongoing processes of change in society.

11 *James Spillius. 1957. Natural Disaster and Political Crisis in a Polynesian Society: An Exploration of Operational Research II*

The first reading in this section concerns the extreme weather patterns called "typhoons," "cyclones," or "hurricanes," which have long drawn anthropological attention (Marshall 1979; Dove and Khan 1995; see also Chapters 15, 16, and 17, this volume). James Spillius (1922–), a British anthropologist, was involved in one of the most famous anthropological studies of hurricanes of all time, which was initiated as part of Sir Raymond Firth's re-study of Tikopia in the Solomon Islands in 1952–1953. Seven weeks before their arrival in March 1952, a severe hurricane struck the island. One year later, while Spillius was still there, a second severe hurricane struck the island. Firth (1959) published an extensive account of the ability of Tikopian social structure to "weather" the impact of the hurricane. From Spillius (1957) we have a more detailed description and analysis of the way that he and Firth were drawn by natives and the colonial British government alike into playing a role in post-disaster assistance. Part One, subtitled "Emerging Research Roles in a Crisis Situation," is a narrative account of their involvement; Part Two, subtitled "Some Principles of Operational Research," draws out the lessons of this involvement for anthropology. The latter part is reprinted here, and the first part will be extensively quoted from.

Spillius' circumstances raised many prescient questions for him, which came increasingly to dominate debates within anthropology during the second half of the twentieth century: "Why do administrators make so little use of anthropological writings? Is it possible to do research that is simultaneously of theoretical and practical importance? Can social change be studied in process, and should anthropologists try to affect the course of such change? . . . What are the anthropologist's responsibilities to the people he studies and works with?" (p. 158).

Spillius represents their situation on Tikopia as one that was, at least for that era, a novel one: "He [the anthropologist] usually takes an active part in the social life of the indigenous community, but he does not try to direct the course of political and social events" (1957: 3–4). Doing so led to a fundamental epistemological challenge, a social science equivalent to physics' indeterminacy principle. ". . . I realized that I was studying social processes not only as they were going on but as I was affecting them" (1957: 4). This led Spillius to question many of the premises of traditional anthropology, especially its ideal of a disinterested stance. He notes that an "interested" stance, far from being alien to anthropological research, is an inevitable byproduct of the core methodology of anthropology, participant observation, with the intense social engagement that this entails, which leads the anthropologist to identify with the community and its well-being. Such emotional identification can be heightened when the community suffers a calamity like a hurricane. As Spillius writes, "No anthropologist, however set he was on sticking to the role of observer, would have done nothing in this situation" (p. 161). Even a non-anthropologist, "even an entomologist," as Spillius (p. 161) avers, would have intervened. In justifying the goal of helping to bring about desired changes in society, Spillius notes that change is taking place in such societies anyway.

In an unusual and prescient recommendation, Spillius says that anthropologists must make an effort to understand state actors, something that they seem predisposed not to do: "It is curious that sometimes anthropologists embarked on 'objective' studies display compassion and understanding for the indigenous community but find it much more difficult to take a dispassionate view of Government as a social system" (p. 166). This, Spillius (p. 166) argues, is a prerequisite to any hope of achieving the goals of operational research: "The practical aims of research can only be achieved if the social anthropologist shows some sensitivity to the social structure of the government he is dealing with."

Spillius observes that he found his period of operational research on Tikopia to be unusually productive. This was in part due to the fact that the hurricane-related crisis was revelatory: "[S]everal aspects of Tikopia social organization were clarified because I watched them changing in response to crisis" (p. 164). In part it was due to the conception of a new research subject, consisting of the relations between community and state and the then-radical view of them as constituting a single social system. Finally, the productivity of Spillius' research was due in part to his creating a role of "engaged" ethnographer: "At the time I did not fully realize how much I was learning about the workings of the society by taking an active part in it" (1957: 25).

Spillius leaves the reader with a statement about the ethics of research, which has perhaps not been bettered in the half-century since: "[H]aving accepted the fact that his mere presence has some effect on the situation, the anthropologist must constantly make judgements about the course of action most likely to achieve the long-term goal of leaving the people studied better able to cope with problems" (p. 165).

12 *Jacqueline S. Solway. 1994. Drought as "Revelatory Crisis": An Exploration of Shifting Entitlements and Hierarchies in the Kalahari, Botswana*

There is a growing literature that suggests that extreme climatic events can act as catalysts to social change. Anthony F.C. Wallace (1957) laid the foundations for this work with his mid-twentieth century analyses of how disaster undermines cultural identity, which leads to cultural revitalization movements. The article by Jacqueline Solway reprinted here focuses on drought, which is the subject of much of the anthropological work on climate (West 2009). Solway is an anthropologist in the Departments of Anthropology and International Development Studies at Trent University in Ontario, whose research and publishing has focused on ethnicity, politics, and the "bushman question" in southern Africa (1998, 2003, 2009).

Solway's analysis focuses on the effects of a recurring drought during the years 1979–1987 in the Kalahari region of Botswana, a semi-arid area receiving 300–350 mm rainfall/year, where the local economy is based on pastoralism, supplemented by rain-fed agriculture. This drought was a "watershed event," Solway says (p. 170), both because it was unusually severe, and because it occurred at a "critical historical juncture" in Botswana, when commercialization and class formation were proceeding at a rapid pace and the nation's economy was growing dramatically. Like some of the other authors in this volume (e.g., Cruikshank, Chapter 19, this volume), Solway studies not just the extreme climatic event, but also its co-occurrence with extreme social events. Together these events produce what Agamben (1998) calls a "state of exception," a state of such disruption of conventional routine that, as Solway writes, "actors are given license to innovate with social and moral ideological and behavioural codes" (p. 170), even to the point of sanctioning what would have previously been seen as anti-social behavior. This exception applies not only to the local community but also to its relations with the central state. Following Ferguson (1990), Solway argues that "The drought was thus an opportunity and provided a point of entry for the state to insert itself in the lives of citizens in new and expanded ways . . ." (p. 170). Since drought conditions tend by their nature to be extended over a long period of time, compared to a hurricane, for example, this opportunity is also extended in time – which perhaps makes drought an inherently state-friendly phenomenon – and raises questions about a similar impact from contemporary global climate change.

Solway begins her study by citing Sahlins' (1972) description of the way that a "revelatory crisis" can expose the contradictions in the domestic mode production, the example of which he draws from Firth's work on typhoons and hurricanes in Tikopia (see Chapter 11, this volume). In her study in Botswana, Solway makes clear, these contradictions were pre-existing ones. The drought revealed but was not responsible for a systemic deterioration in rural conditions and a crisis of social reproduction. Because the drought affords unique opportunities for social change, it "offers a perfect lens for viewing the dialectic between structure and agency" (p. 170).

Whereas Solway is not alone in seeing disasters as revelatory crises, she is one of the few scholars to look at the other side of this coin: "However, in a paradoxical fashion, while a crisis such as a drought reveals and exposes contradictions and deteriorating conditions, it also allows them to be concealed and mystified" (p. 170). A disaster like the Botswana drought draws attention away from other matters. The disaster response of the state, cloaked in benignity, can actually exacerbate these problems. Drawing on Sen's (1982) method of entitlement analysis, Solway argues that the expansion during the drought of social security entitlements from the central government compensated for and thus masked the countervailing contraction in community-based entitlements.

Part III: Vulnerability and Control

This section is concerned with what parts of society are most vulnerable to climatic perturbation, why, and what steps they take to manage their vulnerability, studies of which (not all from anthropologists) include Adger (1999) on coastal Vietnam, Ribot, Magalhães, and Panagides (1995) on the global semi-arid tropics, and Brondizio and Moran (2008) on the Amazon.

Culture and Control of Climate

A great deal of attention is currently being devoted to efforts to mitigate the forces producing climate change and to adapt to those outcomes that are inevitable. There is a history of engagement with such issues. Ratzel and Meggers (Chapters 7 and 8, this volume) suggest that there are but limited means to escape the determinate effects of climate. McGovern (Chapter 9, this volume) sees the potential for both human agency and the failure to exert it. Weiss and Bradley (Chapter 10, this volume) emphasize the inevitability of social collapse in the event of extreme climate events but they also suggest that this is adaptive, whereas Solway (Chapter 12, this volume) sees non-collapse as maladaptive. This next set of papers examines both ideological and material means of mitigating undesirable climatic phenomena and adapting to their effects.

13 Elizabeth Colson. 1957. Rain-Shrines of the Plateau Tonga of Northern Rhodesia

One of the most ubiquitous traditional cultural methods of coping with extreme climate events involves ritual (Vogt 1952; Orlove 1979; Sillitoe 1993). Well known within anthropology are studies of ritual regarding thunder and lightning in the Indo-Malay region (Needham 1964; Freeman 1968). A further topic involves community-based rituals of rain-making, which is the subject of the current paper by Colson (cf. Chapter 20, this volume).

Elizabeth Colson (1917–) spent most of her career at the Rhodes-Livingstone Institute in what was then Northern Rhodesia, now Zambia, and taught at Boston University and the University of California, Berkeley, where she is now Professor Emerita. She began work among the Plateau Tonga in Rhodesia in 1946. In later years, she carried out one of the first in-depth ethnographic studies of the impacts of a development project, a study of the dam-displaced Gwembe Tonga (Colson 1971).

The subject of Colson's paper reprinted here is the rain-shrines of the Tonga. A shrine can be either a natural feature of the land or a man-made hut, in either case thought to be inhabited by a spirit. These spirits can afflict communities with drought, cattle epidemics, or epidemic disease; and they are appealed to on any occasion of community-wide disaster. But their principal concerns are two-fold: first, ensuring the proper rainfall; and second, combating the cultural changes resulting from European contact. In practice, they have an additional concern, concerning socio-political integration.

Colson describes the Tonga as "culturally a have-not group" (p. 192), lacking much of what constitutes society elsewhere in Africa, e.g., an army, an organized state, age-grade sets, secret societies, and social stratification. "Into this anarchy," Colson (p. 193) writes, "some semblance of order is infused by the rain-rituals, which effectively organize small groups of villages for corporate activity . . . ." Colson quotes leaders at one shrine urging all the people and not just the kin-group which controls the shrines to dance and pray for rain: "The rain falls on your fields as well as on ours. You must all dance" (p. 199). In short, there is spatial congruence between climatic events (rain) and social organization (followers of a rain-shrine).

The joint responsibility to carry out annual rites for the upkeep of the rain-shrines obliged villagers to live more or less in peace with one another. Only by fighting against their centrifugal social forces, and by honoring and upholding the tradition of the shrines, are the Tonga able to control the rain. In summary, Colson writes, "It is only in the rain-rituals and their associated shrines that the Tonga show a half-hearted grouping towards the establishment of a larger community than that which existed in the village or in the ties of kinship" (p. 192).

In the absence of other forms of supra-community organization, this allegiance to shrines had political implications, which were reflected in the historic opposition of shrines to the forces of modernization throughout both the colonial and post-colonial eras (cf. Chapter 20, this volume). At the time of Colson's writing in the mid-twentieth century, however, the importance of the shrines was waning, owing to the forces of modernization, in particular opposition by the Christian missions.

14 *Richard L. Burger. 2003. El Niño, Early Peruvian Civilization, and Human Agency: Some Thoughts from the Lurin Valley*

Whereas Colson looked at how people self-organize to pray for rain, Burger looks at how people self-organize to cope with the effects of too much rain, specifically landslides and debris flows precipitated by El Niño events. The El Niño–Southern Oscillation (ENSO) meteorological pattern has drawn increasing academic attention, including by social scientists and historians (Grove and Chappell 2000; Davis 2001; Sandweiss and Quilter 2008). Many studies in this field have looked at ENSO-related drought and forest fires, as in Indonesia in 1997 and 1998 (Harwell 2000). More recently, some scholars, mainly prehistorians, have begun to focus not simply on the short-term role of ENSO in disrupting society, but also on its long-term role in shaping the evolution of society (Billman and Huckleberry 2008; Richardson and Sandweiss 2008; Roscoe 2008). Interest in ENSO is based in part on its perceived utility as a proxy for contemporary climate change. Personal experience of severe El Niño events in recent decades has also led to what Burger calls a "predisposition to take El Niño seriously in the archaeological modeling of civilizational trajectories in the distant past" (p. 202).

Richard L. Burger is a leading archaeologist of pre-Hispanic Peru and directed the repatriation of much of Yale's Hiram Bingham collection of Machu Picchu artifacts (Burger and Salazar 2004). In the study reprinted here, Burger asks "whether the peoples of pre-Hispanic Peru anticipated the dangers posed by El Niño events and whether they were able to develop strategies to mitigate them" (p. 203). The danger consisted of floods, landslides, and debris flows in the deeply cut ravines or *quebradas* of the coast.

With the benefit of the *longue durée* vision of the archaeologist, Burger views ENSO events not as stochastic perturbations but as a recurring "normal" feature of the environment. In this respect he follows Hewitt's (1983) pioneering work in seeing events like ENSO as "characteristic rather than accidental features of the place and societies where they occur" (p. 204). Burger's analysis of the archaeological record at Manchay Bajo on the Peruvian coast shows an impressive adaptive response to the ENSO threat: (i) the threat of ENSO-related landslides was correctly identified; (ii) a solution – a massive stone dam – was devised using available technology and materials; (iii) labor was mobilized to construct a five-meter high dam three-quarters of a kilometer in length; and (iv) labor was mobilized as needed over a period of six centuries to maintain the dam and renovate it on at least two occasions.

This response clearly illustrates the exercise of human agency, which is surprising in the absence of state-like political structures. As Burger notes, all modern disasters elicit an extra-local, state response: "aid comes from 'outside'" (p. 203). But as he continues, "[W]hat disaster strategies were employed prior to the emergence of such over-arching social and political structures?" In the absence of an over-arching state structure at Manchay Bajo, families or small social units developed links with extra-local communities unlikely to be affected by the same calamity at the same time. Such linkages likely crossed vertical zones, a resource use strategy of the Inca famously characterized as the "vertical archipelago" by Murra (1985). The development of extra-local networks as a safety net for coping with localized climatic perturbation has been documented elsewhere in the world as well (Waddell 2008).

Burger argues that more complex, state-like formations might have proved less resilient in the face of extreme ENSO events. As he writes, "[I]t is worth considering whether the pre-state societies of the Initial Period may have been as well as or, perhaps, even better equipped to deal with mega-El Niños than the more fragile complex societies of later times" (p. 214). This point is borne out by Weiss and Bradley in Chapter 10, this volume, who argue that complex state structures collapsed to adapt to extreme climatic events. This counter-intuitive view of the state as not strong but fragile leads to a different reading of some other studies, like Meggers' in Chapter 8, this volume. Whereas Meggers argued that the Amazon was not rich enough to support state formation, Burger's analysis leads us to ask whether the Amazon was too challenging an environment to sustain fragile state systems.

Climatic Disasters and Social Marginalization

For a generation now scholars have emphasized the socio-economic determinants of "natural" disasters. Groups that are socio-economically marginal are more vulnerable to natural perturbations, like extreme climatic events; and the relative position of such groups within society is often, indeed, worsened by such events. Wisner (1993: 128) writes that "Disasters *bring to the surface* the poverty which characterizes the lives of so many inhabitants." As the historian Marc Bloch (1961) wrote, "Just as the progress of a disease shows a doctor the secret life of a body, so to the historian the progress of a great calamity yields valuable information about the nature of the society so stricken." Such revelations are politically sensitive and are typically contested. Hewitt (1983) writes of the effort by modern governments to represent calamity as essentially "accidental," an "unplanned side-effect," to see disasters as "an *archipelago* of isolated misfortune."

15 *Nancy Scheper-Hughes. 2005. The Disaster and its Doubles*

The first reading in this subsection concerns Hurricane Katrina, which devastated the U.S. gulf coast in late August 2005 (Paredes 2006). Scheper-Hughes is a medical anthropologist at the University of California, Berkeley, who has published widely on the socio-cultural dimensions of health and sickness (e.g., 1979, 1993). Her brief analysis of Katrina reprinted here, which was written two months after the storm, posits that the hurricane had a "double" impact, referring to the social and political responses to it that proved to be as injurious as the storm itself (p. 222, n.1).

Scheper-Hughes, like most contemporary scholars of disaster, rejects the idea that the poor are vulnerable to such events because of geography or climate, arguing instead that the problem lies with "political lassitude, racism and entrenched poverty" (p. 221). According to this view, Katrina was so devastating because it impacted a population in New Orleans that was already greatly disadvantaged. As then-Senator Obama said, "The people of New Orleans weren't just abandoned during the hurricane, they were abandoned long ago" (p. 221). Social scientists have theorized such abandonment in terms of a politics of human "disposability" (Giroux 2006), which builds on Agamben's (1998) concept of "homo sacer," the people who are placed outside of the protection of the law by the modern state itself.

The contest over the existential status of Katrina's victims was made manifest in a debate about the propriety of using the term "refugees" to refer to them. Scheper-Hughes writes, "The term 'refugees' implies that there are American-born Americans without a symbolic passport, without a president, without protection, who live and die outside the political circle of trust and care" (p. 222) (cf. Masquelier 2006). Because of this unflattering implication, the use of the term "refugees" became highly contentious. William Safire (2005) argued instead for using the term "Katrina survivors" or "flood victims." The contradictions in the status of Katrina's victims came to a head in one of the disaster's most infamous moments, in which armed sheriffs blocked evacuees from crossing the Greater New Orleans Bridge to safety. In blocking their flight, the sheriffs denied them even the status of refugees.

Scheper-Hughes and other analysts argue that the response to Katrina is an example of "disaster capitalism," referring to the way that state management of disasters creates opportunities for the deployment of capital that would not otherwise have existed (Adams, Van Hattum, and English 2009). Scheper-Hughes refers to the "'golden opportunity' afforded to developers by the destruction of New Orleans" (p. 221), which exemplifies the way that disparate political and economic resources create opportunities for some, and mis-opportunities for others, in the midst of the same, extreme climatic event. Whereas the balance between public and private interests is typically tested in disasters, the outcome is not certain: Katrina tipped the balance toward the private sector; but the 2011 Fukushima nuclear disaster weakened the private sector in Japan; and the disasters described by Spillius and Solway in Chapters 11 and 12, respectively, also seemed to strengthen the public sector over the private.

16 *Rosalind Shaw. 1992. "Nature", "Culture" and Disasters: Floods and Gender in Bangladesh*

Flooding in Bangladesh has been the subject of considerable academic study. Interesting work has been done on the way that elites benefit from floods and, more generally, the way that the impact of disasters in Bangladesh is socially inflected (Zaman 1991; Dove and Khan 1995). The gendered dimension of such disasters, which is the focus of this reading, has been studied in coastal erosion in Papua New Guinea (Lipset 2011) and in rainmaking rituals in Tanzania (Chapter 20, this volume).

Rosalind Shaw is an anthropologist at Tufts University who currently works on issues of religion, violence, and justice in West Africa. Her premise in the current study is that a Cartesian nature/culture divide within anthropology has led to most research on natural hazards being hobbled by two related polarities: "that of the natural and human worlds . . . [and] that of 'normal' versus 'abnormal' events . . . ." (pp. 225–6). She examines this thesis with respect to a devastating flood that struck Bangladesh in the summer of 1988, submerging three-quarters of Dhaka and much of the surrounding countryside.

Shaw begins her analysis with a view of the "duality" of floods in Bangladesh as both resource and hazard, which is reflected linguistically in Bengali as the difference between *borsha* or normal flood and *bonna* or abnormal flood (cf. Zaman 1993). The alluvial silt borne by annual flood waters permits bountiful harvests under normal conditions, based on the cropping of the flood-tolerant *aman* rice variety.

Shaw's thesis is that the experience of the annual floods is socially inflected. Flooding can simultaneously submerge some lands but also create new ones, from accretion, but it is the wealthy and powerful who take control of the new lands, not the landless (cf. Zaman 1991). Not only are the worse-off segments of society less able to benefit from floods as resources, but they are also more vulnerable to floods as hazards, and this vulnerability has been increasing since colonial times, even if the actual incidence of floods has not. As a result, "[T]he 'hazardous' nature of the 1988 flood was differently constituted for men and women, for urban and rural dwellers and for poor and wealthy" (p. 226). Shaw delves most deeply into women's experience of floods. The pressure on women to observe *pardah* or seclusion and the cultural rules applying to female "pollution" make flooding especially burdensome for them. Indeed, flooding can permanently undermine the sustainability of their everyday lives.

Because the experience of floods is thus socially mediated, the same flood may be seen as both a good/normal/*borsha* flood and a bad/abnormal/*bonna* flood, depending upon the gender and class of the actor. Like Burger's and Scheper-Hughes' in Chapters 14 and 15, Shaw's research thereby problematizes the divide between normal and abnormal events. Consequently, it problematizes the divide between natural and cultural climatologies as well. As Shaw says, "This chapter shows that floods not only have varying consequences for rich and poor, men and women, and rural and urban dwellers, but that their very nature as hazards is *constituted* by these and other forms of human social differences" (p. 232). This finding has implications for the orthodox deforestation theory of disastrous floods and the solutions of afforestation and embankments. The most effective flood intervention might simply be to improve the lot of the most vulnerable groups by "attacking poverty and empowering women" (p. 233).

Part IV: Knowledge and its Circulation

How do social, cultural, economic, and political variables affect the way that extreme climatic events are interpreted? And how does that interpretation influence, in turn, the way that the event is experienced?

Emic Views of Climatic Perturbation/Disaster

The first pair of papers concern *emic* views of extreme climatic events. Ethno-climatological analyses were referenced in the earlier discussion of Chapters 5 and 6. These studies concern local, native views not just of climate but also of climate-related disasters. A pioneering study of the local experience of disaster was Erikson's (1976) analysis of the 1972 Buffalo Creek flood in West Virginia. Both of the studies here explicitly compare Western and non-Western experiences of disaster. Firth's (1959) study of typhoons in Tikopia (cf. Chapter 11, this volume) is the canonical effort to define disaster in a pre-industrialized non-Western society. A more recent effort is Dove's (2007) analysis of local interpretations of volcanic hazard in Central Java.

17 David M. Schneider. 1957. Typhoons on Yap

The subject of the first paper is again typhoons (or cyclones or hurricanes). Schneider examines four typhoons that struck the island of Yap, in the Western Carolines of Micronesia, between November 1947 and January 1948. Yap is a "high" island, with rolling hills well above sea level. At the time of Schneider's study, it also had relatively low population/resource pressure. Accordingly, Schneider's thesis is that "the measure of the disaster of a typhoon on Yap is not in its casualty rate, which is low or non-existent, but rather in terms of the disruption of social and emotional relationships which it entails" (p. 241). He sought to answer the question, "[W]hat does a typhoon 'mean' on Yap?" (p. 243). Schneider (1918–1995) was an American anthropologist, who spent most of his career at the University of Chicago.

During the 1947–1948 typhoons, 60 percent of all houses were destroyed, and 100 percent of the remainder were seriously damaged. But shelter was not the focus of Yap concern in their wake, it was food. As Schneider put it, "There is, however, one very notable reaction to the typhoons, and that is the highly vocal, clearly reiterated affirmation that there is no more food left and that starvation is their fate" (p. 243). So adamant were the Yap people on this point that the local civil administrator radioed Guam to send a shipment of emergency food supplies. In fact, the impact of the typhoons on food supplies did not threaten either immediate or future starvation.

The Yap see typhoons as the product of supernatural forces, which are invoked by powerful Yap leaders when they feel neglected by their followers. The occurrence of a typhoon reflects, therefore, a breakdown in socio-political relations. When these relations are intact and functioning properly, this is expressed in ubiquitous exchanges of food. But when socio-political relations break down, these exchanges cease, which is socially articulated as "no food." As Schneider writes, "The exchange of food is the symbol of good relations; the typhoon means that relations have deteriorated, and where social relations deteriorate, food is absent. Hence a typhoon means 'no food' in the sense that the omnipresent symbol of good relations is absent when relations are bad" (p. 246).

Schneider makes a rarely heard distinction between the meaning of acute and chronic disasters: "The unique catastrophe is responded to in terms of the socially structured motives of individuals. The chronic threat takes on common meanings for a wide population" (p. 247). That is, "[T]he chronic threat, the catastrophe that is long-awaited, takes on distinct meanings and provides a focus for long-standing anxieties, guilts, fears, and hostilities . . . " (p. 247). The fact that recurring typhoons are interpreted as, in effect, a need to repair relations between chiefs and commoners suggests that these relations are chronically problematic.

Schneider sees the Yap case as relevant to the U.S., but not with respect to weather-related disasters. As he writes about the U.S., "[T]here is every indication that the speed with which the rescue and clean-up operations are conducted is growing with each disaster" (p. 246). Rather, he thought that the Yap case held lessons for the threatening disaster of nuclear war and the likelihood that Americans would "project" other concerns onto it (p. 247). Anthropologists, like scholars in other disciplines, often seek to link their studies to the overweening concerns of the day. In the mid-twentieth century, that was the threat of global nuclear war; today it is global climate change.

From today's perspective, Schneider was overly confident in thinking that natural disasters would recede in importance in the U.S. because of the speed and efficiency of relief operations, a sanguine view belied by the case of Hurricane Katrina, for example. Nor have disasters in the U.S. shed the sort of social and political meaning that Schneider found in Yap. Many popular as well as academic explanations of Katrina's disproportionate impact on a poor, minority population attributed this to ill-will on the part of the government.

There is an ironic legacy of the nuclear discourse of the mid-twentieth century. One of the chief arguments that was developed against the theoretical use of nuclear weapons was the prediction that their use would trigger a global "nuclear winter," which would devastate all countries, without differentiation. Some nuclear physicists fiercely contested this prediction, which threatened the strategic value of their weapons research; and extant members of this group are among the foremost academic critics today of predictions of global warming, which represents a similar vision of global-scale anthropogenic change (Lahsen 2008).

18 Mark Carey. 2008. The Politics of Place: Inhabiting and Defending Glacier Hazard Zones in Peru's Cordillera Blanca

The focus of the second paper in this section is deglaciation, a subject of increasing academic interest, as the latest successor in the environmental discourse of "loss" (Orlove, Wiegandt, and Luckman 2008; Orlove 2009). One of the best-studied parts of the world in this regard is the Andes (Bolin 2009), where the proximity of densely populated areas to numerous glaciers makes changes in glacial resources and hazards a topic of keen interest, including to Carey (2005, 2008, 2010), an environmental historian at the University of Oregon.

According to Carey, rising global temperatures since the end of the Little Ice Age have led to an increasing incidence of glacial hazards, including glacial lake outburst floods and landslides caused by the deterioration of glacial ice. One of the most devastating examples of the latter occurred on May 31, 1970 in the Peruvian Andes, when an earthquake measuring 7.7 on the Richter scale destabilized a glacier, precipitating an avalanche that buried the highland town of Yungay. The avalanche killed as many as 15,000 people, making this one of the most deadly glacial disasters in history (cf. Oliver-Smith 1986).

Carey's study reprinted here deals with the post-disaster response to the May 1970 glacial avalanche, in particular the zoning of the affected area into inhabitable and uninhabitable zones based on scientific assessment of likely future threats, and government efforts to relocate the remaining population from the latter to the former. For the "revolutionary government" of Peruvian President General Juan Velasco, the near-erasure of Yungay offered a clean slate on which he could rebuild highland society according to his egalitarian ideals (p. 252). Disasters often open up a brief "window of opportunity," during which radical social change – for good or ill – that would otherwise be impossible can take place (cf. Chapters 12 and 15, this volume).

President Velasco's vision was not shared by the affected population, most of whom were adamant about remaining in the hazardous zones. For them, Carey points out, recovery from the May 1970 disaster meant retrieving, not moving away from, the locally embedded way of life that was disturbed by the avalanche. For them, there were risks associated not only with staying in the hazardous zone, but also in trying to establish a completely new life away from it – something that governments intent on relocation routinely overlook (cf. Dove 2007). As Carey writes, "The risks of further losses of social status, economic security, political power, and cultural beliefs were far more important and pressing than the risk of a glacier avalanche or an outburst flood" (p. 255).

There is nearly always a power dynamic at play during and following a natural disaster: in some cases the perturbation benefits local elites (Chapter 16, this volume), whereas in other cases it benefits extra-local elites (Chapters 12 and 15, this volume). In this case, the central government sought to take advantage of the Yungay disaster to impose an egalitarian vision on the affected area: it distributed aid equally to elites and non-elites, and it sought to implement a new agrarian reform program – all of which threatened the privileged status of the local Yungay elites, who took the lead in organizing resistance to the government relocation program.

Carey suggests that planners are prone to overlooking such dynamics because they overemphasize the political forces that oblige people to live in hazardous areas and underemphasize the historical agency that they exercise in living there.

Co-production of Knowledge in Climatic and Social Histories

This set of papers places climate knowledge in historic perspective. Local knowledge and perspective on climate and climatic hazards are not static and they do not lie outside history. They are produced at the intersection of environmental and social processes. A weakness in some studies of climate and society is that they focus on the impact of climate alone, implicitly holding society as a constant. In fact, society typically is changing at the same time that climate is changing.

19 *Julie Cruikshank. 2007. Melting Glaciers and Emerging Histories in the Saint Elias Mountains*

One of the most studied regions of the world, for emic views of climatic perturbation, is the sub-Arctic, especially northwestern America (Berkes and Jolly 2001; Henshaw 2003). The foremost scholar of native views of climate in this region is Julie Cruikshank (1981, 2001, 2005), emeritus professor of anthropology at the University of British Columbia. Her paper reprinted here summarizes her three decades of work on local knowledge and histories of glaciers.

Cruikshank defines "local knowledge" as the "tacit knowledge embodied in life experiences and reproduced in everyday behavior and speech" (p. 264). She rejects the idea that such knowledge is timeless and static, saying rather that it "is continuously made in situations of human encounter: between coastal and interior neighbors, between colonial visitors and residents, and among contemporary scientists, managers, environmentalists, and First Nations" (p. 264). Cruikshank cautions against the idea that such local knowledge is easily "transferable" to climate scientists (p. 270), as exemplified by the myths that she gathered from native informants about glaciers emitting heat and driving people to submerge themselves in glacial rivers to cool off (pp. 266–7). Cruikshank convincingly interprets these puzzling myths as metaphorical references to the smallpox epidemic that devastated the north Pacific coast during the years 1835–1840. As this example suggests, local knowledge of climate and environment differs from scientific knowledge in being more entangled with local history.

One of the attractions that local knowledge of glaciers and climate holds for Cruikshank is that it lays bare our "normalized understanding" of nature (p. 274). Native views of glaciers "encompass both the materiality of the biophysical world and the agency of the nonhuman, and draw on traditions of thought quite different from the conventional framing of nature as a redeemable object to be 'saved'" (p. 274). For example, the histories Cruikshank collected regarding people and glaciers warn against the hubristic notion that we *can* save nature, that we are in control.

Cruikshank's study of glaciers and people on the northwest Pacific coast began with what she calls an ethnographic "puzzle" (p. 263): the unexpected prominence of glaciers in the life histories that she began collecting in the late 1970s. This puzzle pointed toward what proved to be one of the most important theoretical findings of her work: changes in glaciers and climate co-occur with what are often equally momentous changes in human society. The climatic changes that were dramatically affecting glaciers in this region toward the end of the Little Ice Age in the late eighteenth century coincided with equally dramatic social changes stemming from the exploding global fur trade and European exploration. The changes in the glacial landscape provided the "imaginative grist" (p. 267) for comprehending and articulating the attendant social changes. This raises a question about the role that contemporary climate change may play in forging the identity and world view of climate change believers and deniers alike.

20 Todd Sanders. 2008. The Making and Unmaking of Rains and Reigns

There is a growing literature on the colonial politics of climate (Grove 1997; Endfield and Nash 2002). This is the subject of the chapter reprinted here from Sanders, an anthropologist at the University of Toronto, who has published extensively on issues of knowledge and epistemology in Tanzania, including several studies of rainmaking (2000, 2003). Other studies of native rainmaking in Africa include Schapera (1971) and Lan (1985) (cf. Chapter 13, this volume).[10]

The Ihanzu region of Iramba District in northern Tanzania is semi-arid, with just 20–30 inches of annual rainfall. Rainfall or its lack is thus a matter of central concern to those making a livelihood from agriculture and animal husbandry. For centuries, the ability to make rain has been the most important qualification of the royal chiefs of Ihanzu. Indeed, the royal line of Ihanzu was historically famed for possessing unique talents of rainmaking, such that the Ihanzu people played a role in what Sanders calls the "regional rainmaking economy," making this region into a "precolonial focal point" (p. 280). Thus, "[T]hroughout a tumultuous history, a period that witnessed sweeping political, economic, and social changes of every imaginable sort, Ihanzu rainmaking and rainmakers have remained remarkably important" (p. 278).

The powers of a rainmaking chief constituted a threat to colonial power. The first colonial power in the region, the Germans, responded with a strategy of routinely hanging rainmakers. As Sanders points out, this strategy was actually well informed, since rainmakers were "uniquely situated to unite people against colonial forces, both locally and across vast expanses" (p. 282). The British, who succeeded the Germans, tried to turn rainmakers from a "source of mischief" to "useful instruments" (p. 287), but they were no less suspicious of their subversive potential. The Lutheran missionaries who followed the British into the region were openly hostile toward the rainmakers. The post-independence Tanzanian state has been equally hostile as it sought to build a nation in which "modernity destroys tradition" (p. 294).

The Ihanzu villagers maintain that rainmaking is something that they still need. Sanders asks why, given that the rainmaking tradition "provides no obvious material benefits, no privileged access to scarce resources" (p. 294). He suggests that rainmaking may "establish meaningful historical connections with their past" (p. 294) (cf. Chapter 13, this volume); but equally important, he avers, is that "for the women and men of Ihanzu, rainmaking is first and foremost crucial because it brings rain." Without rainmakers and rainmaking, "There would be no rain. There would be no harvest, food, or beer. There would be no animals. There would be no villagers" (p. 295).

"Friction" in the Global Circulation of Climate Knowledge

The final pair of papers concern problems with the idea of the "global" in thinking about climate change. Earlier emphases on the unimpeded global flow of ideas and materials, as in Appadurai (1996), have given way to ideas of "friction" (Tsing 2005), of regions bypassed by globalization (Ferguson 2006), and a belief that global "assemblages" require exertion and agency (Ong and Collier 2005). This new direction challenges the idea of global environmental knowledge. Climate science has become a prominent focus of this critique, with papers like those reprinted here suggesting that the framing of climate change as a global object of scientific study is only one way of understanding it (cf. Adger et al. 2001).

21 *Myanna Lahsen. 2004. Transnational Locals: Brazilian Experiences of the Climate Regime*

The challenges posed to science and policy by climate change include the paradox of local-level adaptation to global-level change (Nelson, West, and Finan 2009), and governance issues involved in relations between local and global levels (Jasanoff and Martello 2004) and between the global North and South (Roberts and Parks 2007). The study by Lahsen reprinted here deals with the "consumption" of Northern climate science by Southern scholars and policymakers. Myanna Lahsen, an American anthropologist working at the Brazilian Institute for Space Research, is one of the leading scholars worldwide of this subject (2007, 2009, 2010).

An influential intellectual construct in recent global environmental governance has been the idea of the "epistemic community," one of the most recent iterations of the idea that scientific knowledge is context-free and universal in character. As originally developed by Haas (1993), this refers to supposed global networks of professionals with shared concerns, epistemologies, and policy agendas. Based on her study of Brazilian scientists and policymakers, Lahsen finds that Brazilian allegiance to the so-called epistemic community that has formed around climate science and policy is "circumscribed, internally fragmented and unstable," reflecting "the continued impact of history, geography and socio-economic realities" (p. 303).

Central to the Brazilian departure from the ideal of the epistemic community is their suspicion of Northern science and scientific institutions as being self-serving. Whereas the Brazilians want to focus on the North's historical contributions to global warming and its high current per capita emissions, they see the North as trying to shift attention to current levels of emissions and national totals, which puts countries like Brazil in a worse light. They say that the North likes to talk about cookstoves but not SUVs, whereas they, the Brazilians, would prefer to focus on "luxury versus survival" emissions (p. 309). Brazilian scientists and policymakers are not monolithic, however; notably, they disagree over the impact on climate change of Amazonian deforestation. Thus, there are fractures in the epistemic community at not only the international but also the national levels. In short, Lahsen says, "Cognitive differences among climate scientists and policy makers at the international level are rooted in national and political identities, cultural memories, and other aspects of personal experience" (p. 306).

The temptation to gloss over such differences and unresolved tensions in the interest of consolidating support for a politically compelling "vision of social reality and social change" (p. 304) can be considerable, Lahsen acknowledges, but it bears a cost. Universalizing discourses like the epistemic climate community simplify reality, in particular by erasing its political character. They thereby "avoid the need to analyze concrete inequalities, such as the distribution of power, costs, profits, and responsibility" (p. 301). The result may be the unintended consequence of "aggravating North–South relations around global environmental problems" (p. 304).

22 *Kenneth Broad and Ben Orlove. 2007. Channeling Globality: The 1997–98 El Niño Climate Event in Peru*

The final chapter has to do with the circulation of a global climate discourse concerning the 1997–1998 El Niño event, one of the most powerful and extensively studied of the century (Harwell 2000). The study reprinted here, which concerns Peru, is by Kenneth Broad and Ben Orlove. Orlove is an anthropologist in the Department of Earth and Environmental Sciences at Columbia University and one of the leading scholars of climate and society within the discipline. Broad is an anthropologist in the School of Marine and Atmospheric Sciences at the University of Miami, whose research and publishing focuses on the interdisciplinary and managerial aspects of climate, hydrology, and fisheries (Broad 2000; Pulwarty, Broad, and Finan 2004; Peterson and Broad 2008).

Broad and Orlove observe that the great scale of El Niño events, and the attendant need to coordinate responses over large areas, creates a "kind of affinity" (p. 325) for such events in national governments. Indeed, in advance of the 1997–1998 El Niño in Peru, then-President Fujimori pushed through the Peruvian Congress funding for an extensive program of mitigation efforts. Referring to the way that elements from beyond Peru (the El Niño event itself as well as the global science concerning it) were linked up with elements from within Peru (the actions of the national media and government), Broad and Orlove observe that "The scale and nature of these connections bring to mind the word *global*" (p. 316). They acknowledge that if they had tried to describe the event at the time, they "might have drawn on earlier models of globalization and emphasized the importance of flow" (p. 333). But with the benefit of a decade's distance from the event, they see it as an example, not of how traditional models of globalization work, but of how they do not work.

Central to Broad and Orlove's revisionist analysis of globality is their replacement of the concept of "flow" with their own concept of "channeling." Through their use of the term "channeling globality," they emphasize "the active nature of efforts that are made to establish connections at a global scale" (p. 333). In contrast to the metaphor of "flow," they use "channeling" to suggest "an intentional directing of distant or new elements along certain paths" (p. 317).

Broad and Orlove (p. 326) cite as an example of a "failure to channel globality" the October 1997 "Climate Outlook Forum," sponsored by Peru, four other South American nations, and the U.S. National Oceanic and Atmospheric Administration. The attending media questioned the purportedly apolitical character of the information presented at the forum by international climate scientists, and they greeted its final product, a hand-drawn map of the consensus El Niño forecasts, with derision. Broad and Orlove contrast this failure with the success of Peru's oldest and largest television network, America TV, in "satisfying the public thirst for high-tech images of the El Niño event" (p. 328). Broad and Orlove break new ground in identifying "public attention" – and the success or failure of the media and others to capture it – as a subject worthy of systematic study.

The capacity of states to deploy their resources to successfully capture and mold public attention on such subjects varies: whereas the Fujimori administration succeeded in the case of the 1997–1998 El Niño, the Bush administration failed with Hurricane Katrina in 2005. Broad and Orlove conclude, "It remains to be seen how public attention and state institutions will respond, in the midst of other pressing conjunctures, to the much slower, but much more powerful, planetary warming that is bearing down on us all" (p. 334).

NOTES

1 Glacken (1967: 121) suggests that this asymmetry has had long-lasting consequences: "If Plato in the Laws had noted that men change their environments through long settlement and that soil erosion and deforestation are parts of cultural history, he could have introduced at an early time these vital ideas into cultural history and changed the course of speculation regarding both man and environment."
2 Hippocrates (1923: 93–5) froze and thawed water to demonstrate that waters from snow and ice are different from, and worse for human health, than other types of water.
3 Hegel is reputed to have similarly said, "Where the Greeks once lived, the Turks now live; and that's an end on the matter."
4 Cf. Zimmermann (1980) on the relationship between the seasonal cycle and human health in traditional Indian culture.
5 A similar sign is reported in *Concerning Weather Signs*: "In the Crab are two stars called the Asses, and the nebulous space between them is called the Manger; if this appears dark, it is a sign of rain" (p. 86, this volume). One of the earliest in-depth anthropological studies in ethno-astronomy and ethno-meteorology, also involving the Pleiades, was Lévi-Strauss (1969: 216–39).
6 This is a companion piece to Janković (2006) on air, medicine, and the construction of the indoors/outdoors distinction in eighteenth-century Britain.
7 The continued thriving of the century-old "Fresh Air Fund" for the children of the urban poor in the U.S. (Muchnick 2010) suggests that the nineteenth-century concept of the therapeutic value of fresh air has not entirely disappeared (cf. p. 92, this volume). Elsewhere around the world, beliefs in the medical importance of wind are still robust. One of the most common labels for illness in the contemporary Indo-Malay world is *masuk angin*, literally "wind entered," its usage commonly applying to colds and flu.
8 The Western European beliefs studied by Janković (p. 89) were themselves affected by colonial-era confrontations with the alien peoples and climates of the tropics.
9 This distinction is not a racial one: "There are peoples of all races who have not yet become civilized, or have, but degraded; civilization has nothing to do with 'mental endowments'" (Ratzel, Division/Volume 1: 19).
10 Cf. Snyder-Reinke (2009) and Elvin (1998) on state rainmaking and responsibility for weather in late imperial China.

REFERENCES

Adams, Vincanne, Taslim Van Hattum, and Diana English. 2009. Chronic Disaster Syndrome: Displacement, Disaster Capitalism, and the Eviction of the Poor from New Orleans. *American Ethnologist* 36(4): 615–636.

Adger, W. Neil. 1999. Social Vulnerability to Climate Change and Extremes in Coastal Vietnam. *World Development* 27(2): 249–269.

Adger, W. Neil et al. 2001. Advancing a Political Ecology of Global Environmental Discourses. *Development and Change* 32: 681–715.

Agamben, Giorgio. 1998. *Homo Sacer: Sovereign Power and Bare Life*. Daniel Heller-Roazen, trans. Stanford: Stanford University Press.

Appadurai, Arjun. 1998. Comment on Francis Zimmermann, "The Jungle and the Aroma of Meats: An Ecological Theme in Hindu Medicine." *Social Science & Medicine* 27(3): 206–207.

Appadurai, Arjun. 1996. *Modernity at Large: Cultural Dimensions of Globalization.* Minneapolis, MN: University of Minnesota Press.

Ayers, Jessica and Tim Forsyth. 2009. Community-Based Adaptation to Climate Change: Strengthening Resilience through Development. *Environment* 51(4): 22–31.

Bawden, G. and R. Reycraft, eds. 2000. Environmental Disaster and the Archaeology of Human Response. University of New Mexico, *Anthropological Papers 7.*

Berkes, Fikret and Dyanna Jolly. 2001. Adapting to Climate Change: Social-Ecological Resilience in a Canadian Western Arctic Community. *Conservation Ecology* 5(2): 18 [online].

Bharara, L.P. 1982. Notes on the Experience of Drought: Perception, Recollection, and Prediction. In *Desertification and Development: Dryland Ecology in Social Perspective.* B. Spooner and H.S. Mann, eds. pp. 351–361. London: Academic Press.

Billman, Brian R. and Gary Huckleberry. 2008. Deciphering the Politics of Prehistoric El Niño Events on the North Coast of Peru. In *El Niño, Catastrophism, and Culture Change in Ancient America.* Daniel H. Sandweiss and Jeffrey Quilter, eds. pp. 101–128. Cambridge, MA: Harvard University Press for Dumbarton Oaks Research Library and Collection.

Bloch, Marc L.B. 1961. *Feudal Society.* L.A. Manyon, trans. Chicago: University of Chicago Press.

Bolin, Inge. 2009. The Glaciers of the Andes are Melting: Indigenous and Anthropological Knowledge Merge in Restoring Water Resources. In *Anthropology and Climate Change: From Encounters to Actions.* Susan A. Crate and Mark Nuttall, eds. pp. 228–239. Walnut Creek, CA: Left Coast Press.

Broad, K. 2000. El Niño and the Anthropological Opportunity. *Practicing Anthropology* 22(4): 20–23.

Brondizio, Eduardo S. and Emilio F. Moran. 2008. Human Dimensions of Climate Change: The Vulnerability of Small Farmers in the Amazon. *Philosophical Transactions of the Royal Society B: Biological Sciences* 363(1498): 1803–1809.

Bulliet, Richard W. 2009. *Cotton, Climate, and Camels in Early Islamic Iran: A Moment in World History.* New York: Columbia University Press.

Burger, Richard L. and Lucy C. Salazar, eds. 2004. *Machu Picchu: Unveiling the Mystery of the Incas.* New Haven, CT: Yale University Press.

Carey, Mark. 2005. Living and Dying With Glaciers: People's Historical Vulnerability to Avalanches and Outburst Floods in Peru. *Global and Planetary Change* 47(2–4): 122–134.

Carey, Mark. 2008. Disasters, Development, and Glacial Lake Control in Twentieth-Century Peru. In *Mountains: Sources of Water, Sources of Knowledge.* Ellen Wiegandt, ed. pp. 181–196. The Netherlands: Springer.

Carey, Mark. 2010. *In the Shadow of Melting Glaciers: Climate Change and Andean Society.* Oxford: Oxford University Press.

Carneiro, Robert L. 2008 [1960]. Slash-and-Burn Agriculture: A Closer Look at its Implications for Settlement Patterns. In *Environmental Anthropology: A Historical Reader.* M.R. Dove and C. Carpenter, eds. pp. 249–253. Malden, MA: Blackwell Publishing.

Casimir, Michael J., ed. 2008. *Culture and the Changing Environment: Uncertainty, Cognition and Risk Management in Cross-Cultural Perspective.* New York: Berghahn Books.

Cohler, Anne M. 1989. Editor's Introduction. *Montesquieu: The Spirit of the Laws.* Cambridge: Cambridge University Press.

Colson, Elizabeth. 1971. The Social Consequences of Resettlement: The Impact of the Kariba Resettlement upon the Gwembe Tonga. *Kariba Studies*, No. 4. Published on behalf of the Institute for African Studies, University of Zambia. Manchester: Manchester University Press.

Conklin, Harold C. 1957. *Hanunóo Agriculture: A Report on an Integral System of Shifting Cultivation in the Philippines.* Rome: Food and Agriculture Organization of the United Nations.

Crate, Susan A. 2011. Climate and Culture: Anthropology in the Era of Contemporary Climate Change. *Annual Review of Anthropology* 40: 175–194.

Crate, Susan A. and Mark Nuttall, eds. 2009. *Anthropology and Climate Change: From Encounters to Actions.* Walnut Creek, CA: Left Coast Press.

Cruikshank, Julie. 1981. Legends and Landscape: Convergence of Oral and Scientific Traditions in the Yukon Territory. *Arctic Anthropology* 18(2): 67–93.

Cruikshank, Julie. 2001. Glaciers and Climate Change: Perspectives from Oral Tradition. *Arctic* 54(4): 377–393.

Cruikshank, Julie. 2005. *Do Glaciers Listen? Local Knowledge, Colonial Encounters, and Social Imagination*. Vancouver: UBC Press.

Davis, Mike. 2001. *Late Victorian Holocausts: El Niño Famines and the Making of the Third World*. London: Verso.

Demeritt, David. 2001. The Construction of Global Warming and the Politics of Science. *Annals of the Association of American Geographers* 91(2): 307–337.

Diamond, Jared. 1997. *Guns, Germs, and Steel: The Fates of Human Societies*. New York: W.W. Norton.

Diamond, Jared. 2005. *Collapse: How Societies Choose to Fail or Succeed*. New York: Viking.

Diemberger, Hildegard, et al. 2012. Communicating Climate Knowledge. *Current Anthropology* 53(2): 226–244.

Doolittle, Amity A. 2010. The Politics of Indigeneity: Indigenous Strategies for Inclusion in Climate Change Negotiations. *Conservation and Society* 8(4): 286–291.

Dove, Michael R. 1992. The Dialectical History of "Jungle" in Pakistan. *Journal of Anthropological Research* 48(3): 231–253.

Dove, Michael R. 2007. Volcanic Eruption as Metaphor of Social Integration: A Political Ecological Study of Mount Merapi, Central Java. In *Environment, Development and Change in Rural Asia-Pacific: Between Local and Global*. J. Connell and E. Waddell, eds. pp. 16–37. London: Routledge.

Dove, Michael R. 2011. *The Banana Tree at the Gate: The History of Marginal Peoples and Global Markets in Borneo*. New Haven, CT: Yale University Press.

Dove, Michael R. and Muhammad H. Khan. 1995. Competing Constructions of Calamity: The Case of the May 1991 Bangladesh Cyclone. *Population and Environment* 16(5): 445–471.

Dumond, D.E. 1961. Swidden Agriculture and the Rise of Maya Civilization. *Southwestern Journal of Anthropology* 17: 301–316.

Eakin, Hallie. 2006. *Weathering Risk in Rural Mexico: Climatic, Institutional, and Economic Change*. Tucson, AZ: University of Arizona Press.

Elvin, M. 1998. Who was Responsible for the Weather? Moral Meteorology in Late Imperial China. *Osiris* 13: 213–237.

Endfield, Georgina H. and David J. Nash. 2002. Missionaries and Morals: Climatic Discourse in Nineteenth-Century Central Southern Africa. *Annals of the Association of American Geographers* 92(4): 727–742.

Erikson, Kai T. 1976. *Everything in its Path: Destruction of Community in the Buffalo Creek Flood*. New York: Simon and Schuster.

Evans-Pritchard, E.E. 1940. *The Nuer: A Description of the Modes of Livelihood and Political Institutions of a Nilotic People*. New York: Oxford University Press.

Ferguson, James. 1990. *The Anti-Politics Machine: "Development", Depoliticization and Bureaucratic Power in Lesotho*. Cambridge: Cambridge University Press.

Ferguson, James. 2006. *Global Shadows: Africa in the Neoliberal World Order*. Durham, NC: Duke University Press.

Finan, Timothy J. and Donald R. Nelson. 2001. Making Rain, Making Roads, Making Do: Public and Private Adaptations to Drought in Ceará, Northeast Brazil. *Climate Research* 19: 97–108.

Firth, Raymond. 1959. *Social Change in Tikopia: Re-Study of a Polynesian Community After a Generation*. London: Allen and Unwin.

Fleming, James R. and Vladimir Janković, eds. 2011. Klima. *Osiris* 26.

Fleming, James Rodger, Vladimir Janković, and Deborah R. Coen, eds. 2006. *Intimate Universality: Local and Global Themes in the History of Weather and Climate*. Sagamore Beach, MA: Science History Publications.

Freeman, Derek. 1955 [1970]. *Report on the Iban*. New York: Athlone Press.

Freeman, Derek. 1968. Thunder, Blood, and the Nicknaming of God's Creatures. *Psychoanalytic Quarterly* 37: 353–399.

Freidel, David and Justine Shaw. 2000. The Lowland Maya Civilization: Historical Consciousness and Environment. In *The Way the Wind Blows: Climate, History and Human Action*. Roderick J. McIntosh, Joseph A. Tainter, and Susan K. McIntosh, eds. pp. 271–300. New York: Columbia University Press.
Gates, Warren E, 1967. The Spread of Ibn Khaldun's Ideas on Climate and Culture. *Journal of the History of Ideas* 28(3): 415–422.
Geertz, Clifford. 1973. *The Interpretation of Cultures: Selected Essays*. New York: Basic Books.
Giroux, Henry A. 2006. *Stormy Weather: Katrina and the Politics of Disposability*. Boulder, CO: Paradigm.
Glacken, Clarence J. 1967. *Traces on the Rhodian Shore: Nature and Culture in Western Thought from Ancient Times to the End of the Eighteenth Century*. Berkeley, CA: University of California Press.
Gladwin, Thomas. 1947. Climate and Anthropology. *American Anthropologist* 49(4): 601–611.
Gladwin, Thomas. 1970. *East is a Big Bird: Navigation and Logic on Puluwat Atoll*. Cambridge, MA: Harvard University Press.
Golinski, Jan. 2007. *British Weather and the Climate of Enlightenment*. Chicago: University of Chicago Press.
Grove, Richard H. 1995. *Green Imperialism: Colonial Expansion, Tropical Island Edens, and the Origins of Environmentalism, 1600–1860*. Cambridge: Cambridge University Press.
Grove, Richard H. 1997. *Ecology, Climate and Empire: Colonialism and Global Environmental History, 1400–1940*. Cambridge: White Horse Press.
Grove, Richard H. and John Chappell, eds. 2000. *El Niño: History and Crisis: Studies from the Asia-Pacific Region*. Cambridge: White Horse.
Haas, Peter M. 1993. Epistemic Community and the Dynamics of International Environmental Cooperation. In *Regime Theory in International Relations*. Volker Rittberger, ed. pp. 168–201. Oxford: Clarendon Press.
Harwell, Emily E. 2000. Remote Sensibilities: Discourses of Technology and the Making of Indonesia's Natural Disaster. *Development and Change* 31: 307–340.
Henshaw, Anne. 2003. Climate and Culture in the North: The Interface of Archaeology, Paleoenvironmental Science and Oral History. In *Weather, Climate, Culture*. Sarah Strauss and Ben Orlove, eds. pp. 217–231. Oxford: Berg.
Hesiod. 1914. *Works and Days*. G. Evelyn-White, trans. Gloucester, UK: Dodo Press.
Hewitt, K. 1983. The Idea of Calamity in a Technocratic Age. In *Interpretations of Calamity, from the Viewpoint of Human Ecology*. K. Hewitt, ed. pp. 3–32. Boston, MA: Allen and Unwin.
Hulme, Mike. 2009. *Why We Disagree About Climate Change: Understanding Controversy, Inaction and Opportunity*. Cambridge: Cambridge University Press.
Janković, Vladimir. 2000. *Reading the Skies: A Cultural History of English Weather, 1650–1820*. Chicago: University of Chicago Press.
Janković, Vladimir. 2006. Intimate Climates, from Skins to Streets, Soirées to Societies. In *Intimate Universality: Local and Global Themes in the History of Weather and Climate*. James Rodger Fleming, Vladimir Janković, and Deborah R. Coen, eds. pp. 1–33. Sagamore Beach, MA: Science History Publications.
Jasanoff, Sheila and Marybeth Long Martello, eds. 2004. *Earthly Politics: Local and Global in Environmental Governance*. Cambridge, MA: MIT Press.
Jones, W.H.S. 1923. General Introduction. *Hippocrates, Volume I*. Cambridge, MA: Harvard University Press.
Kriesel, Karl M. 1968. Montesquieu: Possibilistic Political Geographer. *Annals of the Association of American Geographers* 58(3): 557–574.
Kroeber, A.L. 1947. *Cultural and Natural Areas of Native North America*. Berkeley, CA: University of California Press.
Lahsen, Myanna. 2007. Trust Through Participation? Problems of Knowledge in Climate Decision Making. In *The Social Construction of Climate Change: Power, Knowledge, Norms, Discourses*. M. E. Pettenger, ed. pp. 173–196. London: Ashgate.
Lahsen, Myanna. 2008. Experiences of Modernity in the Greenhouse: A Cultural Analysis of a Physicist "Trio" Supporting the Backlash Against Global Warming. *Global Environmental Change* 18: 204–219.

Lahsen, Myanna. 2009. A Science–Policy Interface in the Global South: The Politics of Carbon Sinks and Science in Brazil. *Climatic Change* 97(3): 339–372.

Lahsen, Myanna. 2010. The Social Status of Climate Change Knowledge: An Editorial Essay. *Wiley Interdisciplinary Reviews: Climate Change* 1(2): 162–171.

Lan, David. 1985. *Guns and Rain: Guerrillas and Spirit Mediums in Zimbabwe*. London: J. Currey.

Launay, Robert. 2010. *Foundations of Anthropological Theory: From Classical Antiquity to Early Modern Europe*. Malden, MA: Wiley-Blackwell.

Lawrence, Bruce B. 2005. Introduction. *Ibn Khaldûn, The Muqaddimah: An Introduction to History*. Franz Rosenthal, trans. and introd., N.J. Dawood, abr. and ed. Princeton, NJ: Princeton University Press.

Lévi-Strauss, Claude. 1969. *The Raw and the Cooked*. John and Doreen Weightman, trans. New York: Harper & Row.

Levin, Lawrence Meyer. 1936. *The Political Doctrine of Montesquieu's Esprit des Lois: Its Classical Background*. New York: Institute of French Studies, Columbia University.

Lipset, David. 2011. The Tides: Masculinity and Climate Change in Coastal Papua New Guinea. *Journal of the Royal Anthropological Institute* 17(1): 20–43.

Little, Peter D. and Michael M. Horowitz, eds. 1987. *Lands at Risk in the Third World: Local-Level Perspectives*. Boulder, CO: Westview Press.

Low, Chris and Elisabeth Hsu. 2007. Introduction to Special Issue. *Journal of the Royal Anthropological Institute* (N.S.): S1–S17.

Magistro, John and Carla Roncoli. 2001. Anthropological Perspective and Policy Implications of Climate Change Research. *Climate Research* 19: 91–96.

Marshall, Mac. 1979. Natural and Unnatural Disaster in the Mortlock Islands of Micronesia. *Human Organization* 38(3): 265–272.

Masquelier, Adeline. 2006. Why Katrina's Victims Aren't *Refugees*: Musings on a "Dirty" Word. *American Anthropologist* 108(4): 735–743.

Mathews, Andrew. 2005. Power/Knowledge, Power/Ignorance: Forest Fires and the State in Mexico. *Human Ecology* 33(6): 795–820.

Mauss, Marcel, with Henri Beuchat. 1979 [1950]. *Seasonal Variations of the Eskimo: A Study in Social Morphology*. James J. Fox, trans. London: Routledge & Kegan Paul.

McAnany, Patricia A. and Norman Yoffee. 2010. *Questioning Collapse: Human Resilience, Ecological Vulnerability, and the Aftermath of Empire*. New York: Cambridge University Press.

McGovern, Thomas H. 1981. The Economics of Extinction in Norse Greenland. In *Climate and History: Studies in Past Climates and Their Impact on Man*. T.M.L. Wigley, M.J. Ingram, and G. Farmer, eds. pp. 404–433. Cambridge: Cambridge University Press.

McGovern, Thomas H. 1991. Climate, Correlation, and Causation in Norse Greenland. *Arctic Anthropology* 28(2): 77–100.

McGovern, Thomas H., Gerald Bigelow, Thomas Amorosi, and Daniel Russell. 1988. Northern Islands, Human Error, and Environmental Degradation: A View of Social and Ecological Change in the Medieval North Atlantic. *Human Ecology* 16(3): 225–270.

McIntosh, Roderick. 2000. Social Memory in Mande. In *The Way the Wind Blows: Climate, History and Human Action*. Roderick J. McIntosh et al., eds. pp. 141–180. New York: Columbia University Press.

Mead, Margaret. 1977. Preface. In *The Atmosphere: Endangered and Endangering*. William W. Kellogg and Margaret Mead, eds. pp. xix–xxiv. Bethesda, MD: Department of Health, Education, and Welfare, Public Health Service, National Institutes of Health.

Meggers, Betty J. 1971. *Amazonia: Man and Culture in a Counterfeit Paradise*. Chicago: Aldine, Atherton.

Mentz, S. 2010. Strange Weather in *King Lear*. *Shakespeare* 6(2): 139–152.

Mills, C.A. 1942. Climatic Effects on Growth and Development with Particular Reference to the Effects of Tropical Residence. *American Anthropologist* 44(1): 1–13.

Moore, Frances C. 2010. "Doing Adaptation": The Construction of Adaptive Capacity and its Function in the International Climate Negotiations. *St. Antony's International Review* 5(2): 66–88.

Muchnick, Barry. 2011. Nature's Republic: Fresh Air Reform and the Moral Ecology of Citizenship in Turn of the Century America. Ph.D. dissertation, Yale University History Department and School of Forestry and Environmental Studies.

Murra, J.V. 1985. The Limits and Limitations of the "Vertical Archipelago" in the Andes. *Andean Ecology and Civilization: An Interdisciplinary Perspective on Andean Ecological Complementarity*. Shozo Masuda, Izumi Shimada, and Craig Morris, eds. pp. 15–20. Tokyo: University of Tokyo Press.

Needham, Rodney. 1964. Blood, Thunder, and the Mockery of Animals. *Sociologus* 14(2): 136–149.

Nelson, Donald R. and Timothy J. Finan. 2000. The Emergence of a Climate Anthropology in Northeast Brazil. *Practicing Anthropology* 22(4): 6–10.

Nelson, Donald R., Colin Thor West, and Timothy J. Finan. 2009. Introduction to: *In Focus: Global Change and Adaptation in Local Places. American Anthropologist* 111(3): 271–274.

Neumann, Franz. 1949. Introduction. *The Spirit of the Laws*, by Baron de Montesquieu. Thomas Nugent, trans. New York: Harper.

Nugent, Thomas. 1752. (Foreword from Translator). *The Spirit of Laws*, by Charles de Montesquieu. London: J. Nourse and P. Vaillant.

Nunn, Patrick D., et al. 2007. Times of Plenty, Times of Less: Last-Millennium Societal Disruption in the Pacific Basin. *Human Ecology* 35: 385–401.

Oliver-Smith, Anthony. 1986. *The Martyred City: Death and Rebirth in the Andes*. Albuquerque: University of New Mexico Press.

Oliver-Smith, Anthony. 1996. Anthropological Research on Hazards and Disasters. *Annual Review of Anthropology* 25: 303–328.

Ong, Aihwa and Stephen J. Collier, eds. 2005. *Global Assemblages: Technology, Politics, and Ethics as Anthropological Problems*. Malden, MA: Blackwell.

O'Reilly, Jessica. 2012. The Rapid Disintegration of Projections: The West Antarctic Ice Sheet and the Intergovernmental Panel on Climate Change. *Social Studies of Science* 42(5): 709–731.

Orlove, Ben S. 1979. Two Rituals and Three Hypotheses: An Examination of Solstice Divination in Southern Highland Peru. *Anthropological Quarterly* 52(2): 86–98.

Orlove, Ben S. 2009. Glacier Retreat: Reviewing the Limits of Human Adaptation to Climate Change. *Environment* 51(3): 22–34.

Orlove, Ben S., John C.H. Chiang, and Mark A. Crane. 2002. Ethnoclimatology in the Andes. *American Scientist* 90: 428–435.

Orlove, Ben, Ellen Wiegandt, and Brian H. Luckman. 2008. *Darkening Peaks: Glacier Retreat, Science, and Society*. Berkeley, CA: University of California Press.

Paredes, J. Anthony. 2006. Introduction to: The Impact of the Hurricanes of 2005 on New Orleans and the Gulf Coast of the United States. *American Anthropologist* 108(4): 637–642.

Peterson, N. and K. Broad. 2008. Climate and Weather Discourse in Anthropology: From Determinism to Uncertain Futures. In *Anthropology and Climate Change: From Encounters to Actions*. S. Crate and M. Nuttall, eds. pp. 70–86. Walnut Creek, CA: Left Coast Press.

Pulwarty, R., K. Broad, and T. Finan. 2004. Science, Vulnerability and the Search for Equity: El Niño Events, Forecasts and Decision-Making in Peru and Brazil. In *Vulnerability: Disasters, Development and People*. G. Bankoff, G. Frerks, and D. Hilhorst, eds. pp. 83–98. London: Earthscan.

Ratzel, Friedrich. 1882–1891. *Anthropogeographie*. Stuttgart: J. Engelhorn.

Ratzel, Friedrich. 1885–1890. *Völkerkunde*. Leipzig: Bibliographisches Institut.

Rayner, Steve. 2003. Domesticating Nature: Commentary on the Anthropological Study of Weather and Climate Discourse. In *Weather, Climate, Culture*. Sarah Strauss and Ben Orlove, eds. pp. 277–290. Oxford: Berg.

Ribot, Jesse C., Antonio Rocha Magalhães, and Stahis S. Panagides, eds. 1995. *Climate Variability, Climate Change, and Social Vulnerability in the Semi-Arid Tropics*. Cambridge: Cambridge University Press.

Richards, Audrey. 1948. *Hunger and Work in a Savage Tribe: A Functional Study of Nutrition among the Southern Bantu*. Glencoe, IL: Free Press.

Richardson, James B. and Daniel H. Sandweiss. 2008. Climate Change, El Niño, and the Rise of Complex Society on the Peruvian Coast during the Middle Holocene. In *El Niño, Catastrophism and Culture Change in Ancient America*. Daniel H. Sandweiss and Jeffrey Quilter, eds. pp. 59–75. Cambridge, MA: Harvard University Press, for Dumbarton Oaks Research Library and Collection.

Roberts, J. Timmons and Bradley C. Parks. 2007. *A Climate of Injustice: Global Inequality, North-South Politics, and Climate Policy*. Cambridge, MA: MIT Press.
Roncoli, Carla. 2006. Ethnographic and Participatory Approaches to Research on Farmers' Responses to Climate Predictions. *Climate Research* 33: 81–99.
Roncoli, Carla, Todd Crane, and Ben Orlove. 2009. Fielding Climate Change in Cultural Anthropology. In *Anthropology and Climate Change: From Encounters to Actions*. S. Crate and M. Nuttall, eds. pp. 87–115. Walnut Creek, CA: Left Coast Press.
Roscoe, Paul B. 2008. Catastrophe and the Emergence of Political Complexity: A Social Anthropological Model. In *El Niño, Catastrophism, and Culture Change in Ancient America*. Daniel H. Sandweiss and Jeffrey Quilter, eds. pp. 77–100. Cambridge, MA: Harvard University Press for Dumbarton Oaks Research Library and Collection.
Rosenthal, Franz. 1958. Translator's Introduction. In Ibn Khaldûn. *The Muqaddimah: An Introduction to History*. 3 vols. New York: Bollingen Foundation.
Safire, William. 2005. On Language: Katrina Words. *New York Times Magazine*, September 18.
Sahlins, Marshall. 1972. *Stone Age Economics*. Chicago: Aldine.
Sanders, Todd. 2000. Rains Gone Bad, Women Gone Mad: Rethinking Gender, Rituals of Rebellion and Patriarchy. *Journal of the Royal Anthropological Institute* (N.S.) 6: 469–486.
Sanders, Todd. 2003. (En)Gendering the Weather: Rainmaking and Reproduction in Tanzania. In *Weather, Climate, Culture*. Sarah Strauss and Ben Orlove, eds. pp. 83–102. Oxford: Berg.
Sandweiss, Daniel H. and Jeffrey Quilter, eds. 2008. *El Niño, Catastrophism, and Culture Change in Ancient America*. Cambridge, MA: Harvard University Press for Dumbarton Oaks Research Library and Collection.
Schapera, I. 1971. *Rainmaking Rites of Tswana Tribes*. Cambridge: African Studies Centre.
Scheper-Hughes, Nancy. 1979. *Saints, Scholars and Schizophrenics: Mental Illness in Rural Ireland*. Berkeley: University of California Press.
Scheper-Hughes, Nancy. 1993. *Death without Weeping: The Violence of Everyday Life in Brazil*. 2nd ed. Berkeley, CA: University of California Press.
Schwartzman, S. and P. Moutinho. 2008. Compensated Reductions: Rewarding Developing Countries for Protecting Forest Carbon. In *Climate Change and Forests: Emerging Policy and Market Opportunities*. C. Streck et al., eds. pp. 227–236. Washington, DC: Brookings Institution Press.
Scott, James C. 2009. *The Art of Not Being Governed: An Anarchist History of Upland Southeast Asia*. New Haven, CT: Yale University Press.
Sen, Amartya. 1982. *Poverty and Famines: An Essay on Entitlement and Deprivation*. Oxford: Clarendon Press.
Sider, David and Carl Wolfram Brunschön, eds. 2007. *Theophrastus of Ereseus: On Weather Signs*. Leiden: Brill.
Sillitoe, Paul. 1993. A Ritual Response to Climatic Perturbations in the Highlands of Papua New Guinea. *Ethnology* 32(2): 169–185.
Sillitoe, Paul. 1994. Whether Rain or Shine: Weather Regimes from a New Guinea Perspective. *Oceania* 64: 246–270.
Smith, Nicholas and Anthony Leiserowitz. 2012. The Rise of Global Warming Skepticism: Exploring Affective Image Associations in the United States Over Time. *Risk Analysis* 32(6): 1021–1032.
Snyder-Reinke, Jeffrey. 2009. *Dry Spells: State Rainmaking and Local Governance in Late Imperial China*. Cambridge, MA: Harvard University Asia Center.
Solway, Jacqueline S. 1998. Taking Stock in the Kalahari: Accumulation and Resistance on the Southern African Periphery. *Journal of Southern African Studies* 24(2): 425–441.
Solway, Jacqueline S. 2003. In the Eye of the Storm: The State and Non-Violence in Southern Africa (Botswana). *Anthropological Quarterly* 76(3): 485–495.
Solway, Jacqueline S. 2009. Human Rights and NGO "Wrongs": Conflict Diamonds, Culture Wars and the "Bushman Question." *Africa* 79(3): 321–346.
Spillius, James. 1957. Natural Disaster and Political Crisis in a Polynesian Society: An Exploration of Operational Research II. Some Principles of Operational Research. *Human Relations* X: 113–125.

Spooner, Brian and H.S. Mann, eds. 1982. *Desertification and Development: Dryland Ecology in Social Perspective*. London: Academic Press.

Steward, Julian H. 1955. *Theory of Culture Change: The Methodology of Multilinear Evolution*. Urbana, IL: University of Illinois Press.

Strauss, Sarah and Ben Orlove, eds. 2003. *Weather, Climate, Culture*. Oxford: Berg.

Toynbee, Arnold J. 1935. *A Study of History*. New York: Oxford University Press.

Tsing, Anna L. 2005. *Friction: An Ethnography of Global Connection*. Princeton, NJ: Princeton University Press.

Vayda, Andrew P. and Bonnie J. McCay. 1975. New Directions in Ecology and Ecological Anthropology. *Annual Review of Anthropology* 4: 293–306.

Vogt, Evon Z. 1952. Water Witching: An Interpretation of a Ritual Pattern in a Rural American Community. *Scientific Monthly* 75(3): 175–186.

Waddell, Eric. 2008 [1975]. How the Enga Cope with Frost: Responses to Climatic Perturbations in the Central Highlands. In *Environmental Anthropology: A Historical Reader*. M.R. Dove and C. Carpenter, eds. pp. 223–237. Malden, MA: Blackwell Publishing.

Wallace, Anthony F.C. 1956. *Human Behavior in Extreme Situations: A Survey of the Literature and Suggestions for Further Research*. Washington, DC: National Academy of Sciences/National Research Council.

Wallace, Anthony F.C. 1957. Mazeway Disintegration: The Individual's Perception of Socio-Cultural Disorganization. *Human Organization* 16: 23–27.

Weiss, H. et al. 1993. The Genesis and Collapse of Third Millennium North Mesopotamian Civilization. *Science* 261: 995–1004.

Weiss, Harvey. 1996. Desert Storm. *The Sciences*, May/June: 30–36.

Weiss, Harvey. 2000. Beyond the Younger Dryas: Collapse as Adaptation to Abrupt Climate Change in Ancient West Asia and the Eastern Mediterranean. In *Environmental Disaster and the Archaeology of Human Response*. G. Bawden and R. Reycraft, eds. pp. 75–95. University of New Mexico, Anthropological Papers 7.

West, Colin Thor. 2009. Domestic Transitions, Desiccation, Agricultural Intensification, and Livelihood Diversification among Rural Households on the Central Plateau, Burkina Faso. *American Anthropologist* 111(3): 276–288.

White, Sam. 2011. *The Climate of Rebellion in the Early Ottoman Empire*. Cambridge: Cambridge University Press.

Whiting, John W.M. 1964. Effects of Climate on Certain Cultural Practices. In *Explorations in Cultural Anthropology*. W.H. Goodenough, ed. pp. 511–544. New York: McGraw-Hill.

Wisner, B. 1993. Disaster Vulnerability: Scale, Power and Daily Life. *GeoJournal* 30(2): 127–140.

Zaman, M.Q. 1991. Social Structure and Process in Car Land Settlement in the Brahmaputra-Jamuna Floodplain. *Man* 26(4): 673–690.

Zaman, M.Q. 1993. Rivers of Life: Living with Floods in Bangladesh. *Asian Survey* 33(10): 985–996.

Zimmermann, Francis. 1980. Ṛtu-s-ātmya: The Seasonal Cycle and the Principle of Appropriateness. *Social Science & Medicine. Part B: Medical Anthropology* 14(2): 99–106.

Zimmermann, Francis. 1987a. *The Jungle and the Aroma of Meats: An Ecological Theme in Hindu Medicine*. Berkeley, CA: University of California Press.

Zimmermann, Francis. 1987b. Monsoon in Traditional Culture. In *Monsoons*. Jay S. Fein and Pamela L. Stephens, eds. pp. 51–76. New York: John Wiley & Sons, Inc.

Part I
Continuities

Climate Theory

1

Airs, Waters, Places

Hippocrates

I. WHOEVER wishes to pursue properly the science of medicine must proceed thus. First he ought to consider what effects each season of the year can produce; for the seasons are not at all alike, but differ widely both in themselves and at their changes. The next point is the hot winds and the cold, especially those that are universal, but also those that are peculiar to each particular region. He must also consider the properties of the waters; for as these differ in taste and in weight, so the property of each is far different from that of any other. Therefore, on arrival at a town with which he is unfamiliar, a physician should examine its position with respect to the winds and to the risings of the sun. For a northern, a southern, an eastern, and a western aspect has each its own individual property. He must consider with the greatest care both these things and how the natives are off for water, whether they use marshy, soft waters, or such as are hard and come from rocky heights, or brackish and harsh. The soil too, whether bare and dry or wooded and watered, hollow and hot or high and cold. The mode of life also of the inhabitants that is pleasing to them, whether they are heavy drinkers, taking lunch,[1] and inactive, or athletic, industrious, eating much and drinking little.

[...]

[Books II–XI are not reprinted here]

XII. So much for the changes of the seasons. Now I intend to compare Asia[2] and Europe, and to show how they differ in every respect, and how the nations of the one differ entirely in physique from those of the other. It would take too long to describe them all, so I will set forth my views about the most important and the greatest differences. I hold that Asia differs very widely from Europe in the nature of all its inhabitants and of all its vegetation. For everything in Asia grows to far greater beauty and size; the one region is less wild than the other, the character of the inhabitants is milder and more gentle. The cause of this is the temperate climate, because it lies towards the east midway between the risings[3] of the sun, and farther away than is Europe from the cold. Growth and freedom from wildness are most fostered when nothing is forcibly predominant, but equality in every respect prevails. Asia, however, is not everywhere uniform; the region, however, situated midway between the heat and the cold is very fruitful, very wooded and very mild; it has splendid water, whether from rain or from springs. While it is not burnt up with the heat nor dried up by drought and want of water, it is not oppressed with cold, nor yet damp and wet with excessive rains and snow. Here the harvests are likely to be plentiful, both those from seed and those which the

The Anthropology of Climate Change: An Historical Reader, First Edition. Edited by Michael R. Dove.

earth bestows of her own accord, the fruit of which men use, turning wild to cultivated and transplanting them to a suitable soil. The cattle too reared there are likely to flourish, and especially to bring forth the sturdiest young and rear them to be very fine creatures.[4] The men will be well nourished, of very fine physique and very tall, differing from one another but little either in physique or stature. This region, both in character and in the mildness of its seasons, might fairly be said to bear a close resemblance to spring. Courage, endurance, industry and high spirit could not arise in such conditions either among the natives or among immigrants,[5] but pleasure must be supreme . . .[6] wherefore in the beasts they are of many shapes.

XIII. Such in my opinion is the condition of the Egyptians and Libyans. As to the dwellers on the right of the summer risings of the sun up to Lake Maeotis, which is the boundary between Europe and Asia, their condition is as follows. These nations are less homogeneous than those I have described, because of the changes of the seasons and the character of the region. The land is affected by them exactly as human beings in general are affected. For where the seasons experience the most violent and the most frequent changes,[7] the land too is very wild and very uneven; you will find there many wooded mountains, plains and meadows. But where the seasons do not alter much, the land is very even. So it is too with the inhabitants, if you will examine the matter. Some physiques resemble wooded, well-watered mountains, others light, dry land, others marshy meadows, others a plain of bare, parched earth. For the seasons which modify a physical frame differ; if the differences be great, the more too are the differences in the shapes.

XIV. The races that differ but little from one another I will omit, and describe the condition only of those which differ greatly, whether it be through nature or through custom. I will begin with the Longheads.[8] There is no other race at all with heads like theirs. Originally custom was chiefly responsible for the length of the head, but now custom is reinforced by nature. Those that have the longest heads they consider the noblest, and their custom is as follows. As soon as a child is born they remodel its head with their hands, while it is still soft and the body tender, and force it to increase in length by applying bandages and suitable appliances, which spoil the roundness of the head and increase its length. Custom originally so acted that through force such a nature came into being; but as time went on the process became natural, so that custom no longer exercised compulsion. For the seed comes from all parts of the body, healthy seed from healthy parts, diseased seed from diseased parts. If, therefore, bald parents have for the most part bald children, grey-eyed parents grey-eyed children, squinting parents squinting children, and so on with other physical peculiarities, what prevents a long-headed parent having a long-headed child?[9] At the present time long-headedness is less common than it was, for owing to intercourse with other men the custom is less prevalent.

XV. These are my opinions about the Longheads. Now let me turn to the dwellers on the Phasis. Their land is marshy, hot, wet, and wooded; copious violent rains fall there during every season. The inhabitants live in the marshes, and their dwellings are of wood and reeds, built in the water. They make little use of walking in the city and the harbour, but sail up and down in dug-outs made from a single log, for canals are numerous. The waters which they drink are hot and stagnant, putrefied by the sun and swollen by the rains. The Phasis itself is the most stagnant and most sluggish of all rivers. The fruits that grow in this country are all stunted, flabby and imperfect, owing to the excess of water, and for this reason they do not ripen. Much fog from the waters envelops the land. For these causes, therefore, the physique of the Phasians is different from that of other folk. They are tall in stature, and of a gross habit of body, while neither joint nor vein is visible. Their complexion is yellowish, as though they suffered from jaundice. Of all men they have the deepest voice, because the air they breathe is not clear, but moist and turbid. They are by nature disinclined for physical fatigue. There are but slight changes of the seasons, either in respect of heat or of cold. The winds are mostly moist, except one breeze peculiar to the country, called *cenchron*, which sometimes blows strong, violent and hot. The north wind rarely blows, and when it does it is weak and gentle.

XVI. So much for the difference, in nature and in shape, between the inhabitants of Asia and the inhabitants of Europe. With regard to the lack of spirit and of courage among the inhabitants, the chief reason why Asiatics are less warlike and more gentle in character than Europeans is the uniformity of the seasons, which show no violent changes either towards heat or towards cold, but are equable. For there occur no mental shocks nor violent physical change, which are more likely to steel the temper and impart to it a fierce passion than is a monotonous sameness. For it is changes of all things that rouse the temper of man and prevent its stagnation. For these reasons, I think, Asiatics are feeble. Their institutions are a contributory cause, the greater part of Asia being governed by kings. Now where men are not their own, masters and independent, but are ruled by despots, they are not keen on military efficiency but on not appearing warlike. For the risks they run are not similar. Subjects are likely to be forced to undergo military service, fatigue and death, in order to benefit their masters, and to be parted from their wives, their children and their friends. All their worthy, brave deeds merely serve to aggrandize and raise up their lords, while the harvest they themselves reap is danger and death. Moreover, the land of men like these must be desert, owing to their enemies and to their laziness,[10] so that even if a naturally brave and spirited man is born his temper is changed by their institutions. Whereof I can give a clear proof. All the inhabitants of Asia, whether Greek or non-Greek, who are not ruled by despots, but are independent, toiling for their own advantage, are the most warlike of all men. For it is for their own sakes that they run their risks, and in their own persons do they receive the prizes of their valour as likewise the penalty of their cowardice. You will find that Asiatics also differ from one another, some being superior, others inferior. The reason for this, as I have said above, is the changes of the seasons.

XVII. Such is the condition of the inhabitants of Asia. And in Europe is a Scythian race, dwelling round Lake Maeotis, which differs from the other races. Their name is Sauromatae. Their women, so long as they are virgins, ride, shoot, throw the javelin while mounted, and fight with their enemies. They do not lay aside their virginity until they have killed three of their enemies, and they do not marry before they have performed the traditional sacred rites. A woman who takes to herself a husband no longer rides, unless she is compelled to do so by a general expedition. They have no right breast; for while they are yet babies their mothers make red-hot a bronze instrument constructed for this very purpose and apply it to the right breast and cauterise it, so that its growth is arrested, and all its strength and bulk are diverted to the right shoulder and right arm.

XVIII. As to the physique of the other Scythians, in that they are like one another and not at all like others, the same remark applies to them as to the Egyptians, only the latter are distressed by the heat, the former by the cold.[11] What is called the Scythian desert is level grassland, without trees,[12] and fairly well-watered. For there are large rivers which drain the water from the plains. There too live the Scythians who are called Nomads because they have no houses but live in wagons. The smallest have four wheels, others six wheels. They are covered over with felt and are constructed, like houses, sometimes in two compartments and sometimes in three, which are proof against rain, snow and wind. The wagons are drawn by two or by three yoke of hornless oxen. They have no horns because of the cold. Now in these wagons live the women, while the men ride alone on horseback, followed by the sheep they have, their cattle and their horses. They remain in the same place just as long as there is sufficient fodder for their animals; when it gives out they migrate. They themselves eat boiled meats and drink mares' milk. They have a sweetmeat called *hippace*, which is a cheese from the milk of mares (*hippoi*).

XIX. So much for their mode of living and their customs. As to their seasons and their physique, the Scythians are very different from all other men, and, like the Egyptians, are homogeneous; they are the reverse of prolific, and Scythia breeds the smallest and the fewest wild animals. For it lies right close to the north and the Rhipaean mountains, from which blows the north wind. The sun comes nearest to them only at the end of its course, when it reaches the summer solstice, and then it warms them but slightly and for a short time. The winds blowing from hot regions do not reach them, save rarely, and with little force; but from the north there are constantly blowing winds that are chilled by snow, ice, and many waters[13] which, never leaving the mountains, render them uninhabitable. A thick fog envelops by day the plains upon which they live, so that winter is perennial, while summer, which is but feeble, lasts only a few days. For the plains are high and bare, and are not encircled with mountains, though they slope from the north. The wild animals too that are found there are not large, but such as can find shelter under ground. They are stunted owing to the severe climate and the bareness of the land, where there is neither warmth nor shelter. And the changes of the seasons are neither great nor violent, the seasons being uniform and altering but little. Wherefore the men also are like one another in physique, since summer and winter they always use similar food and the same clothing, breathing a moist, thick atmosphere, drinking water from ice and snow, and abstaining from fatigue. For neither bodily nor mental endurance is possible where the changes are not violent. For these causes their physiques are gross, fleshy, showing no joints, moist and flabby, and the lower bowels are as moist as bowels can be. For the belly cannot possibly dry up in a land like this, with such a nature and such a climate, but because of their fat and the smoothness of their flesh their physiques are similar, men's to men's and women's to women's. For as the seasons are alike there takes place no corruption or deterioration in the coagulation of the seed,[14] except through the blow of some violent cause or of some disease.

XX. I will give clear testimony to their moistness. The majority of the Scythians, all that are Nomads, you will find have their shoulders cauterized, as well as their arms, wrists, breast, hips and loins, simply because of the moistness and softness of their constitution. For owing to their moistness and flabbiness they have not the strength either to draw a bow or to throw a javelin from the shoulder. But when they have been cauterized the excess of moisture dries up from their joints, and their bodies become more braced, more nourished and better articulated. Their bodies grow relaxed and squat, firstly because, unlike the Egyptians, they do not use swaddling clothes, of which they have not the habit, for the sake of their riding, that they may sit a horse well; secondly, through their sedentary lives. For the boys, until they can ride, sit the greater part of the time in the wagon, and because of the migrations and wanderings rarely walk on foot; while the girls are wonderfully flabby and torpid in physique. The Scythians are a ruddy race because of the cold, not through any fierceness in the sun's heat. It is the cold that burns their white skin and turns it ruddy.

XXI. A constitution of this kind prevents fertility. The men have no great desire for intercourse because of the moistness of their constitution and the softness and chill of their abdomen, which are the greatest checks on venery. Moreover, the constant jolting on their horses unfits them for intercourse. Such are the causes of barrenness in the men; in the women they are the fatness and moistness of their flesh, which are such that the womb cannot absorb the seed. For neither is their monthly purging as it should be, but scanty and late, while the mouth of the womb is closed by fat and does not admit the seed. They are personally fat and lazy, and their abdomen is cold and soft. These are the causes which make the Scythian race unfertile. A clear proof is afforded by their slave-girls. These, because of their activity and leanness of body, no sooner go to a man than they are with child.

XXII. Moreover, the great majority among the Scythians become impotent, do women's work, live like women and converse accordingly. Such men they call Anaries. Now the natives put the blame on to Heaven, and respect and worship these creatures, each fearing for himself. I too think that these diseases are divine, and so are all others, no one being more divine or more human than any other; all are alike, and all divine. Each of them has a nature of its own, and none arises without its natural cause. How, in my opinion, this disease arises I will explain. The habit of riding causes swellings at the joints, because they are always astride their horses; in severe cases follow lameness and sores on the hips. They cure themselves in the following way. At the beginning of the disease they cut the vein behind each ear. When the blood has ceased to flow faintness comes over them and they sleep. Afterwards they get up, some cured and some not. Now, in my opinion, by this treatment the seed is destroyed. For by the side of the ear are veins, to cut which causes impotence, and I believe that these are the veins which they cut. After this treatment, when the Scythians approach a woman but cannot have intercourse, at first they take no notice and think no more about it. But when two, three or even more attempts are attended with no better success, thinking that they have sinned against Heaven they attribute thereto the cause, and put on women's clothes, holding that they have lost their manhood. So they play the woman, and with the women do the same work as women do.

This affliction affects the rich Scythians because of their riding, not the lower classes but the upper, who possess the most strength; the poor, who do not ride, suffer less. But, if we suppose this disease to be more divine than any other, it ought to have attacked, not the highest and richest classes only of the Scythians, but all classes equally—or rather the poor especially, if indeed the gods are pleased to receive from men respect and worship, and repay these with favours. For naturally the rich, having great wealth, make many sacrifices to the gods, and offer many votive offerings, and honour them, all of which things the poor, owing to their poverty, are less able to do; besides, they blame the gods for not giving them wealth, so that the penalties for such sins are likely to be paid by the poor rather than by the rich. But the truth is, as I said above, these affections are neither more nor less divine than any others, and all and each are natural. Such a disease arises among the Scythians for such a reason as I have stated, and other men too are equally liable to it, for wherever men ride very much and very frequently, there the majority are attacked by swellings at the joints, sciatica and gout, and are sexually very weak. These complaints come upon the Scythians, and they are the most impotent of men, for the reasons I have given, and also because they always wear trousers and spend most of their time on their horses, so that they do not handle the parts, but owing to cold and fatigue forget about sexual passion, losing their virility before any impulse is felt.

XXIII. Such is the condition of the Scythians. The other people of Europe differ from one another both in stature and in shape, because of the changes of the seasons, which are violent and frequent, while there are severe heat waves, severe winters, copious rains and then long droughts, and winds, causing many changes of various kinds. Wherefore it is natural to realize that generation too varies in the coagulation of the seed,[15] and is not the same for the same seed in summer as in winter nor in rain as in drought. It is for this reason, I think, that the physique of Europeans varies more than that of Asiatics, and that their stature differs very widely in each city. For there arise more corruptions in the coagulation of the seed, when the changes of the seasons are frequent than when they are similar or alike. The same reasoning applies also to character. In such a climate arise wildness, unsociability and spirit. For the frequent shocks to the mind impart wildness, destroying tameness and gentleness, For this reason, I think, Europeans are also more courageous than Asiatics. For uniformity engenders slackness, while variation fosters endurance in both body and soul; rest and slackness are food for cowardice, endurance and exertion for bravery. Wherefore Europeans are more warlike, and also because of their institutions, not being under kings as are Asiatics. For, as I said above, where there are kings, there must be the greatest cowards. For men's souls are enslaved, and refuse to run risks readily and recklessly to increase the power of somebody else. But independent people, taking risks on their own behalf and not on behalf of others, are willing and eager to go into danger, for they themselves enjoy the prize of victory. So institutions contribute a great deal to the formation of courageousness.

XXIV. Such, in outline and in general, is the character of Europe and of Asia. In Europe too there are tribes differing one from another in stature, in shape and in courage. The differences are due to the same causes as I mentioned above, which I will now describe more clearly. Inhabitants of a region which is mountainous, rugged, high, and watered, where the changes of the seasons exhibit sharp contrasts, are likely to be of big physique, with a nature well adapted for endurance and courage, and such possess not a little wildness and ferocity. The inhabitants of hollow regions, that are meadowy, stifling, with more hot than cool winds, and where the water used is hot, will be neither tall nor well-made, but inclined to be broad, fleshy, and dark-haired; they themselves are dark rather than fair, less subject to phlegm than to bile. Similar bravery and endurance are not by nature part of their character, but the imposition of law can produce them artificially. Should there be rivers in the land, which drain off from the ground the stagnant water and the rain water, these[16] will be healthy and bright. But if there be no rivers, and the water that the people drink be marshy, stagnant, and fenny, the physique of the people must show protruding bellies and enlarged spleens. Such as dwell in a high land that is level, windy, and watered, will be tall in physique and similar to one another, but rather unmanly and tame in character. As to those that dwell on thin, dry, and bare soil, and where the changes of the seasons exhibit sharp contrasts, it is likely that in such country the people will be hard in physique and well-braced, fair rather than dark, stubborn and independent in character and in temper. For where the changes of the seasons are most frequent and most sharply contrasted, there you will find the greatest diversity in physique, in character, and in constitution.

These are the most important factors that create differences in men's constitutions; next come the land in which a man is reared, and the water. For in general yon will find assimilated to the nature of the land both the physique and the characteristics of the inhabitants. For where the land is rich, soft, and well-watered, and the water is very near the surface, so as to be hot in summer and cold in winter, and if the situation be favourable as regards the seasons, there the inhabitants are fleshy, ill-articulated, moist, lazy, and generally cowardly in character. Slackness and sleepiness can be observed in them, and as far as the arts are concerned they are thick-witted, and neither subtle nor sharp. But where the land is bare, waterless, rough, oppressed by winter's storms and burnt by the sun, there you will see men who are hard, lean, well-articulated, well-braced, and hairy; such natures will be found energetic, vigilant, stubborn and independent in character and in temper, wild rather than tame, of more than average sharpness and intelligence in the arts, and in war of more than average courage. The things also that grow in the earth all assimilate themselves to the earth. Such are the most sharply contrasted natures and physiques. Take these observations as a standard when drawing all other conclusions, and you will make no mistake.

NOTES [FROM TRANSLATOR W.H.S. JONES]

1. That is, taking more than one full meal every day.
2. That is, Asia Minor.
3. That is, the winter rising and the summer rising.
4. Or, "they are very prolific and the best of mothers."
5. The writer is thinking of Asiatic natives and the Greek colonists on the coast of Asia Minor.
6. There is a gap in the [original] text here dealing with the Egyptians and Libyans.
7. Or, more idiomatically, "the variations of climate are most violent and most frequent."
8. Practically nothing more is told us about this race by our other authorities, Pliny, Harpocration and Suidas.
9. Modern biologists hold that acquired characteristics are not inherited.
10. Or, reading "the temper of men like these must be gentle, because they are unwarlike and inactive."
11. Both people are of peculiar physique, and the cause of the peculiarity is in the one case extreme heat, and in the other extreme cold.
12. Or, reading "a plateau."
13. Or, "heavy rains."
14. As a modern physiologist might put it, "abnormal variations in the formation of the embryo."
15. *I. e.* "in the formation of the foetus."
16. The people or the rivers? Probably the former, in which case "bright" will mean "of bright (clear) complexion."

2

On the Laws in Their Relation to the Nature of the Climate

Charles de Secondat Montesquieu

Chapter 1

The general idea

If it is true that the character[a] of the spirit and the passions of the heart are extremely different in the various climates, *laws* should be relative to the differences in these passions and to the differences in these characters.

Chapter 2

How much men differ in the various climates

Cold air[1] contracts the extremities of the body's surface fibers; this increases their spring and favors the return of blood from the extremities of the heart. It shortens these same fibers;[2] therefore, it increases their strength[b] in this way too. Hot air, by contrast, relaxes these extremities of the fibers and lengthens them; therefore, it decreases their strength and their spring.

Therefore, men are more vigorous in cold climates. The action of the heart and the reaction of the extremities of the fibers are in closer accord, the fluids are in a better equilibrium, the blood is pushed harder toward the heart and, reciprocally, the heart has more power. This greater strength should produce many effects: for example, more confidence in oneself, that is, more courage; better knowledge of one's superiority, that is, less desire for vengeance; a higher opinion of one's security, that is, more frankness and fewer suspicions, maneuvers, and tricks. Finally, it should make very different characters. Put a man in a hot, enclosed spot, and he will suffer, for the reasons just stated, a great slackening of heart. If, in the circumstance, one proposes a bold action to him, I believe one will find him little disposed toward it; his present weakness will induce discouragement in his soul; he will fear everything, because he will feel he can do nothing. The peoples in hot countries are timid like old men; those in cold countries are courageous like young men. If we turn our attention to the recent wars,[3] which are the ones we can best observe and in which we can better see certain slight effects that are imperceptible from a distance, we shall certainly feel that the actions of the northern peoples who were sent to southern countries[4] were not as fine as the actions of their compatriots who, fighting in their own climate, enjoyed the whole of their courage.

The strength of the fibers of the northern peoples causes them to draw the thickest juices

The Anthropology of Climate Change: An Historical Reader, First Edition. Edited by Michael R. Dove.

from their food. Two things result from this: first, that the parts of the chyle, or lymph,[c] being broad surfaced, are more apt to be applied to the fibers and to nourish them; and second, that, being coarse, they are less apt to give a certain subtlety to the nervous juice. Therefore, these people will have large bodies and little vivacity.

The nerves, which end in the tissue of our skin, are made of a sheaf of nerves. Ordinarily, it is not the whole nerve that moves, but an infinitely small part of it. In hot countries, where the tissue of the skin is relaxed, the ends of the nerves are open and exposed to the weakest action of the slightest objects. In cold countries, the tissue of the skin is contracted and the papillae compressed. The little bunches are in a way paralyzed; sensation hardly passes to the brain except when it is extremely strong and is of the entire nerve together. But imagination, taste, sensitivity, and vivacity depend on an infinite number of small sensations.

I have observed the place on the surface tissue of a sheep's tongue which appears to the naked eye to be covered with papillae. Through a microscope, I have seen the tiny hairs, or a kind of down, on these papillae; between these papillae were pyramids, forming something like little brushes at the ends. It is very likely that these pyramids are the principal organ of taste.

I had half of the tongue frozen; and, with the naked eye I found the papillae considerably diminished; some of the rows of papillae had even slipped inside their sheaths: I examined the tissue through a microscope; I could no longer see the pyramids. As the tongue thawed, the papillae appeared again to the naked eye, and, under the microscope, the little brushes began to reappear.

This observation confirms what I have said, that, in cold countries, the tufts of nerves are less open; they slip inside their sheaths, where they are protected from the action of external objects. Therefore, sensations are less vivid.

In cold countries, one will have little sensitivity to pleasures; one will have more of it in temperate countries; in hot countries, sensitivity will be extreme. As one distinguishes climates by degrees of latitude, one can also distinguish them by degrees of sensitivity, so to speak. I have seen operas in England and Italy; they are the same plays with the same actors: but the same music produces such different effects in the people of the two nations that it seems inconceivable, the one so calm and the other so transported.

It will be the same for pain; pain is aroused in us by the tearing of some fiber in our body. The author of nature has established that this pain is stronger as the disorder is greater; now it is evident that the large bodies and coarse fibers of the northern peoples are less capable of falling into disorder than the delicate fibers of the peoples of hot countries; therefore, the soul is less sensitive to pain. A Muscovite has to be flayed before he feels anything.

With that delicacy of organs found in hot countries, the soul is sovereignly moved by all that is related to the union of the two sexes; everything leads to this object.

In northern climates, the physical aspect of love has scarcely enough strength to make itself felt; in temperate climates, love, accompanied by a thousand accessories, is made pleasant by things that at first seem to be love but are still not love; in hotter climates, one likes love for itself; it is the sole cause of happiness; it is life.

In southern climates, a delicate, weak, but sensitive machine gives itself up to a love which in a seraglio is constantly aroused and calmed; or else to a love which as it leaves women much more independent is exposed to a thousand troubles. In northern countries, a healthy and well-constituted but heavy machine finds its pleasures in all that can start the spirits in motion again: hunting, travels, war, and wine. You will find in the northern climates peoples who have few vices, enough virtues, and much sincerity and frankness. As you move toward the countries of the south, you will believe you have moved away from morality itself: the liveliest passions will increase crime; each will seek to take from others all the advantages that can favor these same passions. In temperate countries, you will see peoples whose manners, and even their vices and virtues are inconstant; the climate is not sufficiently settled to fix them.

The heat of the climate can be so excessive that the body there will be absolutely without strength. So, prostration will pass even to the spirit; no curiosity, no noble enterprise, no generous sentiment; inclinations will all be passive there; laziness there will be happiness; most chastisements there will be less difficult to bear than the action of the soul, and servitude will be less intolerable than the strength of spirit necessary to guide one's own conduct.

Chapter 3

A contradiction in the characters of certain peoples of the south

Indians[5] are by nature without courage; even the children[6] of Europeans born in the Indies lose the courage of the European climate. But how does this accord with their atrocious actions, their barbaric customs and penitences? Men there suffer unbelievable evils; women burn themselves: this is considerable strength for so much weakness.

Nature, which has given these peoples a weakness that makes them timid, has also given them such a lively imagination that everything strikes them to excess. The same delicacy of organs that makes them fear death serves also to make them dread a thousand things more than death. The same sensitivity makes the Indians both flee all perils and brave them all.

As a good education is more necessary to children than to those of mature spirit, so the peoples of these climates have greater need of a wise legislator than the peoples of our own. The more easily and forcefully one is impressed, the more important it is to be impressed in a suitable manner, to accept no prejudices, and to be led by reason.

In the time of the Romans, the peoples of northern Europe lived without arts, without education, almost without laws, and still, with only the good sense connected with the coarse fibers of these climates, they maintained themselves with remarkable wisdom against the Roman power until they came out of their forests to destroy it.

Chapter 4

The cause of the immutability of religion, mores, manners, and laws in the countries of the east

If you join the weakness of organs that makes the peoples of the East receive the strongest impressions in the world to a certain laziness of the spirit, naturally bound with that of the body, which makes that spirit incapable of any action, any effort, any application, you will understand that the soul can no longer alter impressions once it has received them. This is why laws, mores,[7] and manners, even those that seem not to matter, like the fashion in clothing, remain in the East today as they were a thousand years ago.

Chapter 5

That bad legislators are those who have favored the vices of the climate and good ones are those who have opposed them

Indians believe that rest and nothingness are the foundation of all things and the end to which they lead. Therefore, they consider total inaction as the most perfect state and the object of their desires. They give to the sovereign being[8] the title of the unmoving one. The Siamese believe that the supreme felicity[9] consists in not being obliged to animate a machine or to make a body act.

In these countries where excessive heat enervates and overwhelms, rest is so delicious and movement so painful that this system of metaphysics appears natural; and Foë,[10] legislator of the Indies, followed his feelings when he put men in an extremely passive state; but his doctrine, born of idleness of the climate, favoring it in turn, has caused a thousand ills.

The legislators of China were more sensible when, as they considered men not in terms of the peaceful slate in which they will one day be but in terms of the action proper to making them fulfill the duties of life, they made their religion, philosophy, and laws all practical. The more the physical causes incline men to rest, the more the moral causes should divert them from it.

Chapter 6

On the cultivation of land in hot climates

The cultivation of land is the greatest labor of men. The more their climate inclines them to flee this labor, the more their religion and laws should rouse them to it. Thus, the laws of the Indies, which give the land to the princes take away from individuals the spirit of ownership, increase the bad effects of the climate, that is, natural laziness.

[...]

[Chapters 7–12 are not reprinted here]

Chapter 13

Effects resulting from the climate of England

In a nation whose soul is so affected by an illness of climate that it could carry the repugnance for all things to include that of life, one sees that the most suitable government for people to whom everything can be intolerable would be the one in which they could not be allowed to blame any one person for causing their sorrows, and in which, as laws rather than men would govern, the laws themselves must be overthrown in order to change the state.

For if the same nation had also received from the climate a certain characteristic of impatience that did not permit it to tolerate the same things for long, it can be seen that the government of which we have just spoken would still be the most suitable.

The characteristic of impatience is not serious in itself, but it can become very much so when it is joined to courage.

It is different from fickleness, which makes one undertake things without purpose and abandon them likewise. It is nearer to obstinacy because it comes from a feeling of ills which is so lively that it is not weakened even by the habit of tolerating them.

In a free nation, this characteristic would be one apt to frustrate the projects of tyranny,[25] which is always slow and weak in its beginnings, just as it is prompt and lively at its end, which shows at first only a hand extended in aid, and later oppresses with an infinity of arms.

Servitude always begins with drowsiness. But a people who rest in no situation, who constantly pinch themselves to find the painful spots, could scarcely fall asleep.

Politics is a dull rasp which by slowly grinding away gains its end. Now the men of whom we have just spoken could not support the delays, the details and the coolness of negotiations; they would often succeed in them less well than any other nation, and they would lose by their treaties what they had gained by their weapons.

Chapter 14

Other effects, of the climate

Our fathers, the ancient Germans, lived in a climate where the passions were calm. Their laws found in things only what they saw, and they imagined nothing more. And just as these laws judged insults to men by the size of the wounds, they put no greater refinement in the offenses to women. The Law of the Alemanni[26] on this point is quite singular. If one exposes a woman's head, one will pay a fine of six sous; it is the same for exposing a leg up to the knee; double above the knee. The law, it seems, measured the size of the outrages done a woman's person as one measures a geometric figure; the law did not punish the crime of the imagination, it punished that of the eyes. But when a Germanic nation moved to Spain, the climate required quite different laws. The laws of the Visigoths prohibited doctors from bleeding a *freeborn* woman except in the presence of her father or mother, her brother, her son, or her uncle. The imagination of the peoples was fired, that of the legislators was likewise ignited; the law suspected everything in a people capable of suspecting everything. Therefore, these laws gave an extreme attention to the two sexes. But it seems that in their punishing they thought more of gratifying individual vengeance than of exercising public vengeance. Thus, in most cases they reduced the two guilty ones to the servitude of their relatives or of the offended husband. A freeborn woman[27] who had given herself to a married man was put into the power of his wife, to do with as she wanted. These laws required that slaves[28] bind up and present to the husband his wife whom they had caught in adultery; they permitted her children[29] to accuse her, and they permitted torturing her slaves in order to convict her. Thus these laws were more proper for the excessive refinement of a certain point of honor than for the formation of a good police. And one must not be astonished if Count Julian believed an outrage of this kind required the loss of one's country or of one's king. One should not be surprised that the Moors, whose mores were so similar, found it so easy to establish themselves in Spain, to maintain themselves there, and to delay the fall of their empire.

Chapter 15

On the differing trust the laws have in people according to the climate

The Japanese people have such an atrocious character that their legislators and magistrates have not been able to place any trust in them; they have set before the eyes of the people only judges, threats, and chastisements; they have subjected them at every step to the inquisition of the police. These laws that establish, in every five heads of families, one as magistrate over the other four, these laws that punish a whole family or a whole neighborhood for a single crime, these laws that find no innocent men where there can be a guilty one, are made so that all men distrust one another, so that each scrutinizes the conduct of the other, and so that each is his own inspector, witness, and judge.

On the other hand, the people of the Indies are gentle,[30] tender, and compassionate. Thus, their legislators have put great trust in them. They have established few penalties,[31] and these are not very severe or even strictly executed. They have given nephews to uncles and orphans to guardians, as elsewhere they are given to their fathers; they have regulated inheritance by the recognized merit of the heir. It seems they have thought that each citizen should rely on the natural goodness of the others.

They easily give liberty[32] to their slaves; they marry them; they treat them like their children:[33] happy is the climate that gives birth to candor in mores and produces gentleness in laws!

MONTESQUIEU'S NOTES

1 This is even visible: in the cold, one appears thinner.
2 It is known that it shortens iron.
3 The War of the Spanish Succession.
4 In Spain, for example.
5 [Jean Baptiste] Tavernier says, "A hundred European soldiers would have little difficulty routing a thousand Indian soldiers." [*Travels in India*; bk. 2, ch. 9; 1, 391; 1889 edn.]
6 Even Persians who settle in the Indies take on, in the third generation, the indolence and cowardice of the Indians. See [François] Bernier, *Travels in the Mogul Empire*, vol. I, p. 282 ["Letter to Colbert concerning Hindustan"; p. 209; 1916 edn].
7 In a fragment from Nicholas of Damascus, in the collection of Constantine Porphyrogenitus, one sees that it was an ancient custom in the East to send someone to strangle a governor who displeased; this dates from the times of the Medes. [Nicholas of Damascus, *Fragments*, #66.26; *FGrH;* 2A, 366–367.]
8 Panamanack. See [Athanasius] Kircher [*La Chine illustrée*, pt. 3, chap. 6; p. 2415; 1670 edn].
9 [Simon de] la Loubère, *The Kingdom of Siam*, p. 446 [pt. 3, chap. 22; p. 129; 1969 edn].
10 Foë wants to reduce the heart to pure emptiness. "We have eyes and ears, but perfection consists in not seeing or hearing; perfection, for those members, the mouth, hands, etc., is to be inactive." This is taken from the "Dialog of a Chinese Philosopher," reported by Father [Jean Baptiste] du Halde [*Description de l'Empire de la Chine*], vol. 3 [3, 59 H; 3, 49 P; 3, 268 L].

[Notes 11–24 are in chapters not reprinted here]

25 I take this word to mean the design of upsetting the established power, chiefly democracy. This was its meaning for the Greeks and Romans.
26 [*Leges Alamannorum*] chap. 58, paras. 1–2 [A56.1–2; B58.1–2].
27 *Lex Wisigothorum*, bk. 3, tit. 4, para. 9 [3.4.9].
28 Ibid. [*Lex Wisigothorum*], bk. 3, tit. 4, para. 6 [3.4.6].
29 Ibid. [*Lex Wisigothorum*], bk. 3, tit. 4, para. 13 [3.4.13].
30 See [François] Bernier [*Travels in the Mogul Empire*], vol. 2, p. 140 ["History of the States of the Great Mogul"; p. 99, 1916 edn].

31 See *Lettres édifiantes et curieuses*, vol. 14, p. 403 [Lettre du P. Bouchet, Pontichéry, October 2, 1714; 14, 402–404; 1720 edn] for the principal laws or customs of the peoples of India on the peninsula this side of the Ganges.

32 Vol. 9 of the *Lettres édifiantes et curieuses*, p. 378 [Lettre du P. Bouchet, 9, 37; 1730 edn].

33 I had thought that the gentleness of slavery in the Indies had made Diodorus say that in this country there was neither master nor slave, but Diodorus attributed to the whole of India what, according to Strabo [*Geographica*], bk. 15 [15.1.54], was proper only to a particular nation.

TRANSLATORS' NOTES

a *Caractère* can mean mark or sign, trait, or a habitual way of acting and feeling. When Montesquieu uses *caractère*, he seems to mean a form or shape of the spirit, combining these meanings.

b We translate *force as* both "force" and "strength," as the context is more and less abstract.

c In the eighteenth century these words referred to various body fluids, without the precise denotations they have in modern physiology.

Beyond the Greco-Roman Tradition

3

The Muqaddimah

An Introduction to History

Ibn Khaldûn

Supplementary Note to the Second Prefatory Discussion

The northern quarter of the earth has more civilization than the southern quarter. The reason thereof.

We know from observation and from continuous tradition that the first and the second of the cultivated zones have less civilization than the other zones. The cultivated area in the first and second zones is interspersed with empty waste areas and sandy deserts and has the Indian Ocean to the east. The nations and populations of the first and second zones are not excessively numerous. The same applies to the cities and towns there (Figure 3, cf. Frontispiece).

The third, fourth, and subsequent zones are just the opposite. Waste areas there are few. Sandy deserts also are few or non-existent. The nations and populations are tremendous. Cities and towns are exceedingly numerous. Civilization has its seat between the third and the sixth zones. The south is all emptiness.

Many philosophers have mentioned that this is because of the excessive heat and slightness of the sun's deviation from the zenith in the south. Let us explain and prove this statement. The result will clarify the reason why civilization in the third and fourth zones is so highly developed and extends also to the fifth, <sixth,> and seventh zones.

We say: When the south and north poles (of heaven) are upon the horizon, they constitute a large circle that divides the firmament into two parts. It is the largest circle (in it) and runs from west to east. It is called the equinoctial line. In astronomy, it has been explained in the proper place that the highest sphere moves from east to west in a daily motion by means of which it also forces the spheres enclosed by it to move. This motion is perceptible to the senses. It has also been explained that the stars in their spheres have a motion that is contrary to this motion and is, therefore, a motion from west to east. The periods of this movement differ according to the different speeds of the motions of the stars.

Parallel to the courses of all these stars in their spheres, there runs a large circle which belongs to the highest sphere and divides it into two halves. This is the ecliptic (zodiac).

The Anthropology of Climate Change: An Historical Reader, First Edition. Edited by Michael R. Dove.

KEY TO THE MAP

1 South
2 West
3 North
4 East
5 Empty beyond the equator
because of the heat
6 Equator
7 Lamlam Country
8 Maghzâwah (Maguzawa?)
9 Kanem [Country]
10 Bornu
11 Gawgaw
12 Zaghây
13 at-Tâjuwîn
14 Nubia
15 Abyssinia
16 Ghânah
17 Lamtah
18 as-Sûs
19 Morocco
20 Tangier
21 Ṣinhâjah
22 Dar'ah
23 Ifrîqiyah
24 Fezzan
25 Jarîd
26 Kawâr
27 Desert of Berenice
28 Inner Oases
29 Upper Egypt
30 Egypt
31 Beja
32 Ḥijâz
33 Syria
34 Yemen
35 Yamâmah
36 al-Baṣrah
37 'Irâq
38 ash-Shiḥr
39 Oman
40 Western India
41 Mukrân
42 Kirmân
43 Fârs
44 *al-Bahlûs*
45 Azerbaijan
46 Desert
47 Khurâsân
48 Khuwârizm
49 Eastern India
50 Tashkent
51 Soghd
52 China
53 Tughuzghuz
54 Gascogne
55 Brittany
56 Calabria
57 France
58 Venice
59 Germany (Alamâniyah)
60 Macedonia
61 Bohemia
62 Jathûliyah
63 Jarmâniyah
64 al-Baylaqân
65 Armenia
66 Ṭabaristân
67 Alans
68 Bashqirs
69 Bulgars
70 Pechenegs
71 Stinking Land
72 Waste Country
73 Magog
74 Ghuzz
75 Türgish
76 Adhkish
77 Khallukh
78 Gog
79 Kimäk
80 Empty in the north
because of the cold

Figure 3.1 *Map of the World*
Ibn Khaldûn, The Muqaddimah. ©1958, 1967 by PUP Reprinted by permission of Princeton University Press.

It is divided into twelve "signs." As has been explained in the proper place, the equinoctial line intersects the ecliptic at two opposite points, namely, at the beginning of Aries and at the beginning of Libra. The equinoctial line divides the zodiac into two halves. One of them extends northward from the equinoctial line and includes the signs from the beginning of Aries to the end of Virgo. The other half extends southward from it and includes the signs from the beginning of Libra to the end of Pisces.

When the two poles fall upon the horizon <which takes place in one particular region> among all the regions of the earth, a line is formed upon the surface of the earth that faces the equinoctial line and runs from west to east. This line is called the equator. According to astronomical observation, this line is believed to coincide with the beginning of the first of the seven zones. All civilization is to the north of it.

The north pole gradually ascends on the horizon of the cultivated area (of the earth) until its elevation reaches sixty-four degrees. Here, all civilization ends. This is the end of the seventh zone. When its elevation reaches ninety degrees on the horizon – that is the distance between the pole and the equinoctial line – then it is at its zenith, and the equinoctial line is on the horizon. Six of the signs of the zodiac, the northern ones, remain above the horizon, and six, the southern ones, are below it.

Civilization is impossible in the area between the sixty-fourth and the ninetieth degrees, for no admixture of heat and cold occurs there because of the great time interval between them. Generation (of anything), therefore, does not take place.

The sun is at its zenith on the equator at the beginning of Aries and Libra. It then declines from its zenith down to the beginning of Cancer and Capricorn. Its greatest declination from the equinoctial line is twenty-four degrees.

Now, when the north pole ascends on the horizon, the equinoctial line declines from the zenith in proportion to the elevation of the north pole, and the south pole descends correspondingly, as regards the three (distances constituting geographical latitude). Scholars who calculate the (prayer) times call this the latitude of a place. When the equinoctial line declines from the zenith, the northern signs of the zodiac gradually rise above it, proportionately to its rise, until the beginning of Cancer is reached. Meanwhile, the southern signs of the zodiac correspondingly descend below the horizon until the beginning of Capricorn is reached, because of the inclination of the (two halves of the zodiac) upwards or downwards from the horizon of the equator, as we have stated.

The northern horizon continues to rise, until its northern limit, which is the beginning of Cancer, is in the zenith. This is where the latitude is twenty-four degrees in the Ḥijâz and the territory adjacent. This is the declination from the equinoctial at the horizon of the equator at the beginning of Cancer. With the elevation of the north pole (Cancer) rises, until it attains the zenith. When the pole rises more than twenty-four degrees, the sun descends from the zenith and continues to do so until the elevation of the pole is sixty-four degrees, and the sun's descent from the zenith, as well as the depression of the south pole under the horizon, is the same distance. Then, generation (of anything) stops because of the excessive cold and frost and the long time without any heat.

At and nearing its zenith, the sun sends its rays down upon the earth at right angles. In other positions, it sends them down at obtuse or acute angles. When the rays form right angles, the light is strong and spreads out over a wide area, in contrast to what happens in the case of obtuse and acute angles. Therefore, at and nearing its zenith, the heat is greater than in other positions, because the light (of the sun) is the reason for heat and calefaction. The sun reaches its zenith at the equator twice a year in two points of Aries and Libra. No declination (of the sun) goes very far. The heat hardly begins to become more temperate, when the sun has reached the limit of its declination at the beginning of Cancer or Capricorn and begins to rise again toward the zenith. The perpendicular rays then fall heavily upon the horizon there (in these regions) and hold steady for a long time, if not permanently. The air gets burning hot, even excessively so. The same is true whenever the sun reaches the zenith in the area between the equator and latitude twenty-four degrees, as it does twice a year. The rays exercise almost as much force upon the horizon there (at this latitude) as they do at the equator. The excessive heat causes a parching dryness in the air that prevents (any) generation. As the heat becomes more excessive, water and all kinds of moisture dry up, and (the power of) generation is destroyed in minerals, plants, and animals, because (all) generation depends on moisture.

Now, when the beginning of Cancer declines from the zenith at the latitude of twenty-five degrees and beyond, the sun also declines from its zenith. The heat becomes temperate, or deviates only slightly from (being temperate). Then, generation can take place. This goes on until the cold becomes excessive, due to the lack of light and the obtuse angles of the rays of the sun. Then, (the power of) generation again decreases and is destroyed. However, the destruction caused by great heat is greater than that caused by great cold, because heat brings about desiccation faster than cold brings about freezing.

Therefore, there is little civilization in the first and second zones. There is a medium degree of civilization in the third, fourth, and fifth zones, because the heat there is temperate owing to the decreased amount of light. There is a great deal of civilization in the sixth and seventh zones because of the decreased amount of heat there. At first, cold does not have the same destructive effect upon (the power of) generation as heat; it causes desiccation only when it becomes excessive and thus has dryness added. This is the case beyond the seventh zone. (All) this, then, is the reason why civilization is stronger and more abundant in the northern quarter. And God knows better!

The[1] philosophers concluded from these facts that the region at the equator and beyond it (to the south) was empty. On the strength of observation and continuous tradition, it was argued against them that (to the contrary) it was cultivated. How would it be possible to prove this (contention)? It is obvious that the (philosophers) did not mean to deny entirely the existence of civilization there, but their argumentation led them to (the realization) that (the power of) generation must, to a large degree, be destroyed there because of the excessive heat. Consequently, civilization there would be either impossible, or only minimally possible. This is so. The region at the equator and beyond it (to the south), even if it has civilization as has been reported, has only a very little of it.

Averroes[2] assumed that the equator is in a symmetrical position[3] and that what is beyond the equator to the south corresponds to what is beyond it to the north; consequently, as much of the south would be cultivated as of the north. His assumption is not impossible, so far as (the argument of) the destruction of the power of generation is concerned. However, as to the region south of the equator, it is made impossible by the fact that the element of water covers the face of the earth in the south, where the corresponding area in the north admits of generation. On account of the greater amount of water (in the south), Averroes' assumption of the symmetrical (position of the equator) thus turns out to be impossible. Everything else follows, since civilization progresses gradually and begins its gradual progress where it can exist, not where it cannot exist.

The assumption that civilization cannot exist at the equator is contradicted by continuous tradition. And God knows better!

[Remainder of section, pp. 109–166 in original, not reprinted here]

Third Prefatory Discussion

The temperate and the intemperate zones. The influence of the air upon the color of human beings and upon many (other) aspects of their condition.

We[4] have explained that the cultivated region of that part of the earth which is not covered by water has its center toward the north, because of the excessive heat in the south and the excessive cold in the north. The north and the south represent opposite extremes of cold and heat. It necessarily follows that there must be a gradual decrease from the extremes toward the center, which, thus, is moderate. The fourth zone is the most temperate cultivated region. The bordering third and fifth zones are rather close to being temperate. The sixth and second zones which are adjacent to them are far from temperate, and the first and seventh zones still less so. Therefore, the sciences, the crafts, the buildings, the clothing, the foodstuffs, the fruits, even the animals, and everything that comes into being in the three middle zones are distinguished by their temperate (well-proportioned character). The human inhabitants of these zones are more temperate (well-proportioned) in their bodies, color, character qualities, and (general) conditions.[5] They are found to be extremely moderate in their dwellings, clothing, foodstuffs, and crafts. They use houses that are well constructed of stone and embellished by craftsmanship. They rival each other in production of the very best tools and implements. Among them, one finds the natural minerals, such as gold, silver, iron, copper, lead, and tin. In their business dealings they use the two precious metals (gold and silver). They avoid intemperance quite generally in all their conditions. Such are the inhabitants of the Maghrib, of Syria, the two 'Irâqs, Western India (as-Sind), and China, as well as of Spain; also the European Christians nearby, the Galicians,[6] and all those who live together with these peoples or near them in the three temperate zones. The 'Irâq and Syria are directly in the middle and therefore are the most temperate of all these countries.

The inhabitants of the zones that are far from temperate, such as the first, second, sixth, and seventh zones, are also farther removed from being temperate in all their conditions. Their buildings are of clay and reeds. Their foodstuffs are durra and herbs. Their clothing is the leaves of trees, which they sew together to cover themselves, or animal skins. Most of them go naked. The fruits and seasonings of their countries are strange and inclined to be intemperate. In their business dealings, they do not use the two noble metals, but copper, iron, or skins, upon which they set a value for the purpose of business dealings. Their qualities of character, moreover, are close to those of dumb animals. It has even been reported that most of the Negroes of the first zone dwell in caves and thickets, eat herbs, live in savage isolation and do not congregate, and eat each other. The same applies to the Slavs. The reason for this is that their remoteness from being temperate produces in them a disposition and character similar to those of the dumb animals, and they become correspondingly remote from humanity. The same also applies to their religious conditions. They are ignorant of prophecy and do not have a religious law, except for the small minority that lives near the temperate regions. (This minority includes,) for instance, the Abyssinians, who are neighbors of the Yemenites and have been Christians from pre-Islamic and Islamic times down to the present; and the Mâlî, the Gawgaw, and the Takrûr who live close to the Maghrib and, at this time, are Muslims. They are said to have adopted Islam in the seventh [thirteenth] century. Or, in the north, there are those Slav, European Christian, and Turkish nations that have adopted Christianity. All the other inhabitants of the intemperate zones in the south and in the north are ignorant of all religion. (Religious) scholarship is lacking among them. All their conditions are remote from those of human beings and close to those of wild animals. "And He creates what you do not know."[7]

The (foregoing statement) is not contradicted by the existence of the Yemen, the Ḥaḍramawt, al-Aḥqâf, the Ḥijâz, the Yamâmah, and adjacent regions of the Arabian Peninsula in the first and second zones. As we have mentioned, the Arabian Peninsula is surrounded by the sea on three sides. The humidity of (the sea) influences the humidity in the air of (the Arabian Peninsula). This diminishes the dryness and intemperance that (otherwise) the heat would cause. Because of the humidity from the sea, the Arabian Peninsula is to some degree temperate.

Genealogists who had no knowledge of the true nature of things imagined that Negroes are the children of Ham, the son of Noah, and that they were singled out to be black as the result of Noah's curse, which produced Ham's color and the slavery God inflicted upon his descendants. It is mentioned in the Torah[8] that Noah cursed his son Ham. No reference is made there to blackness. The curse included no more than that Ham's descendants should be the slaves of his brothers' descendants. To attribute the blackness of the Negroes to Ham, reveals disregard of the true nature of heat and cold and of the influence they exercise upon the air (climate) and upon the creatures that come into being in it. The black color (of skin) common to the inhabitants of the first and second zones is the result of the composition of the air in which they live, and which comes about under the influence of the greatly increased heat in the south. The sun is at the zenith there twice a year at short intervals. In (almost) all seasons, the sun is in culmination for a long time. The light of the sun, therefore, is plentiful.[9] People there have (to undergo) a very severe summer, and their skins turn black because of the excessive heat. Something similar happens in the two corresponding zones to the north, the seventh and sixth zones. There, a white color (of skin) is common among the inhabitants, likewise the result of the composition of the air in which they live, and which comes about under the influence of the excessive cold in the north. The sun is always on the horizon within the visual field (of the human observer), or close to it. It never ascends to the zenith, nor even (gets) close to it. The heat, therefore, is weak in this region, and the cold severe in (almost) all seasons. In consequence, the color of the inhabitants is white, and they tend to have little body hair. Further consequences of the excessive cold are blue eyes, freckled skin, and blond hair.

The fifth, fourth, and third zones occupy an intermediate position. They have an abundant share of temperance,[10] which is the golden mean. The fourth zone, being the one most nearly in the center, is as temperate as can be. We have mentioned that before.[11] The physique and character of its inhabitants are temperate to the (high) degree necessitated by the composition of the air in which they live. The third and fifth zones lie on either side of the fourth, but they are less centrally located. They are closer to the hot south beyond the third zone and the cold north beyond the fifth zone. However, they do not become intemperate.

The four other zones are intemperate, and the physique and character of their inhabitants show it. The first and second zones are excessively hot and black, and the sixth and seventh zones cold and white. The inhabitants of the first and second zones in the south are called the Abyssinians, the Zanj, and the Sudanese (Negroes). These are synonyms used to designate the (particular) nation that has turned black. The name "Abyssinians," however, is restricted to those Negroes who live opposite Mecca and the Yemen, and the name "Zanj" is restricted to those who live along the Indian Sea. These names are not given to them because of an (alleged) descent from a black human being, be it Ham or any one else. Negroes from the south who settle in the temperate fourth zone or in the seventh zone that tends toward whiteness, are found to produce descendants whose color gradually turns white in the course of time. Vice versa, inhabitants from the north or from the fourth zone who settle in the south produce descendants whose color turns black. This shows that color is conditioned by the composition of the air. In his *rajaz* poem on medicine, Avicenna said:

> Where the Zanj live is a heat that changes their bodies
> Until their skins are covered all over with black.
> The Slavs acquire whiteness
> Until their skins turn soft.[12]

The inhabitants of the north are not called by their color, because the people who established the conventional meanings of words were themselves white. Thus, whiteness was something usual and common (to them), and they did not see anything sufficiently remarkable in it to cause them to use it as a specific term. Therefore, the inhabitants of the north, the Turks, the Slavs, the Tughuzghuz, the Khazars, the Alans, most of the European Christians, the Gog and Magog are found to be separate nations[13] and numerous races called by a variety of names.

The inhabitants of the middle zones are temperate in their physique and character and in their ways of life. They have all the natural conditions necessary for a civilized life, such as ways of making a living, dwellings, crafts, sciences, political leadership, and royal authority. They thus have had (various manifestations of) prophecy, religious groups, dynasties, religious laws, sciences, countries, cities, buildings, horticulture, splendid crafts, and everything else that is temperate.

Now, among the inhabitants of these zones about whom we have historical information are, for instance, the Arabs, the Byzantines (Rûm), the Persians, the Israelites, the Greeks, the Indians, and the Chinese. When[14] genealogists noted differences between these nations, their distinguishing marks and characteristics, they considered these to be due to their (different) descents. They declared all the Negro inhabitants of the south to be descendants of Ham. They had misgivings about their color and therefore undertook to report the afore-mentioned silly story. They declared all or most of the inhabitants of the north to be the descendants of Japheth, and they declared most of the temperate nations, who inhabit the central regions, who cultivate the sciences and crafts, and who possess religious groups and religious laws as well as political leadership and royal authority, to be the descendants of Shem. Even if the genealogical construction were correct, it would be the result of mere guesswork, not of cogent, logical argumentation. It would merely be a statement of fact. It would not imply that the inhabitants of the south are called "Abyssinians" and "Negroes" because they are descended from "black" Ham. The genealogists were led into this error by their belief that the only reason for differences between nations is in their descent. This is not so. Distinctions between races or nations are in some cases due to a different descent, as in the case of the Arabs, the Israelites, and the Persians. In other cases, they are caused by geographical location and (physical) marks, as in the case of the Zanj (Negroes), the Abyssinians, the Slavs, and the black (Sudanese) Negroes. Again, in other cases, they are caused by custom and distinguishing characteristics, as well as by descent, as in the case of the Arabs. Or, they may be caused by anything else among the conditions, qualities, and features peculiar to the different nations. But to generalize and say that the inhabitants of a specific geographical location in the south or in the north are the descendants of such-and-such a well-known person because they have a common color, trait, or (physical) mark which that (alleged) forefather had, is one of those errors which are caused by disregard, (both) of the true nature of created beings and of geographical facts. (There also is disregard of the fact that the physical circumstances and environment) are subject to changes that affect later generations; they do not necessarily remain unchanged.

This is how God proceeds with His servants.—And verily, you will not be able to change God's way.[15]

Fourth Prefatory Discussion

The influence of the air (climate) upon human character.

We[16] have seen that Negroes are in general characterized by levity, excitability, and great emotionalism. They are found eager to dance whenever they hear a melody.[17] They are everywhere described as stupid. The real reason for these (opinions) is that, as has been shown by philosophers in the proper place, joy and gladness are due to expansion and diffusion of the animal spirit. Sadness is due to the opposite, namely, contraction and concentration of the animal spirit. It has been shown that heat expands and rarefies air and vapors and increases their quantity. A drunken person experiences inexpressible joy and gladness, because the vapor of the spirit in his heart is pervaded by natural heat, which the power of the wine generates in his spirit. The spirit, as a result, expands, and there is joy. Likewise, when those who enjoy a hot bath inhale the air of the bath, so that the heat of the air enters their spirits and makes them hot, they are found to experience joy. It often happens that they start singing, as singing has its origin in gladness.

Now, Negroes live in the hot zone (of the earth). Heat dominates their temperament and formation. Therefore, they have in their spirits an amount of heat corresponding to that in their bodies and that of the zone in which they live. In comparison with the spirits of the inhabitants of the fourth zone, theirs are hotter and, consequently, more expanded. As a result, they are more quickly moved to joy and gladness, and they are merrier. Excitability is the direct consequence.

In the same way, the inhabitants of coastal regions are somewhat similar to the inhabitants of the south. The air in which they live is very much hotter because of the reflection of the light and the rays of (the sun from) the surface of the sea. Therefore, their share in the qualities resulting from heat, that is, joy and levity, is larger than that of the (inhabitants of) cold and hilly or mountainous countries. To a degree, this may be observed in the inhabitants of the Jarîd in the third zone. The heat is abundant in it and in the air there, since it lies south of the coastal plains and hills. Another example is furnished by the Egyptians. Egypt lies at about the same latitude as the Jarîd. The Egyptians are dominated by joyfulness, levity, and disregard for the future. They store no provisions of food, neither for a month nor a year ahead, but purchase most of it (daily) in the market. Fez in the Maghrib, on the other hand, lies inland (and is) surrounded by cold hills. Its inhabitants can be observed to look sad and gloomy and to be too much concerned for the future. Although a man in Fez might have provisions of wheat stored, sufficient to last him for years, he always goes to the market early to buy his food for the day, because he is afraid to consume any of his hoarded food.

If one pays attention to this sort of thing in the various zones and countries, the influence of the varying quality of the air upon the character (of the inhabitants) will become apparent. God is "the Creator, the Knowing One."[18]

Al-Mas'ûdî undertook to investigate the reason for the levity, excitability, and emotionalism in Negroes, and attempted to explain it. However, he did no better than to report, on the authority of Galen and Ya'qûb b. Isḥâq al- Kindî, that the reason is a weakness of their brains which results in a weakness of their intellect.[19] This is an inconclusive and unproven statement. "God guides whomever He wants to guide."[20]

Fifth Prefatory Discussion

Differences with regard to abundance and scarcity of food in the various inhabited regions ('umrân) *and how they affect the human body and character.*

It[21] should be known that not all the temperate zones have an abundance of food, nor do all their inhabitants lead a comfortable life. In some parts, the inhabitants enjoy an abundance of grain, seasonings, wheat, and fruits, because the soil is well balanced and good for plants and there is an abundant civilization. And then, in other parts, the land is strewn with rocks, and no seeds or herbs grow at all. There, the inhabitants have a very hard time. Instances of such people are the inhabitants of the Ḥijâz and the Yemen, or the Veiled Ṣinhâjah who live in the desert of the Maghrib on the fringes of the sandy deserts which lie between the Berbers and the Sudanese Negroes. All of them lack all grain and seasonings. Their nourishment and food is milk and meat. Another such people is the Arabs who roam the waste regions. They may get grain and seasonings from the hills, but this is the case only at certain times and is possible only under the eyes of the militia which protects (the hill country). Whatever they get is little, because they have little money. They obtain no more than the bare necessity, and sometimes less, and in no case enough for a comfortable or abundant life. They are mostly found restricted to milk, which is for them a very good substitute for wheat. In spite of this, the desert people who lack grain and seasonings are found to be healthier in body and better in character than the hill people who have plenty of everything. Their complexions are clearer, their bodies cleaner, their figures more perfect and better, their characters less intemperate, and their minds keener as far as knowledge and perception are concerned. This is attested by experience in all these groups. There is a great difference in this respect between the Arabs and Berbers (on the one hand), and the Veiled (Berbers)[22] and the inhabitants of the hills (on the other). This fact is known to those who have investigated the matter.

As to the reason for it, it may be tentatively suggested that a great amount of food and the moisture it contains generate pernicious superfluous matters in the body, which, in turn, produce a disproportionate widening of the body, as well as many corrupt, putrid humors. The result is a pale complexion and an ugly figure, because the person has too much flesh, as we have stated. When the moisture with its evil vapors ascends to the brain, the mind and the ability to think are dulled. The result is stupidity, carelessness, and a general intemperance. This can be exemplified by comparing the animals of waste regions and barren habitats, such as gazelles, wild cows (*mahâ*), ostriches, giraffes, onagers, and (wild) buffaloes (cows, *baqar*), with their counterparts among the animals that live in hills, coastal plains, and fertile pastures. There is a big difference between them with regard to the glossiness of their coat, their shape and appearance, the proportions of their limbs, and their sharpness of perception. The gazelle is the counterpart of the goat, and the giraffe that of the camel; the onagers and (wild) buffaloes (cows) are identical with (domestic) donkeys and oxen (and cows). Still, there is a wide difference between them. The only reason for it is the fact that the abundance of food in the hills produces pernicious superfluous matters and corrupt humors in the bodies of the domestic animals, the influence of which shows on them. Hunger, on the other hand, may greatly improve the physique and shape of the animals of the waste regions.

The same observations apply to human beings. We find that the inhabitants of fertile zones where the products of agriculture and animal husbandry as well as seasonings and fruits are plentiful, are, as a rule, described as stupid in mind and coarse in body. This is the case with those Berbers who have plenty of seasonings and wheat, as compared with those who lead a frugal life and are restricted to barley or durra, such as the Maṣmûdah Berbers and the inhabitants of as-Sûs and the Ghumârah. The latter are superior both intellectually and physically. The same applies in general to the inhabitants of the Maghrib who have plenty of seasonings and fine wheat, as compared with the inhabitants of Spain in whose country butter is altogether lacking and whose principal food is durra. The Spaniards are found to have a sharpness of intellect, a nimbleness of body, and a receptivity for instruction such as no one else has. The same also applies to the inhabitants of rural regions of the Maghrib as compared with the inhabitants of settled areas and cities. Both use many seasonings and live in abundance, but the town dwellers only use them after they have been prepared and cooked and softened by admixtures. They thus lose their heaviness and become less substantial. Principal foods are the meat of sheep and chickens. They do not use butter because of its tastelessness. Therefore the moisture in their food is small, and it brings only a few pernicious superfluous matters into their bodies. Consequently, the bodies of the urban population are found to be more delicate than those of the inhabitants of the desert who live a hard life. Likewise, those inhabitants of the desert who are used to hunger are found to have in their bodies no superfluous matters, thick or thin.

It should be known that the influence of abundance upon the body is apparent even in matters of religion and divine worship. The frugal inhabitants of the desert and those of settled areas who have accustomed themselves to hunger and to abstinence from pleasures are found to be more religious and more ready for divine worship than people who live in luxury and abundance. Indeed, it can be observed that there are few religious people in towns and cities, in as much as people there are for the most part obdurate and careless, which is connected with the use of much meat, seasonings, and fine wheat. The existence of pious men and ascetics is, therefore, restricted to the desert, whose inhabitants eat frugally. Likewise, the condition of the inhabitants within a single city can be observed to differ according to the different distribution of luxury and abundance.

It can also be noted that those people who, whether they inhabit the desert or settled areas and cities, live a life of abundance and have all the good things to eat, die more quickly than others when a drought or famine comes upon them. This is the case, for instance, with the Berbers of the Maghrib and the inhabitants of the city of Fez and, as we hear, of Egypt (Cairo). It is not so with the Arabs who inhabit waste regions and deserts, or with the inhabitants of regions where the date palm grows and whose principal food is dates, or with the present-day inhabitants of Ifrîqiyah whose principal food is barley and olive oil, or with the inhabitants of Spain whose principal food is durra and olive oil. When a drought or a famine strikes them, it does not kill as many of them as of the other group of people, and few, if any, die of hunger. As a reason for that, it may tentatively be suggested that the stomachs of those who have everything in abundance and are used to seasonings and, in particular, to butter, acquire moisture in addition to their basic constitutional moisture, and (the moisture they are used to) eventually becomes excessive. Then, when (eating) habits are thwarted by small quantities of food, by lack of seasonings, and by the use of coarse food to which it is unaccustomed, the stomach, which is a very weak part of the body and for that reason considered one of the vital parts, soon dries out and contracts. Sickness and sudden death are prompt consequences to the man whose stomach is in this condition. Those who die in famines are victims of their previous habitual state of satiation, not of the hunger that now afflicts them for the first time. In those who are accustomed to thirst[23] and to doing without seasonings and butter, the basic moisture, which is good for all natural foods, always stays within its proper limits and does not increase. Thus, their stomachs are not affected by dryness or intemperance in consequence of a change of nourishment. As a rule, they escape the fate that awaits others on account of the abundance of their food and the great amount of seasonings in it.

The basic thing to know is that foodstuffs, and whether to use or not to use them, are matters of custom. Whoever accustoms himself to a particular type of food that agrees with him becomes used to it. He finds it painful to give it up or to make any changes (in his diet), provided (the type of food) is not something that does not fulfill the (real) purpose of food, such as poison, or alkaloids,[24] or anything excessively intemperate. Whatever can be used as food and is agreeable may be used as customary food. If a man accustoms himself to the use of milk and vegetables instead of wheat, until (the use of them) gets to be his custom, milk and vegetables become for him (his habitual) food, and he definitely has no longer any need for wheat or grains.

The same applies to those who have accustomed themselves to suffer hunger and do without food. Such things are reported about trained (ascetics). We hear remarkable things about men of this type. Those who have no knowledge of things of the sort can scarcely believe them. The explanation lies in custom. Once the soul gets used to something, it becomes part of its make-up and nature, because (the soul) is able to take on many colorings. If through gradual training it has become used to hunger, (hunger) becomes a natural custom of the soul.

The assumption of physicians that hunger causes death is not correct, except when a person is exposed suddenly to hunger and is entirely cut off from food. Then, the stomach is isolated, and contracts an illness that may be fatal. When, however, the amount of food one eats is slowly decreased by gradual training, there is no danger of death. The adepts of Sufism practice (such gradual abstinence from food). Gradualness is also necessary when one gives up the training. Were a person suddenly to return to his original diet, he might die. Therefore, he must end the training as he started it, that is, gradually.

We personally saw a person who had taken no food for forty or more consecutive days. Our *shaykhs* were present at the court of Sultan Abûl-Ḥasan[25] when two women from Algeciras and Ronda were presented to him, who had for years abstained from all food. Their story became known. They were examined, and the matter was found to be correct. The women continued this way until they died. Many persons we used to know restricted themselves to (a diet of) goat's milk. They drank from the udder sometime during the day or at breakfast.[26] This was their only food for fifteen years. There are many others (who live similarly). It should not be considered unlikely.

It should be known that everybody who is able to suffer hunger or eat only little, is physically better off if he stays hungry than if he eats too much. Hunger has a favorable influence on the health and well-being of body and intellect, as we have stated. This may be exemplified by the different influence of various kinds of food upon the body. We observe that those persons who live on the meat of strong, large-bodied animals grow up as a (strong and large-bodied) race. Comparison of the inhabitants of the desert with those of settled areas shows this. The same applies to persons who live on the milk and meat of camels. This influences their character, so that they become patient, persevering, and able to carry loads, as is the case with camels.[27] Their stomachs also grow to be healthy and tough as the stomachs of camels. They are not beset by any feebleness or weakness, nor are they affected by unwholesome food, as others are. They may take strong (alkaloid) cathartics unadulterated to purify their bellies, such as, for instance, unripe colocynths, *Thapsia garganica*, and Euphorbia. Their stomachs do not suffer any harm from them. But if the inhabitants of settled areas, whose stomachs have become delicate because of their soft diet, were to partake of them, death would come to them instantly, because (these cathartics) have poisonous qualities.

An indication of the influence of food upon the body is a fact that has been mentioned by agricultural scholars[28] and observed by men of experience, that when the eggs of chickens which have been fed on grain cooked in camel dung, are set to hatch, the chicks come out as large as can be imagined. One does not even have to cook any grain to feed them; one merely smears camel dung on the eggs set to hatch, and the chickens that come out are extremely large. There are many similar things.

When we observe the various ways in which food exercises an influence upon bodies, there can be no doubt that hunger also exercises an influence upon them, because two opposites follow the same pattern with regard to exercising an influence or not exercising an influence. Hunger influences the body in that it keeps it free from corrupt superfluities and mixed fluids that destroy body and intellect, in the same way that food influenced the (original) existence of the body.

God is omniscient.

NOTES [FROM 1958 TRANSLATOR FRANZ ROSENTHAL]

1 Cf. Issawi, pp. 39 f.

2 Muhammad b. Aḥmad b. Rushd, 520–595 [1126–1198]. Cf. *GAL*, I, 461 f.; *Suppl.*, I, 833 ff.

3 Translation of *mu'tadil* in the usual way by "temperate" would not seem to be correct here. The word must here be translated by "symmetrical," or the like. This becomes clear from the discussion of Averroes' view of the problem found in L. Gauthier, *Ibn Rochd* (Paris, 1948), pp. 84 ff. Averroes argues against the opinion advanced by Ibn Ṭufayl that the region around the equator was temperate. He maintains that Ibn Ṭufayl misunderstood the word *mu'tadil*, which could mean both "uniform" (symmetrical) and "temperate." Averroes further rejects the idea that the southern part of the earth contains habitable areas comparable to those in the north.

This would seem, in effect, the direct opposite of the opinion Ibn Khaldûn here attributes to Averroes. However, the latter came out elsewhere for the theory of a habitable area in the south, which would be in a symmetrical position with relation to that in the north, as we learn from Gauthier, *ibid.*, pp. 87 f. Consequently, Ibn Khaldûn's report on Averroes here is incomplete – in a way, misleading – but it is not incorrect. Cf. also C. Issawi, *Osiris*, X (1952), 114 f.

The idea that the equator has a temperate climate is also mentioned in al-Bîrûnî, *Chronologie orientalischer Völker*, ed. C. E. Sachau (Leipzig, 1878; 1923), p. 258; tr. by the same (London, 1879), p. 249.

4 Cf. Issawi, pp. 42–46.

5 Bulaq adds here: "and religions, even including the various (manifestations of) prophecy that are mostly to be found there, in as much as no historical information about prophetic missions in the southern and northern zones has come to our notice. This is because only those representatives of the (human) species who have the most perfect physique and character are distinguished by prophets and messengers. The Qur'ân says [3.110 (106)], 'You are the best group (ever) produced for mankind.' The purpose of this is to have the divine message of the prophets fully accepted."

The available MSS, including E, do not have this passage, which apparently was deleted by Ibn Khaldûn very early as superfluous, in view of such later remarks as those below, pp. 59 and 60.

6 Bulaq adds: "Romans (Rûm), Greeks. . . ."

7 Qur'ân 16.8 (8).

8 Cf. Gen. 9:25.

9 See pp. 56–57, above.

10 As we can observe throughout this chapter, the same Arabic word is used by Ibn Khaldûn to designate temperateness of climate and living

conditions, and the resulting temperance of moral qualities.

11 See p. 58, above.

12 Cf. the translation of Avicenna's poem by K. Opitz in *Quellen und Studien zur Geschichte der Naturwissenschaften und der Medizin*, VII[2] (1939), 162, vv. 50–51. The same work appears to have been the subject of a study by H. Jahier and A. Noureddine, in *IVe Congrès de la Fédération des Sociétés de Gynécologie et d'Obstétrique* (1952), pp. 50–59, and of a new edition and translation by the same authors, published in 1956.

On the subject of the origin of the black and the white colors of skin, cf. also *Rasâ'il Ikhwân aṣ-ṣafâ'* (Cairo, 1347/1928), I, 233f.

13 Bulaq and B have "names."

14 Cf. Issawi, p. 50.

15 Cf. Qur'ân 33.62 (62); 35.43 (41); 48.23 (23). The last sentence is also often translated, "You will not find any change in God's way." The translation given in the text appears to represent the meaning as intended by the Prophet. It would be difficult to be certain about Ibn Khaldûn's understanding of the passage. Qur'ân commentators, such as al-Bayḍâwî, combine both translations.

16 Cf. Issawi, pp. 46f.

17 Cf. R. Dozy in *Journal asiatique*, XIV[6] (1869), 151, and *Supplément aux dictionnaires arabes*, II, 831*b*. Cf. also A. Mez, *Die Renaissance des Islâms* (Heidelberg, 1922), p. 157. For the theory that expansion and contraction of the animal spirit cause joy or sadness, cf. F. Rosenthal, *Humor in Early Islam*, p. 137.

18 Qur'ân 15.86 (86); 36.81 (81).

19 Cf. al-Mas'ûdî, *Murûj adh-dhahab*, I, 164f. For the famous ninth-century philosopher al-Kindî, see *GAL*, I, 209f.; *Suppl.*, I, 372ff. From among the many recent publications concerning him, we may mention M. 'A. Abû Rîdah, *Rasâ'il al-Kindî al-falsafîyah* (Cairo, 1369–72/1950–53). Cf. also R. Walzer, "The Rise of Islamic Philosophy," *Oriens*, III (1950), 1–19.

20 Qur'ân 2.142 (136), 213 (209), etc.

21 Cf. Issawi, pp. 47–49.

22 Ibn Khaldûn has just mentioned them as belonging to the former group. Cf. A. Schimmel, *Ibn Chaldun*, p. 26 (n. 9)

23 *'Aymah* means, in particular, "thirsting after milk."

24 Cf. Bombaci, p. 444. *Yattû'* may be specifically Euphorbia, but below, p. 183, it is used as a general term for alkaloids taken as cathartics.

25 The Merinid of Fez who ruled from 1331 to 1351 and was the predecessor of Abû 'Inân, under whom Ibn Khaldûn came to Fez.

26 Or, "when breaking their fast." This may be the preferable translation, even though Ibn Khaldûn does not seem to think of ascetics in this passage.

27 This remark occurs in an appendix to L. Mercier's translation of Ibn Hudhayl, *La Parure des cavaliers* (Paris, 1924), p. 355. The autjor of the appendix, however, is not the fourteenth-century Ibn Hudhayl, or any other old author, but the modern Muḥammad Pasha. Cf. *GAL Suppl*, II, 887.

28 That is, people familiar with works on agriculture such as the *Falâḥah an-Nabaṭîyah*.

4

The Jungle and the Aroma of Meats

An Ecological Theme in Hindu Medicine

Francis Zimmermann

The inquiry reported in my 1987 book *The Jungle and the Aroma of Meats* began with a few pages in a Sanskrit medical treatise (the *Suśrutasaṃhitā*), a catalogue of meats where animals were classified in two groups: *jāṅgala*, 'those of the dry lands', and *ānūpa*, 'those of the marshy lands'. The meats of the jungle were light and astringent, those of the marshy lands heavy, unctuous and liable to provoke fluxes. There was no zoology in ancient India, only catalogues of meats. The division of the various branches of knowledge favored utilitarian or anthropocentric disciplines such as medicine, and devalued the pure sciences: the rudiments of zoology and botany were literally dissolved into pharmacy. But that pharmacy in turn presupposed a cosmic physiology: the world seen as a sequence of foods, and a series of cooking operations or digestions at the end of which nourishing essences derived from the soil were exhaled in the medicinal aroma of meats.

The polarity between the dry lands and the marshy ones occurred in the inventory of flora and fauna: acacias and coconut palms, and so on (Figure 4.1). It reappeared in therapeutics as a polarity between a basic savor and one of the three humors: unctuosity to be applied against the disorders of wind, bitterness against the acridity of bile, and astringency to counteract a superabundance of phlegm. Eventually, medical doctrine fitted the general Brahminic tradition with its polarities of Agni (the Sun) and Soma (the Moon), fire and water. The aridity of the jungles scorched by the sun could be at one point associated with and at another point opposed to the nourishing unctuosity of the rain. My research describes this curious wrapping of different layers of knowledge in each other.

Since ancient science was interested in the dietetic and therapeutic virtues of the jungle and its fauna, I pursued the register of pharmacy by analysing the language used to describe the aroma, savor or bouquet of a particular soil and of its inhabitants, taken as potential remedies. The use of remedies led to the superior

The Anthropology of Climate Change: An Historical Reader, First Edition. Edited by Michael R. Dove.

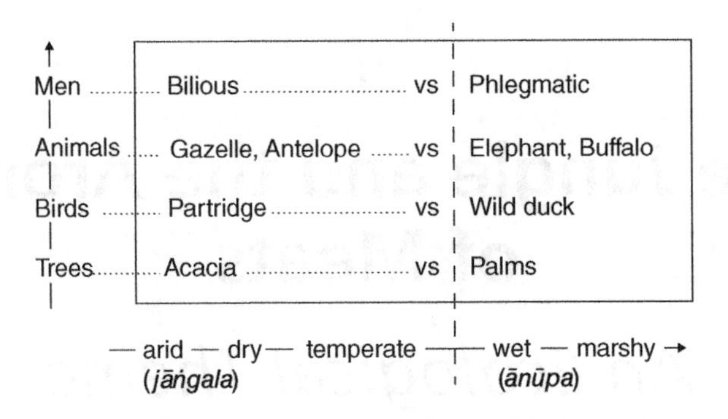

Figure 4.1

register of physiology which, in the ancient sense, governed the circulation of fluids in the environment, including the saps in plants, the aromas in cooking different kinds of meat, and the interplay of humors within the human body.

The Great Chain of Being

The Seers of Vedic times quite literally conceived of the universe as a kitchen in which a kind of chemistry of savors fed now one and then another domain of Nature: stars, waters, the earth, plants, and finally, animals. The cold rays of moonlight increased the unctuosity of waters which impregnated the earth with *rasa*, or nourishing essences. The *rasa* were in the first place six fundamental savors: sweet, sour, salty, acrid, bitter and astringent. Savors were invisible, as sugar and salt are in water, but they were present everywhere in the sap of plants and the juice of meats. They were not mere qualities of things, but essences which circulate to the depths of the landscape and are diffused through the great chain of being.

What Europeans called 'the great chain of being' was in India a sequence of foods. At the end of the sequence meats were cooked and the cooking was the last predigestion of foods before they were consumed. Animals in this sequence were eaten eaters, at once flesh and meat, whose ambiguous position revealed the natural violence of life. "Immobile beings are the food of those which are mobile, those without teeth are the food of those with teeth, those without hands are the food of those with hands, and the coward is the food of the brave".[1] To account for the subordination of some classes of beings to others, it was not enough to say that animals ate plants and human beings ate meat. In the animal kingdom and then the human one, the dialectic of the eaten eater introduced further divisions between the strong and the weak, the predator and his prey, the carnivore and the vegetarian.

Vegetarianism, a Brahminic ideal and a common social practice in India, precisely called into question the fateful dialectic in which every class of being fed upon another. The prohibition of flesh, which became increasingly strict among Brahmins, was to break the chain of alimentary violence, and to affirm that it was not necessary to kill to eat. A new opposition between human beings was emphasized. Rather than being a matter of courage and fear, domination and servitude, the opposition was phrased as one between the pure and impure in a ritual hierarchy of castes. Abstention from eating meat became a criterion of purity.

Since the organizing principles of Hindu tradition were essentially juridical and religious, I am interested in the way that medicine tended to contradict orthodoxy and ritualism in confronting physical distress and emergency. Ayurvedic physicians superimposed upon the religious principles of purity and hierarchy other principles of a different order. Starting from the observation that animals composed a series of hostile couples: herbivores vs carnivores, game vs beasts of prey, the antelope vs the lion, and the partridge vs the crow, physicians used natural violence for therapeutic ends. Flesh nourished flesh, and the nourishment that was the most restorative for an anaemic or consumptive person was not just meat, but the meat of animals that ate meat. The state of the patient might render such medication indispensable, even if a vegetarian patient had to be lied to in order to overcome his disgust.

Physicians could not escape violence. This fact, within a civilization that invented and highly valued the concept of *ahiṃsā* (absence of the desire to kill), lowered the physician's social status. And this occurred despite the Brahminic character of classic medical knowledge.

Cognitive Approach

My inquiry borrows illustrations from contemporary tropical geography, botany and zoology, but its point of departure is the study of texts. I am concerned with a corpus of traditional notions into which it is nevertheless possible to subsume observations from the natural sciences to discover what is at stake in Hindu scholasticism—the demands, conflicts, and vision to which the Sanskrit texts bear witness.

A complex of arguments, keys, classificatory grids and cartographic procedures make it possible for a modern scholar to situate the principles of traditional Hindu ecology and the typology of soils laid down in the Sanskrit medical treatises. My 1987 book contains maps, but here I only refer to the most salient division within the Indo-Gangetic plains where in the west an open vegetation of arid bushes, sand and rock formed the *jāṅgala*, and to the east the forests and marshes of Bengal formed the *ānūpa*. In between these extremes the dry tropical forest of sparsely scattered trees which lost their leaves in the summer merged with the tree-bearing savannah, and this constituted the *sādhāraṇa*, or middle region. This polarity was expressed in the food (wheat to the west, rice to the east), the pharmacy and the bodily techniques of human communities, giving rise to people of thin, dry and bilious temperament at one extreme, and at the other to rotund people who were susceptible to the disorders of phlegm. Consumption was a characteristic malady of the *jāṅgala*, and elephantiasis of the *ānūpa*.

The traditional geography of malaria also displays this polarity, with the epidemic form in the east and the endemic form in the west. The phytogeographical frontier between the thorny formations to the west and the eastern forest assumed a capital importance for me when I noticed that it corresponded almost exactly to the ecology of malaria in India right down to 1947, when campaigns to eradicate malaria disturbed this historical continuity.[2] Nevertheless, the traditional polarity between the dry *jāṅgala* and marshy *ānūpa* lands took on its full meaning for me in reference to the long-standing contrast between regions of endemic and epidemic malaria.

In the classical texts the polarity was obscured by a plethora of adjectives. In *La Formation de l'Esprit Scientifique*, Gaston Bachelard described the semantic proliferation which infested the whole of ancient scientific literature. Every medicament was covered in adjectives, and it came as no surprise to find in Ayurveda a luxuriance of synonyms, a mixture of wordy empiricism and phantasmagoria, an omnipresent dialectic between contrary qualities. This was the stuff of all old medical treatises and in different societies reflected similar unconscious reveries of pre-scientific thought. At first, then, the doctrine that the meats of dry lands are astringent and those of wet lands provoke fluxes struck me as pure verbiage. In my tradition, after all, Bachelard related how, in 1669, the Académie Française dissected a civet-cat in order to compare it with a beaver. Their conclusions ran as follows: The strong and unpleasant smell of the *castoreum* came "from the cold wetness of the beaver, which is half-fish"; the fluid secreted by the civet-cat, in contrast, was sweet-smelling because the animal "is of a hot and dry temperament, drinks little and usually lives in the sands of Africa".[3] Wet, cold habitats went with the stench of fermenting substances; hot, dry habitats with ethereal fragrances. No need, one might say, to go seeking in the Hindu texts for principles so well formulated by the French 'Immortals' of 1669!

Yet these texts may also be considered from a slightly different angle. For example: *The gazelle is light, the buffalo indigestible.* Let us determine the medical, religious, political and social context of such a statement in the Ayurvedic texts. A whole world looms up in the background: the jungle and the anthropological structures of space.

Garlands of Names

Modern works devoted to Indian medicine generally classify the doctrine under headings to which we are accustomed in the Greek and Latin tradition: anatomy, physiology, pathology, and so on. Interpreters consequently suppose the ordering of the material to have been similar to that adopted in our own ancient medicine. But the analogy should not be taken for granted. One difference of capital importance which does not emerge from this mode of presentation affects the very nature of the reasoning of ancient Sanskrit texts and the rules of literary composition. Once a field, or subject of study has been determined, we heirs of Greco-Latin logic proceed by way of description and argument. Each heading announces the account of a particular line of research: osteo-logy, physio-logy, patho-logy. The suffix itself indicates an objective approach, a distinction drawn on principle between the researcher and his field of study. The situation in the Ayurvedic treatises is quite different. In ancient India, the primary material of knowledge was constituted by the recitation of series of words of a more or less stereotyped nature. For example, if one was concerned with the fauna and materia of animal origin then it would be quite wrong to give the name of zoology to what in Sanskrit literature is at the most a bestiary. There was no zoology in the minds of the Indian scholars, nor any osteology or physiology. Instead, there were endless 'garlands of names' (*nāmamālā*), the image is a Sanskrit one, and onto these name lists an amazing combinative system of qualities and savors (*rasa*) was grafted.

A single animal could possess a dozen or more names. The particular connotations of each name added to knowledge, and thus, to describe was to name. Take, for example, the cuckoo. Aristotle mentioned a well-known fact in describing the behavior of these birds, "The cuckoo makes no nest but lays its eggs in the nests of other birds … It lays only one egg but does not itself hatch it. It is the bird in whose nest the egg has been laid that breaks it to hatch it out and that feeds the young bird".[4] Ayurvedic treatises contain nothing of this kind—no account, no story, only the equivalence between synonyms. One of the names of the *kokila*, the Indian cuckoo, is in effect *parabhṛta*, 'the-one-which-was-fed-by-another'. This is the name in the Catalogue of Meats in the great medical text, the *Suśrutasaṃhitā*. Where Greek and Latin naturalists observed and registered curious details, and invented the model of knowledge which we call natural history, Indian scholars followed a different path of reasoning and created a thesaurus of names, and of semantic equivalences. My concern was the manner in which therapeutic savors and qualities were grafted upon this animal nomenclature. Or, to put it another way, I sought the logical and poetic rules which determined the use of hundreds of adjectives and the nominal forms of verbs in medical discourse.

A Multi-Faceted Knowledge

All Sanskrit medical treatises contain a Catalogue of Meats, more or less clearly based upon the polarity between *jāṅgala* and *ānūpa*. Suśruta's list began as follows:

> And now we shall enumerate the different kinds of meats, namely: aquatic, *ānūpa*, domesticated, carnivorous, whole-hoofed and *jāṅgala*. That makes six kinds of meats, which are listed in order of increasing excellence.

A hierarchy was thus established, for which the criteria were given in the conclusion of the catalogue:

> Aquatic, *ānūpa*, domesticated, carnivorous and whole-hoofed, and those which tear at their food, those which have a burrow, those called *jaṅghāla* ("which have legs"), those which peck and scatter grain (in that order) precisely—these are (meats), each one of which is lighter than the one before and produces far fewer fluxes than the one which precedes it (in the list), and vice versa.[5]

I will skip the catalogue of about two hundred meats to which medicinal properties were ascribed in the form of stereotyped phrases, such as:

1. astringent, sweet, cordial,
2. calming bile-blood and disorders of phlegm,
3. constricting, appetising, fortifying
4. is the black antelope, also febrifuge.

The catalogue was versified, and each of the lines printed above corresponded to an octosyllabic Sanskrit hemistich or quarter of a distich. These sets of adjectives and therapeutic indications constituted a phraseology in the sense that the same groupings recurred time and again throughout the corpus of medical texts. Moreover, some sort of synonymy existed between 'astringent–sweet–cordial' and 'calming bile-blood and phlegm'. These were mutually substitutable formulas which belonged to the same glossarial paradigm. An opposite paradigm would be composed of formulas such as: 'unctuous–hot–calms wind', 'unctuous–virilifying–calms wind', and so on, which again corresponded to octosyllabic Sanskrit hemistiches describing meats of opposite medicinal properties. The interplay of these octosyllabic sets of words (a few thousand hemistiches) is the main tool the doctor could use to assess, compose and prescribe remedies. And this is still true today of any learned Ayurvedic practitioner.

What were the rules which governed this rotation of the lexical stock, this interplay of hemistiches, this combinative system of both taxonomy and prescriptions? Let us start from a classic problem in the natural and social sciences: the criteria used to classify objects also require classification. Here, the objects were medicinal substances and the criteria for ordering them were their savors, qualities and therapeutic or pathogenic effects. The simple method of ordering them was a linear one that gave rise to hierarchies of technical terms. For example, the six categories of meats provided by the different animals: aquatic–marshy–domesticated–carnivorous–whole-hoofed–*jāṅgala*. To arrange these terms on the same scale all that was needed was to select a variable, in this case lightness, for an operator which would indicate increase or decrease: 'they are progressively (*yathottaram*) light', each successive term is lighter than the one which precedes it. Or, again, 'less prone to produce fluxes', where each successive term is less productive of fluxes than the one which precedes it. Here, then, was a series of names; r<s<t<u<v<... which, in this order, were more and more or less and less. The operators often used were: *yathottaram*, 'in an increasing order, each a degree more than the one before' and its converse, *yathāpūrvam*, 'in a decreasing order, each a degree less than the one before'.

But Ayurvedic doctors also practiced another infinitely more complex typological operation which assumed combinations of variables. This time the variables were not linear but dichotomous. Thus: *jāṅgala* meats which were light–dry–astringent–constrictive were contrasted with *ānūpa* meats, which were heavy–unctuous–emollient–diaphoretic. These were homogeneous constellations of savors, qualities and actions. To discover the method for such grouping I distinguish three different kinds of elements in the technical language:

{ the *operators*
{ the objects to be classed { the *names* of substances
{ the medical *criteria*

(1) The *operators*. These were the logical means of a combinative system: 'which excites', 'which calms', 'good for', 'not too much', 'more and more', 'generally', 'specially', and so on.
(2) The *names*. The names of plants, animals and other medicinal substances (milk, fats, minerals, etc.) constituted the basic material for typological operations. The elements 'antelope' and 'partridge' had the same logical value as the categories in which they were included. Whether specific or generic, names were the basis for the combinative system of medicinal properties.
(3) The *criteria*. These consisted of all the humors, savors and other properties distributed over a number of dimensions within the same logical space, forming a single property space divided into a number of facets. Each substance or category of substances could be defined by its action upon the humors, or by the savors of which it is composed, and so on. It thus presented multiple facets each of which corresponded to one of the dimensions of an *n*-dimensional space.

The three humors, the six savors and a stereotyped list of ten couples of contrary qualities constituted the most important facets or dimensions

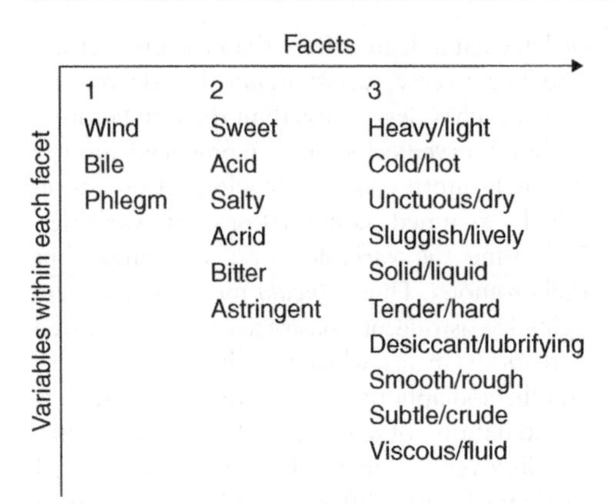

Facets
Variables within each facet
1
Wind
Bile
Phlegm
2
Sweet
Acid
Salty
Acrid
Bitter
Astringent
3
Heavy/light
Cold/hot
Unctuous/dry
Sluggish/lively
Solid/liquid
Tender/hard
Desiccant/lubrifying
Smooth/rough
Subtle/crude
Viscous/fluid

Figure 4.2

Facets →

...	4	5	...
	Earth	Depletive	
	Water	Nutritive	
	Fire	Drying	
	Air	Hydrating	
	Ether	Sudorific	
		Styptic	

Figure 4.3

in this combinative system (Figure 4.2). However, it was always possible to increase the number of dimensions of this ideal space in which each substance was situated by combining the characteristics attributed to it. This produced further facets (Figure 4.3).

Depending upon which of the five cosmic elements predominated within it, each substance was essentially earthy, watery, fiery, airy or ethereal; dependent upon which of the six modes of therapeutic action it most favored, each drug was depletive, nutritive, drying, and so on. Thus adjectives, whether simple (*laghu*, 'light'), derivative (*pittala*, 'bilious', *pārthiva*, 'earthy') or compound (*māruta-nāśana*, 'which calms wind'), and verbal nouns such as *laṅghana*, 'fast, slimming cure, depletive action', and *bṛṃhaṇa*, 'feeding-up, nutritive action', were multiplied through a kind of word inflation, or a mania for enumerations, recitations, seriations and verbigerations.

Certain homonymies may cause confusion, as in the case of *vāyu*, which means 'wind' (the humor) and 'air' (the element), or *soma*, which means 'moon', and by extension, 'water' (the element), and sometimes also 'phlegm' (the humor). This was a consequence, in the domain of logic, of the religious and metaphysical analogy between the microcosm of the human body and the macrocosm of the surrounding world.

From Images to Concepts

The system drew upon familiar images with which the notion of vital functions was commonly associated. For example, digestion was the work of fire and bile, a fiery fluid. The humors had a double reality. They were a set of images of saps, winds, and even fire, as fluids circulating outside and inside human bodies. In medical reasoning they were also abstract functions, the morbid factors that could accumulate or become disordered and thus engender disease.

The point of interest upon which the entire weight of my analysis was concentrated was the articulation between the image of a fluid and the concept of a pathological factor, for this was the point at which the Ayurvedic doctor launched a conceptual system. At that moment Wind ceased to mean wind or flatulence and became the principle of consumption and rheumatism, or, similarly, Phlegm was no longer simply an image of excessive serosity or unctuosity, but became the abstract principle of elephantiasis. The leap that this abstraction involved was very different from the one made in the Greek and Latin tradition, and which Westerners still tend to make. Abstraction for us means generalization, a reduction of specifications. For Ayurvedic scholars abstraction meant and still means over-determination, multiplying adjectives and points of view. The difference is enormous.

The expressions I have used to describe this system, the notions of a 'property space' and a 'facet design', were borrowed from American sociologists who in the line of Paul Lazarsfeld studied how typologies were constructed and deconstructed.[6] A prolix series was turned into a concise one by selection, fusion and reduction of the variables; in short, by a process of subsumption. Thus, unctuous, tender, emollient, and nourishing were subsumed under the genus sweet; and cold, dry, light, desiccant were subsumed under the concept of wind. But the converse typological operation, which Paul Lazarsfeld called substruction, required also to be described. The problem was to work back from a typology to the combinative system which produced it. Wind and sweet resulted from a simplification; we must reconstruct the complexity from which they emerged, spell out all the variables that they summarize. In the Ayurvedic system, this substruction, or deconstruction, to keep in fashion, was of primary importance. It determined the very formulation of the hemistiches or quarter distiches. For example, let us select unctuous from among the 20 qualities (*guṇa*), hot from among the two energies (*vīrya*), sweet from among the six savors (*rasa*), and virilifying from among the countless effects of drugs, so as to compose the hemistich 'unctuous–hot–sweet–virilifying' to describe a particular drug which calms wind. Each drug thus is ascribed a particular constellation of properties, and reciprocally, the different facets of the combinative system of properties overlap at one of the variables from which each one is composed, since the unctuousness, heat and sweetness of a drug which calms wind entail its being also virilifying.

Better still, the substruction of humors and savors was systematically explained in the Sanskrit treatises. For example: "Wind is characterized by the qualities of coldness, dryness, lightness, desiccation, constriction; the astringent savor is homogeneous to it".[7] The qualities could be repressed by drugs of opposite qualities. When the attributes of each humor were enumerated it was not simply a matter of deconstructing the existing categories to qualify and refine them, but of pressing on to reconstruct the system upon other bases, multiplying all the possible points of view. What was involved was not the division of a genus into species, which could be represented by a genealogical tree, nor the organization of an overall view by means of inclusion, which might be represented by a trellis design.

We are a long way away from the Western logical tradition and would do better to think of China. In contrast to the Chinese treatises, in which the very writing, an emblematic writing, conditioned the existence of a many-faceted type of thought, the Sanskrit texts were quite devoid of any means of iconographic expression. All the same, the linear appearance of the writing should not mask the spatial quality of the thought, for each link in the chain of writing implicitly referred to one or another of its facets. The writing was linear, but the thought was combinative.

Deconstruction: An Example

The concept of energy (Sanskrit *vīrya*) and digestion, a technical term, Sanskrit *vipāka*, to denote the alteration of a savor through digestion, seem to have been forged to facilitate reductions within a complex typology. All digestion (*vipāka*) was either acrid or sweet. All energy was either cold or hot. Savors and qualities were thus reduced to two pairs of contraries: acrid/sweet, cold/hot. However, the possibilities for reducing dimensions complicate the system of evaluating and classifying substances by superinducing background factors. Thus, acrid and sweet are savors (*rasa*), but by inference and with hindsight they are alterations in primary savors due to digestion (*vipāka*). Cold and hot are qualities (*guṇa*), and, by inference, energies (*vīrya*) resulting from a particular configuration of savors and qualities.

The Catalogue of Meats was an excellent example of the multiplication of points of view. As a whole, meats were of a sweet savor, sweet digestion and cold energy; a set of properties linked to the production of phlegm, unctuosity and fluxes. Within that whole, *jāṅgala* meats were sweet but also astringent, and were distinguished from *ānūpa* meats in which only traces of astringency were as subsidiary savor (*anurasa*). This is outlined in Figure 4.4. At the same time another point of view surreptitiously impinged. Some substances of sweet savor and digestion unexpectedly develop a contrary hot energy. Among these exceptions were the *ānūpa* meats.[8] Hence, Figure 4.5 is quite different. This is why, in the catalogue of Suśruta's early text meats which excited phlegm through their unctuosity, a quality usually associated with a cold energy, also excited

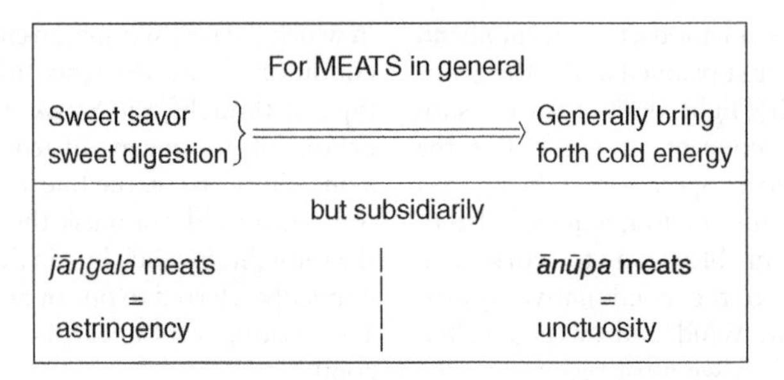

Figure 4.4

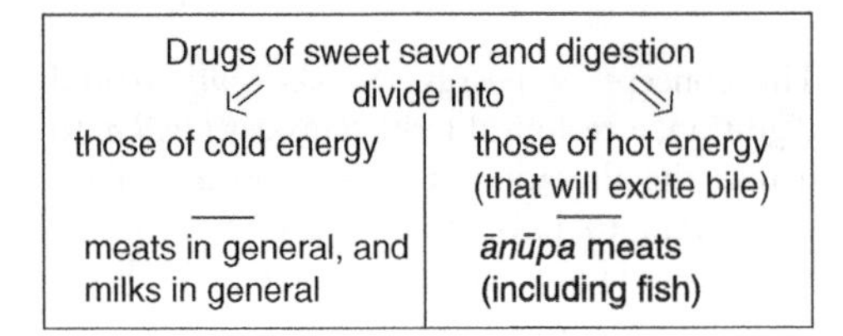

Figure 4.5

bile through their hot energy. This was the exceptional consequence of a sweet savor and a sweet digestion. The excitation of bile thus appeared on the side of the marshy meats, whereas one would have expected it to be on the side of the dry lands.

A possible justification for an important dietetic principle emerges: the incompatibility within a single dietetic regimen between milk, which in general was of sweet savor, sweet digestion and cold energy,[9] and the domesticated, marshy and aquatic meats, which were of hot energy.[10] Foods and remedies were evaluated and classified by means of dichotomies which gave the whole discourse a scholastic and artificial air. In all probability, however, this sophisticated system of humors, savors and qualities, digestions and energies, in most cases simply systematized the facts of traditional observation or the rules for popular dietetic practices. For example, among the causes of leprosy the Ayurvedic treatises mention the harmfulness of a diet that associates incompatible foodstuffs such as milk and fish. Ethnographic evidence of this traditional belief exists,[11] and the 1790 issue of *Asiatick Researches* included a letter from a Muslim medical practitioner 'On the Cure of Elephantiasis', which denoted a form of leprosy, the 'elephantiasis' of the Greeks. This text from At'har Ali' Kha'n of Delhi, ran as follows:

> A common cause of this distemper is the unwholesome diet of the natives, many of whom are accustomed, after eating a quantity of fish, to swallow copious draughts of milk, which fail not to cause an accumulation of yellow and black bile, which mingles itself with the blood and corrupts it.[12]

The incompatibility which the Sanskrit texts expressed in logical form was here described as a curious fact in the language of trial-and-error experience. The bile mentioned was not that excited by the hot energy of fish, which in Ayurveda was a specific pathogenic factor in a rigorously logical system. For this eighteenth century Muslim it was the yellow and black bile of the Hippocratics. In either case, a popular belief that the mixture of milk and fish provokes leprosy was justified by appeal to a scholarly notion. The difference lies in the fact that black and yellow bile was an imported vulgarized notion that was no longer co-ordinated within an overall etiological system, and did not refer to any background. Quite the opposite of the intact classical thought of Ayurveda. One may also find in modern times similar vulgarization of this system.

The Poetics of Medicine

The Sanskrit treatises constitute an erudite or technical literature, but above all they were thought to be transcriptions of an authoritative word. They were normative texts that transcribed authentic, orthodox knowledge. And this property was obtained through poetry. The phraseology, the combinative interplay of formulas, the way in which the hemistiches recur in the versified sequences, the rhythmic and syntactic parallelism of opposite formulas constituted a poetic.

Roman Jakobson and the whole school of linguists, folklorists and theoreticians of literature who founded the famous Society for the Study of the Poetic Language in Russia in 1916–17 systematically extended the study of poetic language beyond the limits of poetry. Jakobson showed Sanskrit didactic poetry was quite relevant to that approach:

> Mnemonic lines like *Thirty days hath September*, modern advertising jingles, and versified medieval laws .... or finally Sanskrit scientific treatises in verse which in Indic tradition are strictly distinguished from true poetry ... all these metrical texts make use of poetic function without, however, assigning to this function the coercing, determining role it carries in poetry.[13]

It is sometimes said that the versification of scientific texts had the utilitarian purpose of making them easier to memorize. But it was not so much a matter of aiding the memory as of the promotion, prescription and scansion of knowledge.

The fundamental problem of poetry, Jakobson went on to say, was that of parallelism. The word evokes a mass of contemporary studies on folklore and oral poetry, but the principle was noted by Gerard Manley Hopkins as early as 1865:

> The artificial part of poetry, perhaps we shall be right to say all artifice, reduces itself to the principle of parallelism .... But parallelism is of two kinds .... Only the first kind, that of marked parallelism, is concerned with the structure of verse—in rhythm, the recurrence of a certain sequence of syllables .... Now the force of this recurrence is to beget a recurrence or parallelism answering to it in the words or thought.[14]

Speech encourages semantic stereotypes, and the formal similarities displayed by writing on the blank page, or palm-leaf in the case of the Indian manuscript, induces a parallelism of meanings within the space to which the text refers.

I shall content myself here with a few allusions to the sequences of octosyllabic hemistiches for which my 1987 book *The Jungle and the Aroma of Meats* offers more complete translations. Recurrence and parallelism are the same as Hopkins noted, whether it be the period repetition of one formula through a gamut of variants such as astringent–sweet–cordial, astringent–sweet–light, astringent–sweet–dry, which are mutually substitutable variants within one and the same paradigm, or a pair of formulas which constitute a negative parallelism, such as:

Few fluxes ...	(= *jāṅgala*)
Many fluxes ...	(= *ānūpa*)
Little water and few trees ...	(= *jāṅgala*)
Much water and many trees ...	(= *ānūpa*)
Gives fire, eliminates Phlegm ...	(= the sun)
Gives Phlegm, eliminates fire ...	(= the moon)

raktapittaharaḥ sí tō
'cold, calming bile-blood'
raktapittakarāś coṣṇā
'hot, provoking bile-blood'
(a parallelism partly based on a pun: the substitution of *k* for *h*, which transforms °*hara* 'calming' into °*kara* 'provoking').

A thousand other examples of this kind could be given. Whether it be a repetition of formulas in oral recitations or a juxtaposition of formulas in a written text, a given sequence of octosyllabic hemistiches constitutes through recurrence and parallelism, a rhythmic and semantic multifaceted system.

Space, Word and Perception

In the vocabulary of the materia medica, we can distinguish between the names of drugs, which denote the most concrete things from the highly abstract language of medicinal qualities. When the text asserted that wind accumulated in summer, bile during the rains and phlegm during winter, the most abstract terms of humoral meta-language were applied to the concrete series, summer–rains–winter.

The cycle of the seasons was in turn used as an abstract framework to classify specific foods and bodily techniques. We thus pass through different strata of language. And again, from the conceptual opposition between dry land and marshy land, we may move either down the ladder of abstraction to the distributional patterns of the flora and fauna over the terrain or else up the ladder of abstraction to language in which theory was formulated. This provided the poetic of recurring hemistiches, which made it possible to formulate the *jāṅgala/ānūpa* polarity in a thousand different ways. The same intellectual structures that appeared most clearly in the parallelism of versified texts also governed instruction about geography and the distribution of the fauna and flora. Thus, traditional space was structured by the language.

The physician, or any other specialist in the Brahminic concept of knowledge, would be by himself incapable of discovering the characteristics or effects of a substance that he was about to use. There was no experimental method, or methodic reasoning other than on the basis of traditional teachings. The model of knowledge was perception, and insofar as perception was imperfect, the word transmitted in the texts aided the physician. Madeleine Biardeau concluded from her masterly researches into Brahminic gnoseology:

> The summit of all knowledge is perception, having reached that peak, thought can rest, in satisfaction .... The desire to know is never gratuitous, it always implies that one expects to benefit, not from knowledge itself but from the object to be known .... Everything that is useful is given, either in perception or in Revelation and the Tradition which completes it.[15]

The medical texts witness this pragmatism and anthropocentricity, as Suśruta testified:

> No need to examine them, no need to reflect on them, they will make themselves known, the remedies that the clear-sighted must prescribe in accordance with tradition.
>
> Their characteristics and effects are self-evident and they make themselves known, the medicinal plants; scholars should in no way quibble about them.
>
> Were there a thousand reasons to do so, the (styptic) *ambaṣṭhā* group will never set itself to purging. The wise must therefore adhere to tradition, without arguing.[16]

Of all that they saw around them, what was best to use? In making that choice, tradition was the law. To the extent that the distribution of things was anthropocentric, the principles of that distribution eluded perception and belonged to the domain of the word.

The Human Body Concept

Since all living beings were eaters soon to be eaten who transmitted one to another the nourishing essences drawn from the earth, the study of pharmacy led to physiology. This account of my 1987 *The Jungle and the Aroma of Meats* is too brief for detailed arguments about the physiology of digestion and the question of violence linked to carnivorous diet. Nevertheless, I must give some account of Hindu physiology.

Where to us physiology is the science of organic functions, precisely the opposite held in Ayurveda, which was a medicine of properties or virtues, a medicine of metamorphoses. The very idea of an organ would make no sense in Ayurvedic physiology, because that physiology was unacquainted with a fundamental distinction which we owe to Aristotle and from which all subsequent Western biology has stemmed. That distinction was between the homoiomeric or uniform parts of the living body (for example, flesh divides up into flesh) and the anhomoiomeric or composite parts (for example, the hand does not divide up into hands, nor the face into faces). The former are the tissues, such as flesh, bone, blood; the latter are the organs. The tissues only exist for the sake of the organs, upon which all physiological functions depend. Nothing of this kind existed in India where all arguments and prescriptions about the human body fell within the framework of a physiology founded upon the concept of *dhātu*, 'tissues' divisible into homoiomeric or uniform parts, and which was ignorant of the idea of an organ. Indeed, the organs of the human body were named, using words which have been habitually and conventionally translated as the heart, the liver, the intestine, and so forth. But when it came to expanding the physiological facts, the parts of the body were mentioned only as reservoirs or receptacles to which fantastical contents were attributed.

The seven *dhātu* or tissues, chyle, blood, flesh, fat, bone, marrow and semen, were changed one into another through a process of successive cooking. Chyle, cooked by a fire which was specially designed for it in the depths of the organism, produced blood; blood through cooking engendered flesh, and so on. Each element was provided with its own particular fire which converted a fraction of this element into its successor in the series. The image of the seven successive fires suggested the chain of metabolic processes by which the living tissues were constructed.

Hence the comparison that Ayurvedic revivalists are today tempted to draw between the organic fires and the enzymes. This was one of the themes dear to the late Dr C. Dwarakanath, who worked on updating Ayurveda by comparing it to modern science, in his capacity as a Secretary at Independence of a famous governmental committee to formulate a policy on indigenous medicine for the Ministry of Health, and, later, as a professor at Jamnagar, and an advisor in New Delhi on indigenous medicine.[17] One of his pupils, Dr Bhagwan Dash, who similarly held office for a while in the Ministry of Health in New Delhi, has also produced a quite brilliant book on the Ayurvedic physiology of nutrition, which is typical of this integrative trend.[18,19] I skip here the detailed analysis of a few medieval Sanskrit texts quoted by them to substantiate their conclusions, and I only concern myself with the ideological aspects of those conclusions.

Both Dwarakanath and Bhagwan Dash explain that the seven fires "obviously refer to substances which, like enzymes, catalyze the synthesis of seven kinds of nutrient substances required for the use of the seven species of *dhātu*, each fire aiding the conversion of nutrient substances in what may permissibly be called precursor substances of the formed *dhātu* already present in the body".[18,19] Each *dhātu* is present in two successive states: fluid and solid, first as a precursory element, next as a constituted one, depending upon whether the chemical syntheses are in the process of taking place or have already done so. In short, each *dhātu* first exists in a nascent state within hot and nourishing fluids, which then solidify and crystallize into stable tissues. With deceptive allusions to modern biochemistry, this is the modern dressing of an ancient image, the image of a network of nutrient-carrying channels which become increasingly elongated and thin as they are required to supply increasingly re-cooked or sublimated tissues. This image of fluids eventually crystallizing into tissues also entails the description of all physiology as a great chain of cooking. Reciprocally, the analysis of the process of digestion in Ayurveda remains entirely on the level of images. For example, the bones allow semen given off by the marrows to swell up through their pores like a new clay pot which is slightly porous: the nutritive juice accumulates in a plethora in the bodily channels like massing clouds before a rainstorm bursts;[20] and a hundred other images of the kind.

The evolution of the body tissues or *dhātu* was strictly determined by the interplay between the three *doṣa*, the humors or pathogenic entities, as the classic text, the *Carakasaṃhitā* made it known in a chapter deceptively entitled *Analysis of the Body*. In spite of its title, it did not produce an anatomy or physiology. It showed how wind, bile and phlegm acted upon the human body and produced diseases. It was a speculative pathogenesis, a reflection upon balance and imbalance between the humors. The text opened with a definition of the healthy body as *samayogavāhin*, 'a vehicle for congruous junctions'. The junction of an element of the body with foodstuffs of similar qualities produced that element's growth. The physician controlled movement within the body by means of an appropriate diet, and through the combinative system of medicamentous qualities:

> Thus, for all tissues and qualities, growth results from junction with what is identical, and diminution results from the contrary.
>
> Consequently, flesh makes flesh increase much more than the other tissues of the body, and blood makes blood increase.

That was why meat was an important drug, and even blood, raw blood, prescribed in some rare emergencies to counterbalance hemorrhages.

> But when foodstuffs possessing that identity ... are not to hand or when, even if they are available, it is impossible to prescribe them on account of impropriety, repulsion or any other reason, and yet it is nevertheless necessary to provoke growth in the body tissue the qualities of which are identical with those of the forbidden foods, then one must prescribe foods which, although different in nature, possess those identical qualities in abundance.[21]

Medicine was a calculus of imagery. Growth resulted from digestion, and to digest was to concentrate within oneself the active powers carried by the nourishing juices. The central image was an agricultural metaphor of the irrigation of body tissues. This irrigation was described in a set of images in which chyle circulated through channels in the body, carrying juices through a series of cooking. At the more abstract level the humors were ambiguous pathogenic factors in a combinative system of qualities, which as fluids also irrigate the tissues.

Unctuosity Sublimed, and a Conclusion

Physiology, which consisted mainly of a set of basic images, clearly was less elaborate than therapy and pharmacy. Accordingly, we shall advance deeper in the analysis of the Ayurvedic tradition by approaching it from a cognitive rather than a symbolic angle, although I have combined the two approaches to implement the rapprochement advocated some years ago in two special issues of the *American Ethnologist. On* the one side, the Ayurvedic physician both classified and evaluated drugs. To go further into this model taxonomy we have to revise current methodology. On the other side, the basic symbols of Hindu cosmic physiology, Agni (the sun) and Soma (the moon) have to be deconstructed.

Needless to say, the dialectic of dry and marshy lands, the alliance between the jungle and the water's edge, is an expression of the 'Agni-Soma-ness' of the world. What is less commonplace is the architecture of the Ayurvedic doctrine, with its layering of different planes of knowledge. What I have interpreted are specialized concepts that form the basis for an abstract combinative system, and which, lower in the scale of abstraction, were rooted in the ground itself.

Let us follow the links in this sequence from the jungle to the gods Agni and Soma. In the beginning the sun and water were allied. Men built towns in the jungle, "in dry lands, where there was perennial water", as an old text puts it.[22] The implication was that Soma was in the *jāṅgala*. The sun, which ripened the wheat in the granary plains, the *jāṅgala* at its most prized, started a long chain of cooking through which Soma was assimilated. This chain was continued in the household kitchen, and finally in bodily digestion. The lightness of *jāṅgala* meats was a distant consequence of this assimilation of Soma by Agni: their astringency tempered, sublimated and refined the unctuosity that they carried.

The most telling illustration of sublimated unctuosity was the elaboration of milk in the organism. The thick fluidity and whiteness of milk resemble phlegm, semen and *ojas*, the vital fluid. It shared their properties: heavy, sweet, cold, unctuous, and yet the milk of the cow, goat and camel produced few fluxes. It was therefore fortifying, aphrodisiac, and an elixir of youth. But of all its eminent qualities, its purity was stressed. I do not mean simply that in Brahminic ritual cow's milk exercised purifying functions, although these functions were mentioned in the Ayurvedic treatises—*pāpmāpaha*, "it wards off sin" [23]—but that its functions were reflected in its nature, in its physical purity. Charles Malamoud noted the Vedic paradox of the cooked milk in the raw cows. It was cooked by nature. "Whereas the cow is raw, the milk in her is cooked", says the *Śatapatha Brāhmaṇa*, "for it is the semen of Agni".[24] Expressed in medical terms, it was the pure essence of all *rasa*, the result of a sublimation of all saps, juices or savors. Suśruta taught that the herbivores' milk which man consumed "is the pure essence of the *rasa* of all plants".[23] It was as if the organism of the cow, goat or camel performed an initial cooking, sublimating the nourishing juices that they derived from the great herbarium of the world.

However brilliant a calculator he may have been, and this aspect of the Hindu person does legitimate our resorting to the methods of cognitive anthropology, we must admit that the Ayurvedic physician was first of all a man of the soil. He sought to procure a well-tempered unctuosity to the patient's body. Nutritive treatments (*bṛṃhaṇa*) would increase unctuosity, while the depletive treatments (*laṅghana*) would temper it with soothing drugs, or eliminate it through evacuant drugs. To describe this nutritive and depletive action, the classic authority, Caraka used agricultural metaphors of irrigating and draining the body.[25] Grave disorders of the humors were cured by elimination. How could a flooded rice paddy be drained unless the ridge of earth surrounding it was breached? The same truth held for the patient's body in profuse dyscrasia. A moderate dyscrasia was cured by fasting and by soothing remedies, "as one dries out a draining well by tossing sand and ashes to complete the action of the sun and wind". A very slight excess in a humor was eliminated spontaneously by fasting "just as a little puddle of water evaporates in the sun and wind".

This essay will introduce the curious reader to my 1987 book *The Jungle and the Aroma of Meats*, but it cannot do full justice to the arguments developed there. As the title itself suggests, readers will be plunged into a colorful world of collective imagery. Yet the book is nothing short of a theoretical disquisition on the place of humoral medicine in traditional society and the structural features of medical humoralism. But theory is developed through the workings of style. One aspect of the book which the present summary conveys most imperfectly is the systematic use of stylistic devices. My work, in so far as it revolves around the syntactic structures of medical discourse and Sanskrit didactic poetry, tends to cast itself in the mould of convoluted, redundant, animistic patterns of 'emic' description, which are currently used by French essayists but do not go down easily with English, I would like to mention here not only the names of our Fathers, Lévi-Strauss and Dumézil, but also the work of younger scouts on the track of a new structuralism in anthropology, like Marcel Détienne's *Gardens of Adonis* and Tzvetan Todorov's *Conquest of America*. They strive to construe the unsaid, the unconscious, the margins, emotion and affect, thus reaching out to American and British studies. Also they throw a bridge between classical studies and anthropology. The study of Ayurvedic medicine in India is a privileged domain for such an attempt to combine the humanities (here, Sanskrit studies) with anthropology, since Ayurveda has traditionally been practiced in the context of a *diglossia*, that is, a linguistic situation where the learned language is superimposed on the vernacular in medical practice itself. In my 1987 book, dwelling upon Sanskrit texts, I followed up the great chain of being all throughout from botany to medicine to cosmology. But, one can ask, is this Medical Anthropology? My book questions the status of 'applied' discipline currently ascribed to Medical Anthropology. It blows up the so-called 'medical system', to reach upstream the more basic categories of collective thought, and downstream the more basic facts of ecology.

NOTES/REFERENCES

1 *Laws of Manu*. Vol. 29.
2 Spate O. H. K. and Learmonth A. T. A. *India and Pakistan*, 3rd edn, p. 139. Methuen, London, 1967.
3 Conclusions of the Académie cited by Bachelard G. *La Formation de l'Esprit Scientifique*, p. 125. Vrin, Paris, 1938.
4 Aristotle. *Historia Animalium* IX, 29, 618a8.
5 Suśruta. *sūtra* XLVI, 53, 135–136.
6 See for example: Barton A. H. The concept of property-space in social research. In *The Language of Social Research* (Edited by Lazarsfeld P. F. and Rosenberg M.), pp. 40–53. Free Press, Glencoe, 1955. Especially p. 50 (substruction of the property-space of a typology).
7 Suśruta. *sūtra* XLII, 8(1).
8 Suśruta. *sūtra* XL, 5; Caraka. *sūtra* XXVI, 48d.
9 Substances of a sweet savor and sweet digestion are normally of cold energy, and the same goes for milk: Caraka. *sūtra* XXVI, 45–47. Milk 'in general' is of cold energy: Caraka. *sūtra* I, 107ab, and Cakra. ad Caraka. *sūtra* XXVII, 217–224.
10 Suśruta. *sūtra* XX, 13; Caraka. *sūtra* XXVI, 82–84.
11 Jolly J. *Medicin*, Vol. 68, p. 4. Trübner, Strassburg, 1901. *Indian Medicine*, 2nd edn, 117.

Munshiram Manoharlal, New Delhi, 1977. *Census of India*, 1891, Vol. 23, p. 366.

12 *Asiatick Researches or Transactions of the Society Instituted in Bengal for Inquiring into the History and Antiquities, the Arts, Sciences, and Literature of Asia*, Vol. 2. pp. 149–158, especially p. 156. Calcutta, 1790.

13 Jakobson R. Linguistics and poetics. In *Style in Language*, (Edited by Sebeok T. A.), p. 359. MIT Press, Cambridge, Mass., 1960.

14 Hopkins G. M. *The Journals and Papers* (Edited by House H.), p. 84. Oxford University Press, London, 1959.

15 Biardeau M. *L'Hindouisme*, p. 84. Flammarion, Paris, 1981.

16 Suśruta. *sūtra* XL, 19–21.

17 Jamnagar (Gujarat) is, with Benares, one of the major centres for Ayurvedic research in India. A planning commission was set up under the chairmanship of R. N. Chopra to elaborate the policy to be followed in the newly independent India with regard to indigenous systems of medicine. The Report of the Chopra Committee published in 1948 thus determined the policy followed during the fifties. The aim was to combine the Ayurvedic system with the allopathic system (Western medicine) to produce an 'integrated system', at the cost of an extraordinary telescoping of the Sanskrit texts into the facts of modern biology. A policy subsequently abandoned, at least on the practical level; the double qualification of practitioners was scrapped and those of each of the two systems were thereafter inscribed upon two separate professional registers. But the books of C. Dwarakanath are a typical product of this attempt at an integration on a doctrinal level.

18 Dash B. *Concept of Agni in Ayurveda*. Chowkhamba, Varanasi, 1971.

19 Dwarakanath C. *Digestion and Metabolism in Ayurveda*. Baidyanath, Calcutta, 1967.

20 Caraka. *cikitsā* XV, 1–40.

21 Caraka. *śārīra* VI, 10–11.

22 *Vāyu Purāṇa* VIII, 94.

23 Suśruta. *sūtra* XLV, 48–49.

24 III, 2, 4, 15, cited by Malamoud C., Cuire le monde. In *Puruṣārtha*, Vol. I, p. 104. C.E.I.A.S., Paris 1975.

25 Caraka. *vimāna* III, 43–44.

Ethno-climatology

5

Concerning Weather Signs

Theophrastus

Introductory: General Principles

I. The signs of rain wind storm and fair weather we have described so far as was attainable, partly from our own observation, partly from the information of persons of credit.

Now those signs which belong to the setting or rising of the heavenly bodies must be learnt from astronomy.[1] Their settings are twofold, since they may be said to have set when they become invisible. And this occurs when the star sets along with the sun, and also when it sets at sunrise. In like manner their risings are twofold: there is the morning rising, when the star rises before the sun, and there is the rising at nightfall, when it rises at sunset.

Now what are called the risings of Arcturus occur at both times, his winter rising being at nightfall and his autumn rising at dawn. But the rising of most of the familiar constellations is at dawn, for instance, the Pleiad Orion and the Dog.

Of the remaining signs some belong specially to all such lands as contain high mountains and valleys, specially where such mountains extend down to the sea: for, when the winds begin to blow, the clouds are thrown against such places, and, when the winds change, the clouds also change and take a contrary direction, and, as they become laden with moisture, they settle down in the hollows because of their weight. Wherefore good heed must be taken to the local conditions of the region in which one is placed. It is indeed always possible to find such an observer, and the signs learnt from such persons are the most trustworthy.

Thus in some parts have been found good astronomers: for instance,[2] Matriketas at Methymna observed the solstices from Mount Lepetymnos, Cleostratus[3] in Tenedos from Mount Ida, Phaeinos at Athens from Mount Lycabettus: Meton, who made the cycle[4] of nineteen years, was the pupil of the last-named. Phaeinos was a resident alien at Athens, while Meton was an Athenian. Others also have made astronomical observations in like manner.

Again there are other signs which are taken from domestic animals or from certain other quarters and happenings. Most important of all are the signs taken from the sun and moon: for the moon is as it were a nocturnal sun. Wherefore also the meetings of the months are stormy, because the moon's light fails from the fourth day from the end of one month to the fourth day from the beginning of the next: there is therefore a failure of the moon corresponding to the failure of the sun. Wherefore anyone who desires to forecast the weather must pay especial heed to the character of the risings and settings of these luminaries.

Now the first point to be seized is that the various periods are all divided in half, so that

The Anthropology of Climate Change: An Historical Reader, First Edition. Edited by Michael R. Dove.

one's study of the year the month or the day should take account of these divisions. The year is divided in half by the setting and rising of the Pleiad[5]: for from the setting to the rising is a half year. So that to begin with the whole period is divided into halves: and a like division is effected by the solstices and equinoxes. From which it follows that, whatever is the condition of the atmosphere when the Pleiad sets, that it continues in general to be till the winter solstice, and, if it does change, the change only takes place after the solstice: while, if it does not change, it continues the same till the spring equinox: the same principle holds good from that time to the rising of the Pleiad, from that again to the summer solstice, from that again to the autumnal equinox, and from that to the setting of the Pleiad.

So too is it with each month; the full moon and the eighth[6] and the fourth days make divisions into halves, so that one should make the new moon the starting-point of one's survey. A change most often takes place on the fourth day, or, failing that, on the eighth, or, failing that, at the full moon; after that the periods are from the full moon to the eighth day from the end of the month, from that to the fourth day from the end, and from that to the new moon.

The divisions of the day follow in general the same principle: there is the sunrise, the mid-morning, noon, mid-afternoon, and sunset; and the corresponding divisions of the night have like effects in the matter of winds storms and fair weather; that is to say, if there is to be a change, it will generally occur at one of these divisions. In general therefore one should observe the periods in the way indicated, though as to particular signs we must follow the accepted method.

The Signs of Rain

Now the signs of rain appear to be as follows: most unmistakable is that which occurs at dawn, when the sky has a reddish appearance before sunrise; for this usually indicates rain within three days, if not on that very day. Other signs point the same way: thus a red sky at sunset indicates rain within three days, if not before, though less certainly than a red sky at dawn.

Again, if the sun sets in a cloud in winter or spring, this generally indicates rain within three days. So too, if there are streaks of light from the south, while, if these are seen in the north, it is a less certain sign. Again, if the sun when it rises has a black mark, or if it rises out of clouds, it is a sign of rain; while, if at sunrise there are rays[7] shooting out before the actual rising, it is a sign of rain and also of wind. Again if, as the sun sinks, a cloud forms below it and this breaks up its rays, it is a sign of stormy weather. Again, if it sets or rises with a burning heat, and there is no wind, it is a sign of rain.

Moonrise gives similar indications, at the time of full moon: they are less certain when the moon is not full. If the moon looks fiery, it indicates breezy weather for that month, if dusky, wet weather; and, whatever indications the crescent moon gives, are given when it is three days old.

Many shooting stars are a sign of rain or wind, and the wind or rain will come from that quarter from which they appear. Again, if at sunrise or sunset the sun's rays appear massed together, it is a sign of rain. Also it is a sign of rain when at sunrise[8] the rays are coloured as in an eclipse; and also when there are clouds[9] like a fleece of wool. The rising of bubbles in large numbers on the surface of rivers is a sign of abundant rain. And in general, when a rainbow[10] is seen round or through a lamp, it signifies rain from the south.

Again, if the wind is from the south, the snuff of the lamp-wick indicates rain; it also indicates wind in proportion to its bulk and size: while if the snuff is small, like millet-seed,[11] and of bright colour, it indicates rain as well as wind. Again, when in winter the lamp rejects[12] the flame but catches, as it were, here and there in spurts, it is a sign of rain: so also is it, if the rays of light leap up on the lamp, or if there are sparks.

It is a sign of rain or storm when birds which are not aquatic take a bath. It is a sign of rain when a toad takes a bath, and still more so when frogs are vocal. So too is the appearance of the lizard known as 'salamander,'[13] and still more the chirruping of the green frog in a tree. It is a sign of rain when swallows[14] hit the water of the lakes with their belly. It is a sign of storm or rain when the ox licks his fore-hoof; if he puts[15] his head up towards the sky and snuffs the air, it is a sign of rain.

It is a sign of rain when a crow puts back its head on a rock which is washed by waves, or when it often dives or hovers over the water. It is a sign of rain if the raven, who is accustomed to make many different sounds, repeats one of these twice quickly and makes a whirring sound and shakes his wings. So too if, during a rainy season, he utters many different sounds, or if he searches for lice perched on an olive-tree. And if, whether in fair or wet weather, he imitates, as it were, with his voice falling drops, it is a sign of rain. So too is it if ravens or jackdaws fly high and scream[16] like hawks. And, if a raven in fair weather does not utter his accustomed note and makes a whirring with his wings, it is a sign of rain.

It is a sign of rain if a hawk perches on a tree, flies right into it and proceeds to search for lice: also, when in summer a number of birds living on an island pack together: if a moderate number collect, it is a good sign for goats and flocks, while if the number is exceedingly large, it portends a severe drought. And in general it is a sign of rain when cocks and hens search for lice; as also when they make a noise like that of falling rain.

Again it is a sign of rain when a tame duck gets under the eaves and flaps its wings. Also it is a sign of rain when jackdaws and fowls flap their wings whether on a lake or on the sea—like the duck. It is a sign of wind or rain when a heron utters his note at early morning: if, as he flies towards the sea, he utters his cry, it is a sign of rain rather than of wind, and in general, if he makes a loud cry, it portends wind.

It is a sign of rain or storm if a chaffinch kept in the house utters its note at dawn. It is also a sign if any pot filled with water causes sparks to fly when it is put on the fire. It is also a sign of rain when a number of millepedes are seen crawling up a wall. A dolphin[17] diving near land and frequently reappearing indicates rain or storm.

If the lesser Mount Hymettus, which is called the Dry Hill, has cloud in its hollows, it is a sign of rain: so also is it, if the greater Hymettus has clouds in summer on the top and on the sides: or if the Dry Hymettus has white clouds on the top and on the sides; also if the south-west wind[18] blows at the equinox.

Thunder in winter and at dawn indicates wind rather than rain; thunder in summer at midday or in the evening is a sign of rain. If lightning is seen from all sides, it will be a sign of rain or wind, and also if it occurs in the evening. Again, if when the south wind[19] is blowing at early dawn, there is lightning from the same quarter, it indicates rain or wind. When the west wind is accompanied by lightning from the north, it indicates either storm or rain. Lightning in the evening in summer time indicates rain within three days, if not immediately. Lightning from the north in late summer is a sign of rain.

[20]When Euboea has a girdle about it up to the waist, there will be rain in a short space. If cloud clings about Mount Pelion, it is an indication of rain or wind from the quarter to which it clings. When a rainbow appears, it is an indication of rain; if many rainbows appear, it is an indication of long-continued rain. So too is it often when the sun appears[21] suddenly out of cloud. It is a sign of rain, if ants[22] in a hollow place carry their eggs up from the ant-hill to the high ground, a sign of fair weather if they carry them down. If two mock suns appear, one to the south, the other to the north, and there is at the same time a halo, these indicate that it will shortly rain. A dark halo round the sun indicates rain, especially if it occurs in the afternoon.

In the Crab are two stars called the Asses, and the nebulous space between them is called the Manger; if this appears dark, it is a sign of rain. If there is no rain at the rising of the Dog or of Arcturus, there will generally be rain or wind towards the equinox. Also the popular saying about flies is true; when they bite excessively, it is a sign of rain. If a chaffinch[23] utters its note at dawn, it is a sign of rain or storm, if in the afternoon, of rain.

When at night a long stretch of white cloud encompasses Hymettus below the peaks, there will generally be rain in a few days. If cloud settles on the temple of Zeus Hellanios[24] in Aegina, usually rain follows. If a great deal of rain falls in winter, the spring is usually dry; if the winter has been dry, the spring is usually wet. When there is much[25] snow in winter, a good season generally follows.

Some say that, if in the embers there is an appearance as of shining hail-stones, it generally prognosticates hail; while, if the appearance is like a number of small shining millet-seeds,[26] it portends fair weather, if there is wind at the time, but, if there is no wind, rain or wind. It is better both for plants and for animals that rain should come from the north before it comes from the south; it must however be fresh and not briny to the taste. And in general a season[27] in which a north wind prevails is better and healthier than one in which southerly winds prevail. It is a sign of a long winter when sheep or goats have a second breeding season.

NOTES [1926 FROM TRANSLATOR A.F. HORT]

1 Or, perhaps, ' from my astronomical works.'
2 Plinius, Naturalis Historia, 5. 140. Of Matriketas nothing is known.
3 Said (Plinius 2. 31) to have first recognised the Ram and the Archer. Athenaeus (7. 278 *b*) connects him with Tenedos.
4 Called 'the great year': *cf.* Aelian. *V.H.* 10. 7.
5 Plinius 18. 280.
6 *cf.* Arat. 73 f.
7 Plinius, 18. 344.
8 Plinius 18. 344.
9 Plinius 18. 356.
10 *cf.* Aristotle *Meteorologica.* 3. 4; Plutarch *Quaest. Nat.* 1. 2.
11 *i.e.* breaks up into small 'grains' (?). *cf.* p. 86, paragraph 1
12 *i.e.* refuses to light properly. The appearance seems to be that described [in] Virgil's *Georgics* 1. 391 (*scintillare oleum*). In the same passage *putres concrescere fungos* perhaps illustrates the comparison of the snuff to millet-seed above.
13 *cf. de igne* 60, where it is explained why the salamander puts fire out.
14 Plinius 18. 363; Vergil 1. 377.
15 Plinius 18. 364; Vergil 1. 375.
16 ? 'hover like hawks.' However, Arat. 231 understood it to refer to the voice: so LS.
17 *cf.* Plinius, 18. 361; Cic. *Div.* 2. 70.
18 *cf.* Aristotle *Probl.* 26. 26.
19 *cf.* Sophocles *Aj.* 257; Aristotle *Probl.* 26. 20.
20 Evidently an Attic saying, of days when only the upper part of the Euboean mountains was visible.
21 *cf. H.P.* 8. 10. 3.
22 Plinius 18. 364; Vergil 1, 379.
23 *cf.* p. 85, paragraph 5, of which this seems to be in part a repetition.
24 So called also by Pind. *Nem.* 5. 19.-
25 *cf. C.P.* 2. 2.
26 *cf.* p. 84, paragraph 8.
27 *cf. C.P.* 2. 2.

6

Gruff Boreas, Deadly Calms

A Medical Perspective on Winds and the Victorians

Vladimir Janković

In this paper I propose to reflect on the medical meanings of the nineteenth-century winds. I am not prepared to say anything about their qualities as meteorological phenomena nor am I equipped to judge on whether Victorian winds differed from the winds of today. Perhaps no amount of weather modelling and past data interpretation would do to complete such a project. My intention is rather to look at the period in which the cultural and medical understanding of wind came to reflect the role of scientific knowledge in shaping both the traditional and scientific understanding of health and environment. The Victorian period in particular witnessed a series of negotiations on the limits, meanings, and purposes of the natural knowledge as it came to be increasingly institutionalized (Levine 2002; MacLeod 1982; Yeo 1993). Scientific authority, utility, and cultural status have been brought to bear on the forms of traditional knowledge and, in particular, on medical practices. Where previously a multiplicity of environmental meanings coexisted in epistemic competition, the nineteenth-century public appreciation of science effected an exclusivist hierarchy of knowledge that opposed rational truth to mere opinion and 'folklore'. Increasingly the public and the state acquiesced to the idea that the 'everyday nature' was a 'scientific nature'. What was once a knowledge available to the majority was appropriated by an expert clique speaking esoteric tongues. This bifurcation, of course, was not a new development. Ever since natural philosophy had gathered institutional momentum in the early eighteenth century, nature had been systematically deprived of emblematics and accessibility, the result of which was a trend culminating in the naturalism of the mid-Victorian period. This redefinition is apparent in the history of atmospheric sciences (Favret 2004; Jankovic 2002; also general discussion in Anderson 2005; Friedman 1993; Reed 1983). In the natural philosophy textbooks, the many sentimental and spatial meanings of the eighteenth-century 'atmosphere' gave way to the notion of a grand chemical 'laboratory' (Adams 1798: 474; Ferriar 2004). The nineteenth-century medical aetiology of diseases pursued an analogous reductionist turn (Hamlin 1992).

What follows is an examination of the wind's medical significance within neo-Hippocratic

The Anthropology of Climate Change: An Historical Reader, First Edition. Edited by Michael R. Dove.

medical thinking and environmentally inflected discourses, including medical topography (Cantor 2002; Harrison 1999; Jordanova 1979; Livingstone 1999; Riley 1987). The examination will scrutinize the meandering meanings of the wind as a public meteor. As with other public meteors – the atmospheric phenomena that could be observed regardless of the status and education of an individual – the wind had often been described as an agency whose qualities included those of breath, omen, fertilizer, or destructor. The wind had also been known as a carrier of smoke, smell, miasma, and death, but also, as a power that propelled ships, scattered seeds, fertilized mares, ventilated rooms, and mixed gases. In literary tradition winds howled, whistled, roared, and moaned. They changed people's opinions, confined them indoors, and taxed their lungs. They altered outcomes of horse races. But despite these roles and presences, wind was also regarded, quite simply and since at least Seneca, as the air in motion. The anemometer and weather-vane measured its speed and direction. Like other public events, winds were shared across communities and discourses, and it is pointless to provide an absolute rationale for isolating its medical meaning on the basis of medical theory only. Indeed, medical meanings of wind emerged from the situational logic of daily lives that comprised travel, economic transactions, exploration, colonial epidemiology, and anecdotal evidence. Yet it should be noted that the creation of medical space in these realms often depended on a 'poetics of pathology' rather than medical maps or experimental knowledge (Wrigley 1996; 2000).

The purpose of this paper, however, is not to engage with the many medical connotations of the wind. Rather, it is to explore the ways in which wind has been constructed as a medical force, the ways whereby it could acquire a 'pathogenic' agency, that is, the ability to act as a cause of disease. I conceive wind's agency as a crossing of the boundary between the non-medical and the medical. Historically, such crossing could take numerous manifestations – from nuisance to pain, from discomfort to debility, from chronic exposure to acute condition – and one of the aims of this paper is to show how such changes took place in relation to medicalization of wind in a period of European history known for its pervasive interest in 'exteriority' and, more generally, the polarization of social space (Pilloud & Louis-Courvoisier 2003). The goal is to examine what might be termed a *meteorological pathogenesis*, that is, the understanding of how the pre-pathological elements of the 'environment' occupied medical attention and public anxiety as a disease-bearing agency. How did winds, rains, and sunlight enter medical research and regimen? As later sections demonstrate, such research involved a deconstruction of the wind and, in effect, the removal of its 'blowing' from medical problematics.

Winds of Risk and Death

The medical meanings of winds have been known since antiquity. Hippocratic tradition put emphasis on the city's exposure to winds; the winds specific to places and seasons represented the elements of an 'atmospheric constitution' that defined the nature and intensity of the prevailing diseases (Miller 1962; Sargent 1982). Yet the Hippocratic opus did not strive to assemble a causal theory of environmental influence; it merely prepared a physician for practice in a new town. A new physician was required to know that, for example, the inhabitants in a city exposed to hot winds suffered from bodily flabbiness, had humid heads, and irritable bowels unsupportive of wine; that women suffered from excessive menstruation, infants from asthma, and other adults from dysentery and haemorrhoids. Whether such conditions stemmed from the hot wind itself or were merely coincidental to it remained outside the physician's purview. Following the writings of the seventeenth-century doctor Thomas Sydenham – called the 'English Hippocrates' for his interest in disease as an environmental imbalance – Enlightenment medics continued to pursue this route and opened up a possibility of thinking about diseases as naturally reoccurring phenomena (see Bisset 1762; Dolan 2002; Huxham 1759; Lee 1973; Murray 1774; Poynter 1973; Rusnock 2002; Rutty 1770; Short 1749; Weindling 1985; Wintringham 1727).

Early modern European culture, however, witnessed an intense cultural exposure to the non-European world. Travellers, colonists, and explorers published accounts about diseases and environments not felt in European lands. Like plants, people, diseases, and topography, foreign winds elicited interest for their odd and unsuspected qualities that captured the European imagination regardless of their relevance for philosophical or other forms of knowledge (Barnes & Mitchell 2002; Knellwolf 2002; Rousseau & Porter 1990). The French scholar Chardin publicized an account of people killed by a sudden blow of the African 'mortifying' *samiel*. They seemed asleep, with limbs separated from their bodies. Volney observed that the victim of the *khamsin* wind remained warm, swollen, and blue for a long time, as if killed by thunder (Robertson 1808: 305). Harmattan, the dry African wind, had attracted attention for its desiccating and healing properties. It killed plants and parched the skin, but cured fevers and the bleeding fatigue; it could put a stop to smallpox and diarrhoea and exert preventive powers against infections, ulcers, and skin eruptions (Dobson 1782–3: 261–2). Yet if it reached Europe, it became pestilential through the 'adventitious' qualities that it soaked up from the underlying soil.

The influence of soil gave some winds a terrifying notoriety. This was the case with the easterly wind in Senegal 'with which those who are suddenly met ... are scorched up as by a blast from a furnace', or of the notorious Falkland Island winds that attacked fowl with cramps and stopped the perspiration of the locals (Adams 1798: 541–2). Known for their suffocating, depressing, and 'relaxing' qualities were the Sirocco and the Levant (Jugo in the Adriatic), which badly affected Minorca and Gibraltar but which originated over the burning expanses of Libya and Egypt. Sirocco in Naples and Palermo stopped digestion and killed over-eaters; city folks shut their houses and stayed in until the end of its onslaught. The identification of the geographical origin of these winds (often African) is an indication that learned Europeans thought of the tropics as a place of 'elementary contention and violence', of luscious vegetation and putrefaction, shaken by the geological convulsions and winds that 'vary their terrors: sometimes involving all things in a suffocating heat, sometimes mixing all the elements ... and sometimes destroying all things in their passage'. Given the mildness of mid-latitude Europe, it seemed necessary that infectious diseases had to be 'imported from some sultry climate' (Brown 1797: 52, 53).

Crucial in these disquisitions was physicians' tacit expectation that a physical place should exhibit a stable homeostasis between its inorganic features, such as winds, soil and precipitation, and human life and health. This reciprocity has been debated in the context of climatological determinism and natural theology (Gates 1967; Glacken 1973). Into the nineteenth century, it even became possible to speak about 'hereditary climates': 'Has not Nature adapted the constitution of man to his hereditary climate?', asked Thomas Burgess. 'Is it consistent with nature's laws that a person born in England and attacked by consumption can be cured in a foreign climate?' (1850: 591). An environmental pathology therefore came about only as a *relative* misadaptation to local conditions, which by and in themselves did not cause disease. Sir Walter Scott's Highlanders were reported to sleep outdoors, unperturbed by the hoarfrost in their hair (1996 [1817]: 76). The notorious quack James Graham claimed to have achieved robust health by keeping his windows open for the maximum wind impact during Scottish storms (1793: 5).

While some of these feats were explained by a sympathy between the body and its native exteriors, no precise formulations had been advanced on the physiological responses to weather and climate:

> We see then that the insalutary impressions of the atmosphere transmitted from the surface to the central parts of our bodies, may be reproduced and transmitted from organ to organ, by means of the sympathies ... [t]ill various and complicated maladies, accompanied by a tribe of obscure and complicated symptoms have arisen, that are as embarrassing to the Physician as they are distressing to the patient (Johnson 1818: 12; see also Vila 1997).

Keeping unperturbed the sympathetic exchanges between the 'in and out' was tantamount to good health.

Such views were the medical orthodoxy during the Enlightenment (Rosenberg 1992). In 1785, Andrew Hamper, the author of a health treatise, phrased it thus: 'All things considered to act immediately upon the Body, or make Impression on it, at a Distance, through the Medium of the Senses, the changes that happen in the Atmosphere which surrounds us, and the Air we breathe, constitute the general, *external causes* that affect Health' (1785: 1).[1] The wording is significant: the external causes of disease – what the Augustan physician John Arbuthnot (1733) had called the 'powers without' – could act even 'at a Distance' and through 'the Medium of the Senses', rendering the body a vital barometer recording the alteration of its surrounding milieu. These associations testified to the common use of 'influence' among eighteenth-century physicians, who often met with patients with extreme atmospheric sensibility (Castle 1995; Golinski 1998; Phelps 1743; Shuttleton 1995). These 'human barometers' suffered from a 'nervous' disease which made them exceptionally attuned to the 'vicissitude' of the weather. They gave emphatic meaning to being 'under the weather'. In the age of sympathetic sensibilities this condition was virtually self-evident and a subject of moral analysis: '[I]t was a damn'd bad thing to have a constitution subject to squalls of weather' (Bage 1796: 88).[2]

As for the medical views, the condition was diagnosed as a result of under-exposure to fresh air. Eighteenth-century sociability (salons, libraries, lectures, card routs, dancing, clubability) was related to environmental improvements designed to provide thermal equability in interiors whose calms and comfort contrasted with the wind-bruised English outdoors. In these environmental technologies, the domestic space was the haven of calms. This accorded a greater value to the cosiness of private space, which favoured a low-metabolism and high-caloried sedentariness, and which were in turn tainted by traditional fears of draughts and the miasmatic night air. The specifically medical reasons for this over-protection were to do with the perceived risks of sudden changes caused not only by the alterations in outside weather or instant exposure to winds, but by a careless night life (Gregory 1815: 40). Thermometric equability and regulated exposure ruled the medical enlightenment, which, as was pointed out by contemporaries, bred a peculiar type of citizens, pale, out of shape, pampered and unreasonably shielded from the bracing outdoors (Johnson 1839).

Anthony Florian Madinger Willich wrote in 1800, for example, that such trends created an acute medical condition. Next to gout, he wrote, there is the 'still more general malady of the times, ... an extreme sensibility to every change of the atmosphere; or, rather, constantly sensible relation to its influence'. Willich noted the sensibility of some of his patients to identify the direction of the wind while even inside their apartments. He was struck by the 'talent so peculiar to our age' and acquired in a climate that defined English health as 'dependent, frail and transitory'. Even the question of whether political strife itself owed to people's 'secret dependence on the weather' was legitimate because 'beings so organized cannot warrant, for a single hour, their state of health, their good-humour, or their physical existence' (Willich 1800: 58–9). For the British, this dependency also became crucial for the colonial project (Curtin 1964).

As the analyses and criticisms of the New World, tropics, and the otherwise popular Mediterranean health resorts amply illustrate, foreign climates were practically equivalent to (and judged in terms of) the harm they wrought on their visitors (Arnold 1996; Bolton-Valencius 2002; Hoolihan 1989; Wrigley 2000). The Mediterranean, for example, came under negative scrutiny increasingly because its local climates violated the hereditary argument. Medical commentators claimed that even though the British enjoyed rather chilly, windy, and rainy weather for the most part of the year, its bleak uniformity none the less spared consumptives from adjustments to the violent changes in temperature known to affect the European South. In some of the Mediterranean resorts, local winds were claimed to be especially morbid: the Sirocco, Liebeccio, and Leste depleted the vitality with blasts followed by copious dews. These were conditions unknown to the unsuspecting British invalid. In Malta, cold winds ('bracing' in the British climate) were too piercing to allow one to venture outdoors: 'I do not remember ever to have felt the sensation of cold so acutely in this country [England] as I have done in Malta during a dry, north-westerly, or north-easterly wind' (Burgess 1850: 592). For contrast, in England, Charles Kingsley paraphrased Shelley's 'Ode to the West Wind' with his own 'Ode to the North-east Wind', to celebrate its hardening qualities suited to Victorian robust moral ambition: 'Welcome, black North-easter! O'er the German foam, the Danish moorlands, thy frozen home. Tired we are of summer, Tired of gaudy glare' (Kingsley 1909–14 [1889]).

In medical theory and topography, however, the fear of Southern, non-European winds was further heightened by their supposed miasmatic qualities (Cipolla 1992; Hannaway 1993; Pelling 1978). 'Miasma' was the generic term used to describe the allegedly material agency that spread epidemics: the 'anticontagionists' contended that epidemics had almost nothing to do with bodily contact between the infected and the healthy, because the magnitude, spread, and speed of the pestilence could be explained only by the wind-blown miasmas (Ackerknecht 1948). Winds were foul insofar as they pushed infectious material across national, even continental, borders. Wind was a vehicle of transmission and a medium of pollution – in a broadest possible sense, a carrier of 'matter out of place' (Douglas 1966; Hamlin 1994). But this also made it a ventilating force, a harbinger of health. Authorities and the medical profession of eighteenth-century Europe grappled with these two meanings in a variety of ways. The environmental medicine assumed that air could become corrupted by subterranean processes, organic decomposition, re-breathing, and urban waste (Hannaway 1972; 1974; Jordanova 1981). It was asserted that in unfavourable conditions (never fully specified) adulterated air caused cholera, fevers and typhoid, and even death itself by the 'mortiferous exhalations' that sometimes spurted out from pits, mines, and coffins (Laqueur 2002; Robertson 1808).

The wind here was a purifier and a polluter. How did such qualifications influence the works of mid-nineteenth-century medics? How were they prompted to investigate the inner workings of wind's purifying or polluting powers? In the remainder of the paper, I show that the answers to these problems also bear on popular versus professional readings of public meteors and on the epistemic status of the almost universal belief in the refreshing powers of wind and pure air generally.

The Winds Electrical; the Winds of Ozone

Eighteenth- and nineteenth-century physicians, geographers, town planners, medical authorities, meteorologists, and common folk often referred to winds as having unique ventilating powers capable of supplying unlimited amounts of fresh air to a space of any scale, from privies to churches, from streets to industrial districts, from the sea coast to the Scottish moors. The litany of fresh air and ventilation established itself as the moral basis of the British public health movement, transfixed as it was by the paradigm of environmental purity at the expense of other socially and economically created vectors of misery and disease: 'What habitation in the British empire can have an excuse for a stagnant atmosphere?', asked a reviewer of Thomas Bateman's article on contagion in 1820. 'If such places there be, they ought not to be permitted to exist' ('Review' 1820; see also Hamlin 1992; Rome 1996; Wohl 1983). The zeal for ambiental purity that such information fomented was recorded in the fact that between 1800 and the 1860s, the estimated need for indoor fresh air soared from four to sixty cubic feet per person (Bruegman 1978: 153)!

This culture of 'fresh air' did not go uncriticized for its exaggerated claims and unreasonably high expectations. One of its most prominent critics was London practising physician and author John Charles Atkinson, the authority on the 'change of air', and a member of the Royal College of Surgeons with long-standing interests in environmental epidemiology and health. The fad of 'fresh air', he claimed, was of the same nature as those to do with Elixirs, Drops, Waters, mesmerism, hydropathy, homeopathy. 'Change of air' is the last of these and 'is forced upon everyone', as the public and the professionals advise visits to spas, places with special healing powers, and those with a good exposure to prevailing winds (Atkinson 1848: 8). Atkinson's early works focused on the role of atmospheric electricity during cholera epidemics, while in his later years he researched the therapeutic aspects of wind, sleep, and fear. Writing in 1849 in the medical journal *The Lancet*, he noted a lack of exact knowledge into the effects of winds on the national morbidity. Of his own research he found no precursors and only a 'domestic and proverbial' knowledge: 'When the wind is in the east / It's neither good for man nor beast'. This he thought was puzzling given the role of winds in Hippocrates' *Airs, waters and places*, where the author urged physicians to pay attention first to the seasons and 'Then the winds, the hot and the cold, especially such as are common to all countries, and then such as are peculiar to each locality' (2004: 3).

Atkinson set to cut the Gordian knot of causality on the premise that most of winds' medical effects owed to imagination and 'common voice' rather than their real action. Neither was the direction nor the action of winds observed properly: telescopes could be used to observe the motion of clouds and thus determine the direction of wind at higher altitude. The action of winds was not properly discussed, because not all patients were affected similarly during the same wind: some had ailments in which the afflicted organ showed more susceptibility to a particular wind's influence regarding the amount of damaged tissue. Furthermore, observing that chemistry could not differentiate between the airs found in different locations – valleys or plains, one country or another, the ratio of its constituents remained the same – Atkinson queried as to 'what can we attribute the variable influences of different winds, if not to something yet undiscovered something which has wholly escaped observation?' (Atkinson 1849*a*: 208).

Patients, in particular, guessed the change in wind when 'closely shut up or imprisoned in a room, not exposed even to the whispering of the wind' because there was something 'in the atmosphere which, by change of wind, causes a simultaneous action on all living substances and structures', this something being 'the actions and decomposition, attraction and repulsion', that is, the properties of electromagnetism (Atkinson 1848: 318). This fact explains the truth of animals' and plants' prognostication of the weather because – like the hypersensitive and the sick – these 'poor-man's barometers' (pimpernel, chickweed, leeches, horses) instinctively respond to the stimuli of atmospheric electricity. As far as the human body was concerned, the changes of wind, temperature, and electricity were sufficient to disturb the body's secretion and excretion and thus cause 'a loss of the balance of the proper and health functioning of life'. But winds in particular determine local conditions. For Atkinson, without a comprehensive anemometry that took into account electrical states of the weather, no full understanding of epidemics was forthcoming (Atkinson 1843: 323). But because such work had not been done, medical writers struggled to determine the exact nature of wind's pathogenicity by other means and considerations. The discussions were structured according to the dichotomy between the wind as a polluter and as a purifier.

The dichotomy was indirectly tackled by Alfred Haviland (1853), a self-confessed Hippocratist educated at St Thomas's Hospital in London and a practising surgeon in Cannington, Somerset. His works give an indication of the shared anxiety with regard to the 'foreignness' of pestilential winds but also the scale of the Victorian quest to assess the medical co-ordinates of a place (Barret 1998; Freeman 1978–9; Harley 2003). Haviland in this respect belongs to the vibrant tradition of medical topography and regional geography that had thrived in the British Isles since the early 1800s, following the work of Leonhard Ludwig Finke, James Lind, James Johnson, and James R. Martin.[3] For our purpose what is important is, first, Haviland's apparent rejection of the wind's miasmatic character in favour of its ozonic content. In this reduction of wind's medical properties to wind's chemical composition, Haviland was, like his contemporaries in other field sciences, effectively putting an end to a common-sensical understanding of 'public nature'. Introducing analytical chemistry to replace empirical diagnostics based on an anecdotal evidence of the wind's properties announced a parting of the ways between mundane and esoteric winds, between shared and analytically circumscribed natures.[4] This is significant in that such parting went directly against the entrenched psychological identification with the environment that had been repeatedly inscribed in the many manifestation of the Victorian medical geography (Bewell 1996).

Haviland's professional involvement with wind started during the 1849 English cholera outbreak when he looked after patients in his native Bridgewater. In four months he witnessed over 200 deaths. Pondering over the circumstances that might have contributed to such mortality, he entertained the possibility that the number of new cases varied according to the wind patterns. This led him to take weather observations, not least important of which were his avant-garde experiments with atmospheric ozone, the results of which were compiled in *Climate, weather and disease* (1855). While the idea to correlate weather with cholera was not uncommon (Atkinson 1849*b*; 1850), the book aimed at a broader if erudite audience by offering a one-of-a-kind pastiche of ancient history-cum-climatology and the statistical data gleaned from the recently available tables of the Royal Observatory at Kew Gardens.[5] Most relevant for our purpose was the discussion on the 'pestilential constitution', a Hippocratic concept defined as a mix of environmental circumstances thought to be conducive to an epidemic (Ackerknecht 1982). The notion served as the framework within which meteorological elements could cause pathogenic effects. The pestilential constitution was thus the notional site of pathogenesis, a site in which the wind could cross the threshold of its medical identity within the learned discourse of Victorian science.

Haviland's method of discussion and analysis involved the use of both meteorological and historical sources. These helped him move from the colloquial and descriptive to what seemed medically validated. No sympathetic theory ever got involved. In this he self-avowedly remained faithful to Hippocrates. His method is illustrated in his dealing with the malignity of the south wind. He opens the discussion by the doxographic reference to the ancients and follows it up by the modern opinions of Fodéré (1813), who, in discussing the wind's origins, relied on plague's Egyptian provenance: 'There only is it engendered; in the other regions it is always alien' (Haviland 1855: 92–3). From then on Haviland plunges into mortality data and the views of the eighteenth-century epidemiologist Joseph Browne. Browne thought that as the plague deaths typically peaked during the periods marked by south winds, those must have been passing over the deserts in southern Egypt, where they became 'loaded with putrid emanations exhaled from the animal and vegetable substances which are decomposed in the lakes formed by the retiring' of the Nile and its surrounding cemeteries. Thus impregnated, the wind 'blasted' the inhabitants of the Mediterranean and their northern neighbours (quoted in Haviland 1855: 122–3). The picture thus emerges of a wind as a bearer of exogenous material, but in a way in which the content and the medium cannot easily be separated.

Haviland takes these views with a grain of salt. The wind as a carrier cannot really explain the geography of the plague. For if the above is correct, every country would have to have its own Egypt! Instead, it might be the case that the south wind itself, regardless of its poisonous content, is responsible for creating a set of conditions in a given place that gives rise to a positively pestilential atmosphere. While any given location can be prone to the plague, it is the south wind that triggers it by its warmth and humidity (Haviland 1855: 123). Indeed, he notes that Galen believed that heat was the active, and humidity the material, cause of putrefaction – a view he thinks amply confirmed in the annals of plague, which show their unfailing concomitance. The annals of influenza, sweating sickness, catarrhs, and cholera indicate a 'southern constitution' of elements such as wet and warm conditions accompanied with 'great gluts of rain' and 'stinking fogs' (Haviland 1855: 125). What remained unclear was the precise nature of the south wind's harmfulness. Is it in its general 'hemispheric' flow, which enables humidity and warmth to generate a specific disease? Is it through its stimulation of putrefaction and miasmas? Is it by bringing them already formed? Or is it by some other means, perhaps entirely mechanical? The language is not straightforward. Sometimes, the south wind is 'favourable' or 'conducive' to cholera, suggesting its predisposing or exciting but not immediate role in its causation. In other places its 'morbific property' as an active agent has to do with some quality like the haziness of air: '[W]hat is this haze?,' Haviland asks. 'In the West of England a hazy day in spring is called a blight' (Haviland 1855: 127).

In other places, to further complicate the picture, Haviland suggests that instead of the south wind, one might need to concentrate on the northerly, or even on no wind at all, but on its absence, the calms. In this he follows the results of the doctor J.A. Hingeston connecting the highest mortality during the 1832 Asiatic cholera to the 'fearful atmospheric stillness and that seemingly interminable moistness', combined with the 'gloomy and cloudy' weather and a certain grey mist which painters call 'scumbling' (Hingeston 1853). The calms might have been deadly simply because they let miasmas hover for longer and instigate the creation of local poisons. Atkinson called these 'disease mists' that dissipated with the wind (1848: 17). In such situations human habitations exercise an influence on the 'supernatant air of a district, especially during the prevalence of calms, when whatever rises from them accumulates and concentrates in such a manner as seriously to affect the health of those who inhale the poisoned atmosphere' (Haviland 1855: 109; see also Allison 1839). With regard to endemic disease, Haviland finds that 'bronchocele' (i.e. goitre or the enlargement of the thyroid gland) reigns in the deep, dark, and humid valleys of Switzerland, 'where the atmosphere is seldom ruffled by a breeze of sufficient power to remove the accumulate poison ... Were this valley beneath a tropical sun, it would be the seat of pestilence and death' (Haviland 1855: 92). The tropics re-emerge as the baseline of morbidity.

This reasoning had implication for climatotherapy. By Haviland's time the practice had all but reigned supreme among those able to take a Mediterranean or Continental *tour de bagnes* to imbibe the qualities of fresh air.[6] Haviland, for instance, considers it inadvisable to shut the windows in a hospital under the pretence of protecting patients from the cold east winds because 'the atmosphere thus pent up gets loaded with animal emanations' and becomes positively poisonous (1855: 118). The 1832 cholera was the most lethal during the calms, the dead calms, what the soldiers call 'cat's paws' – and the disease began to decline only when the wind rose. This view, parenthetically, seems to have been widely spread: many noted the coincidence of highest mortality and the sickly, sticky calms during the periods of an unusually luxuriant growth of plants, greasy appearance of window glass, and numerous dead flies circled by their solidified faeces (Cross 1857). Charlotte Brontë's *Shirley* captured the role of wind in dispelling these morose times:

> So long as the breath of Asiatic deserts parched Caroline's lips and fevered her veins, her physical convalescence could not keep pace with her returning mental tranquillity; but there came a day when the wind ceased to sob at the eastern gable of the Rectory, and at the oriel window of the church. A little cloud like a man's hand arose in the west; gusts from the same quarter drove it on and spread it wide; wet and tempest prevailed a while. When that was over the sun broke out genially, heaven regained its azure, and earth its green: the livid cholera-tint had vanished from the face of nature: the hills rose clear round the horizon, absolved from that pale malaria-haze (1849: 34).

The importance of calms is indicative of Hingeston's and Haviland's interest in considering winds as purifiers, not polluters. But does this purifying happen in virtue of mere ventilation (mechanical removal of filth and miasmas) or in virtue of some positive property inherent to wind? And which wind? Haviland rejects the former possibility and introduces ozone as the agency. Ozone was a newly discovered mid-century gas that was just about beginning to rule among public health officials due to its alleged disinfecting powers (Fox 1873; Schönbein 1845). With ozone, purification was easier to explain: the wind was now imagined to supply a region with fresh amounts of ozone, but when it stopped blowing, the gas dissipated and was replaced by the putrid emanations from the locale, eventually giving rise to epidemics. When somewhat later Thomas Moffat from Hawarden, Flintshire, the early English ozone authority, showed that the cholera-free atmosphere contained ozone only when three concomitant conditions were met (high pressure, low temperature, and wind from the north), it became plausible to imagine that the healthy place can be defined as exposed to a steady (north) wind (Moffat 1860). Haviland asks: 'May not the almost perpetual calms that obtain in lowest valleys of the earth, be powerful encouragers of the endemics which prevail therein?' (Haviland 1855: 130). To say this is to say that the pathogenicity of wind equalled the pathogenicity of its absence, the calm; but also, with the ozone as the measure of a place's healthiness, the wind itself lost its proper medical role except that of a dynamic container of a chemical agency.

The Mucous Membrane and the Ontology of Winds

Victorian men of science and medicine increasingly saw wind as a medium of transfer, not a phenomenon *per se*. The studies undertaken by Atkinson and Haviland showed that the change of environmental scrutiny meant a shift from the notoriously unhealthy south quadrant to the newly discovered north quadrant of bracing purity. Not all would agree with this trend, however. In fact, the north (and the east) wind came under repeated scrutiny for their notorious unhealthiness. Haviland himself admitted that the eastern and northeastern *aspect* of land was unfavourable for British agriculture, referring to a report of a Fellow of the Royal Agricultural Society on how the shepherds on the Wiltshire and Sussex downs exercise caution on the approach of northeasterlies, making sure 'when their flocks are in a situation having easterly aspect, to remove them at once ... to prevent "the shotting of the blood" ' (1855: 93), an inflammation resulting from exposure in an easterly aspect. Similarly, the popularity of health resorts in South Wales, the Isle of Wight, the south Hams of Devon, and the neighbourhood of Penzance almost entirely derived from their situation being sheltered from the 'chilly influence of the northerly winds' (1855: 93).

This would ring true with most contemporaries, who saw the north and northeasterly winds as the epitome of discomfort and disease, but in the epithets different from those they would use to describe the southern winds of lassitude. A medical periodical reported that 'our coldest, north east winds occasion croup, soar throat, swelled glands, pulmonary ailments' (Collactenea 1834: 190). Medical statisticians found that they increased the levels of urban mortality; others noted their prevalence during paralytic attacks (Drummond 1849: 410). The business commentary in *The Times* was replete with reports on shipping disruptions and accidents occasioned by the north wind's gusts. The literary take of its influence ranged in nature and seriousness. In Jane Austen's *Emma* one reads that, 'though you will never own being affected by the weather, I think everybody feels a North-west wind' (1816: 2 58). Non-medical depiction of the north wind as cold, piercing, and snowy was common: Charlotte Brontë's *Jane Eyre* described how it had been 'whistling through the crevices of our bed-room windows all night long, had made us shiver in our beds, and turned the contents of the ewers to ice' (1847: 91). Others experienced distraction or mood changes, like a late Hanoverian novel hero, Baron Phillips, who, on hearing that the wind veered to the northeast, immediately realized why he felt 'not know how-ish' (Bage 1796: vol. 3, 88). The critic Leigh Hunt commented that we cannot answer 'for what a north-east wind or a fall of snow may do to us. I have myself, before now, had a whole host of fine ideas blown away by the one; and have been compelled to retreat from the other, mind and body, with my knees almost into the fire' (Hunt 1817: 11). The northeaster howled, roved, moaned, crept, cut, froze, and did all these things loudly enough to elicit a poet's supplication:

> Silence, oh North-East wind thy saddening cry
> Silence, oh wind thine everlasting moan!
> Is the child Innocence all naked thrown
> Out on the freezing earth, is the great sky
> Now made of lead for ever, nor again
> May the heart cheer up nor sweet lips be
> curled?
> Silence oh deadly wind!
>
> (W.B. Scott 1854)

The more damning judgment still came from medical professionals. Dr C.B. Garrett of the Royal Human Society, a medical officer based in a dispensary in Kingston, Surrey, outlined a ghastly gallery of afflictions that his countrymen suffered during the onslaught of this grand 'Aeolian enemy'. He sketched an image of a fictitious London pedestrian caught in the February northeaster, faltering in step, forced to frequent stoppages to draw full breath, and with a nervous glance:

> Observe how he elevates his shoulders and supports his hands on his side as he stops to inhale, and accomplish a husky raking clearance

> of the windpipe by a cough. This is an endurer of no common an affliction with a sense of impending suffocation, and the most pitiable feebleness, his very look is that of trepidation and supplication for assistance (Garrett 1855: 4).

From Garrett we learn that the wind causes a range of physiological debilities, from private to national: a sense of heaviness, swelling, and tightness in the head that was accompanied by sudden flashes of giddiness. People feel faint, short of breath, without strength and appetite. Some suffer from a short-term memory loss; many have restless sleep wetted by uncontrollable salivating; in time, the whole nation is apt to lose temper, become fidgety, fretful, and excitable, and eventually to plunge into a great depression of spirits (Garrett 1855: 78).

Clearly this was not the miasmatic and sticky south wind packed with infectious particles from African deserts and tropical marshes. The north wind's signature was in its dynamic aggression – sometimes thought of as a 'lazy' wind in going 'through' rather than 'around' the person – combined with its parching dryness that attacked the throat's mucous membrane and from where it triggered further symptoms. Not all people suffered from the condition, however, and Garrett explained this differential in an analysis of digestive function, concluding that the northeaster's victims include those who share the so-called 'lactacidic' constitution, or a form of gastritically induced susceptibility. Garrett identified that it was the thermo-physiometric quality of wind's air that was pathological, not its chemical, electrical, or dynamic composition. Neither was he interested in epidemics; this allowed him to avoid any involvement with miasmas. The pathogenicity arose only through the 'gruff Boreas's' desiccating powers, which worked their way through the mucous membrane of the nose. The latter had already been investigated for its important physiological role as the first interface, with the skin, between the body and the ambient air. Although Garrett does not specify other researchers, he must have been aware of the research on the subject by the French tissue pathologist Xavier Bichat and that of François Magendie in which the nose's membrane appears as an 'air conditioning system' (Proctor 1974: 360). This explains Garrett's discussion on the wind-shielding role of the recently introduced ventilators by Julius Jeffreys (1842; 1850).

As Proctor (1974) explains, by the 1880s, the importance of the mucous membrane in wind-related diseases (including the common cold) gained widespread currency. *The Lancet* published a commentary that exemplified this by explicitly rejecting the impressionistic discourse and putting forward the *real* explanation:

> [That] the 'east wind' has from the earliest times been credited with evil influence is apparent in the fact that Ephraim was described in one of the most ancient of the Jewish traditions as feeding upon it. What a repast! Modern science has determined that what we call the east wind is *really* a current of air from the north, its direction being modified by the rotatory movement of the earth. This is not difficult to understand. The avidity with which the dry cold wind takes moisture, and particularly that from the mucous membrane lining of the air passages, inspissating the mucus, and as it were, gluing together the cilia of the epithelium, is the *real* cause of the malignancy of the 'east wind'. We have repeatedly explained this fact, and pointed to the use of a succulent jujube or something which will give off moisture in the mouth to be carried into the air-passages with the inspired air as the best of practicable measures of protection. We believe that if only the habit of keeping a morsel that gives off aqueous vapour – not one that dissolves by taking up moisture – in the mouth during the time of exposure to the east wind were formed, there would be a very few colds from that cause (*The Lancet* 1886: 559).

The important consequence of this rebuttal of wind's 'popular' ontology was that it invalidated the notion of wind as a medical phenomenon in its own right. It also questioned the credibility of the public understanding of 'nature' insofar as it reduced the health risks of exposure to a 'wind' to the risks of exposure to the qualities of air, regardless of their association with the wind. It is not in the nature of *winds* to be healthy or harmful, but in the nature of the concomitant *conditions* of electricity, ozone, and other meteorological qualities such as rain or humidity. Later investigation took this point seriously and avoided doing anything other than correlating the wind's direction to the prevailing weather condition and the statistics of morbidity (Gordon 1903; 1905; 1910). If the twentieth-century volumes of the pre-eminent *Lancet* are anything to go by, the medical importance of wind has all but lost relevance for the profession. The only researches published on the topic included that on the 'wind of explosives' during the First World War (*The Lancet* 1915), the minor effects of wind in the treatment of tuberculosis (Girdlestone 1925), the (questioned) role of wind in the spread of foot-and-mouth disease (Burne 1969), the Foehn and myocardial infarction (Ambach, Tributsch, Mairinger, Steinacker & Reinegger 1992), wind-related injuries (*The Lancet* 1992), and wind noise when motorcycling (McCombe, Binnington, Donovan & McCombe 1992).

We should not conclude from this declining interest in medical analyses of wind (*qua* motion) that there has been a corresponding decline in interest in the phenomenon on the modern public, spiritual, industrial, energy-sector, or personal level. The above analysis does not contend that these processes brought about anything like a 'redefinition' of the natural world. Instead, the different generic engagements with the wind indicate for us an unstable coexistence within the nineteenth-century medical space, even when a reductionist Hippocratism sought to purge this space of the naïvety of colloquial meteorology. But regardless of the relative merit which this and other weather idioms enjoy in contemporary society – and they can be prophetic, prognostic, theoretical, naval, agricultural, or racial – they all seem to outline something resembling a modern *climatological citizenship* within putative polity boundaries which are fixed by the physical impact of the weather on local people's lifework.

NOTES

1 On smell, see Corbin (1986); on medical police, see Rosen (1957); on eudometry, see Schaffer (1990).
2 On sensibility, see Barker-Benfield (1996), Frye (1990), and Rousseau (1976).
3 Local topographers included Burrows (1814), Coldstream (1833), Hastings (1834), Hooper (1837), Royston (1809), Smith (1816), and Walker (1818). For history, see Jankovic (2000) and Rupke (2000).
4 For uses of chemistry in the analysis of spa waters, see Porter (1990).
5 On meteorology and public health, see Burton (1990).
6 Medical texts on the qualities of foreign and domestic climates were numerous, e.g. Bennet (1861); Bright (1854); Burgess (1850); Clark (1820; 1841); Cullen (1852); Dalrymple (1861); Davis (1807); Farr (1841); Francis (1853). For history, see Turner (1967).

REFERENCES

Ackerknecht, E. 1948. Anticontagionism between 1821 and 1867. *Bulletin of the History of Medicine* 22, 562–93.

—— 1982. Diathesis: the word and the concept in medical history. *Bulletin of the History of Medicine* 56, 317–25.

Adams, G. 1798. *Lectures on natural and experimental philosophy*. London: J. Dillon.

Allison, S.S. 1839. *An inquiry into the propagation of contagious poisons, by the atmosphere; as also into the nature and effects of vitiated air*. Edinburgh: MacLachlan and Stewart.

Ambach, E., W. Tributsch, T. Mairinger, R. Steinacker & G. Reinegger 1992. Fatal myocardial infarction and Tyrolean winds (the Foehn). *The Lancet* 339, 1362–3.

Anderson, K. 2005. *Predicting the weather: Victorians and the science of meteorology*. Chicago: University Press.

Arbuthnot, J. 1733. *An essay concerning the effects of air on human bodies*. London: Thonson.

Arnold, D. (ed.) 1996. *Warm climates and Western medicine: the emergence of tropical medicine, 1500–1900*. Amsterdam: Rodopi.

Atkinson, J.C. 1843. Meteorology and cholera. *The Lancet* 62, 322–3.

——1848. *Change of air: fallacies regarding it*. London: John Ollivier.

——1849*a*. On the effects of different winds on the human constitution. *The Lancet* 53, 207–8, 318, 410, 533–35; 54, 91–3.

——1849*b*. Electricity in cholera. *The Lancet* 54, 50.

——1850. On the connection of meteorological phenomena with cholera, influenza, and other epidemic diseases. *The Lancet* 1, 240.

Austen, J. 1816. *Emma: a novel, vol. 3*. London: J. Murray.

Bage, R. 1796. *Hermsprong; or, man as he is not. A novel. In three volumes. By the author of Man as he is*. London: William Lane.

Barker-Benfield, G.J. 1996. *The culture of sensibility: sex and society in eighteenth-century Britain*. Chicago: University Press.

Barnes, G. & A. Mitchell 2002. Measuring the marvelous: science and the exotic in William Dampier. *Eighteenth-Century Life* 26, 45–7.

Barret, F.A. 1998. Alfred Haviland's nineteenth-century map analysis of the geographical distribution of disease in England and Wales. *Social Science and Medicine* 46, 757–81.

Bennet, H.J. 1861. *Menton and the Riviera as a winter climate*. London.

Bewell, A. 1996. Jane Eyre and Victorian medical geography. *English Literary History* 63, 773–808.

Bisset, C. 1762. *An essay on medical constitution of Great Britain to which are added observations on the weather*. London: A. Millar and Wilson.

Bolton-Valencius, C. 2002. *Health of the country: how American settlers understood themselves and their land*. New York: Basic Books.

Bright, J. 1854. *A practical synopsis of diseases of the chest and air-passages*. (Second edition). London: J. Churchill.

Brontë, C. 1847. *Jane Eyre*. London: Smith, Elder & Co.

——1849. *Shirley*. London: Smith, Elder & Co.

Brown, S. 1797. *An inaugural dissertation on the bilious malignant fever. Read at a public examination, held by the medical professors, before the Rev. Joseph*. Boston.

Bruegman, R. 1978. Central heating and forced ventilation: origins and effects on architectural design. *Journal of the Society of Architectural Historians* 37, 143–60.

Burgess, T. 1850. Inutility of resorting to the Italian climate for the cure of pulmonary consumption. *The Lancet* 1, 591.

Burne, J.C. 1969. Not blowing in the wind. *The Lancet* 294, 901.

Burrows, G.M. 1814. Medical topography of London. *London Medical Repository* 1, 80–90.

Burton, J.M.C. 1990. Meteorology and the public health movement in London during the late nineteenth century. *Weather* 45, 300–7.

Cantor, C. (ed.) 2002. *Reinventing Hippocrates*. Ashgate: Aldershot.

Castle, T. 1995. *The female thermometer: eighteenth-century culture and the invention of the uncanny*. Oxford: University Press.

Cipolla, C.M. 1992. *Miasmas and disease: public health and environment in the pre-industrial age*. New Haven: Yale University Press.

Clark, J. 1820. *Medical notes on climate, disease, hospitals, and medical schools in France, Italy, and Switzerland*. London: T.G. Underwood.

——1841. *The sanative influence of climate*. (Third edition.) London: J. Murray.

Coldstream, J. 1833. An account of the topography, climate, and present state of the town of Torquay (Devonshire). *Edinburgh Medical and Surgical Journal* 40, 351–64.

Collectanea, 1834. The hardening System. *Collectanea. Medical Quarterly Review* 1, 190–3.

Corbin, A. 1986. *The foul and the fragrant: odor and the French social imagination*. Leamington Spa: Berg.

Cross, E. 1857. Cholera and diarrhœa. *The Lancet* 70, 180–1.

Cullen, W.H. 1852. *The climate of Sidmouth*. Sidmouth.

Curtin, P.D. 1964. Promise and the terror of the tropical environment. In *The image of Africa: British ideas in action 1780–1850* (ed.) P.D. Curtin, 58–87. Madison: University of Wisconsin Press.

Dalrymple, D. 1861. *Meteorological and medical observations on the climate of Egypt*. London.

Davis, J.B. 1807. *The ancient and modern history of Nice*. London: Tipper & Richards.

Dobson, M. 1782–3. An account of the Harmattan, a singular African wind. *The British Magazine and Review* 3. 259–84.

Dolan, B. 2002. Conservative politicians, radical philosophers and the aerial remedy for the diseases of civilization. *History of the Human Sciences* 15: 2, 35–54.

Douglas, M. 1966. *Purity and danger*. London: Routledge & Kegan Paul.

Drummond, H. 1849. On the effects of winds etc. in the production of disease. *The Lancet* 53, 410.

Farr, W. 1841. *A medical guide to Nice*. London: John Churchill.

Favret, M.A. 2004. War in the air. *Modern Language Quarterly* 65, 531–59.

Ferriar, D. 2004. The erotics of empiricism. Unpublished paper, University of Leeds.

Fodéré, F.E. 1813. *Traité de médecine légale et d'hygiène publique ou de police de santé adapté aux codes de l'empire français et aux connaissances actuelles*. Paris.

Fox, C.B. 1873. *Ozone and antozone: their history and nature. When, where, why, how is ozone observed in the atmosphere?* London: J. and A. Churchill.

Francis, D.J.T. 1853. *Change of climate considered as a remedy in dyspeptic, pulmonary, and other chronic affections*. London: John Churchill.

Freeman, T.W. 1978–9. Alfred Haviland: 19th-century medical geographer. *Geographical Magazine* 51, 90.

Friedman, F. 1993. *Appropriating the weather: Vilhelm Bjerknes and construction of a modern meteorology*. Ithaca, N.Y.: Cornell University Press.

Frye, N. 1990. Varieties of eighteenth-century sensibility. *Eighteenth-Century Studies* 24, 157–72.

Garrett, C.B. 1855. *East and northeast winds, the nature, treatment, and prevention of their distressing morbid effect on the respiratory and other organs, especially the larynx*. London: Samuel Highley.

Gates, W.E. 1967. The spread of Ibn Khaldun's ideas on climate and culture. *Journal of the History of Ideas* 28: 3, 415–22.

Girdlestone, G.R. 1925. The sun, the wind, and the skin. *The Lancet* 205, 1227–8.

Glacken, C. 1973. *Traces on the Rhodian shore: nature and culture in Western thought from ancient times to the end of the eighteenth century*. Berkeley: University of California Press.

Golinski, J. 1998. The human barometer: weather instruments and the body in eighteenth-century England. Paper at the American Society for Eighteenth-Century Studies Annual Meeting, Notre Dame.

Gordon, W. 1903. The influence of wind on phthisis. *British Medical Journal, 23* May.

——1905. On the influence of rainy winds on phthisis. *The Lancet* 165, 10–16, 77–82.

——1910. *The influence of strong, prevalent rain-bearing winds on the prevalence of phthisis*. London: H.K. Lewis.

Graham, J. 1793. *A new and curious treatise of the nature and effects of simple earth, water, and air*. London: Richardson.

Gregory, J. 1815. *A dissertation on the influence of change of climate in curing diseases* (trans. and notes by W.P.C. Barton). Philadelphia: Thomas Dobson.

Hamlin, C. 1992. Predisposing causes and public health in early nineteenth-century medical thought. *Social History of Medicine* 5, 43–70.

——1994. Environmental sensibility in Edinburgh, 1839–1840: the 'fetid irrrigation' controversy. *Journal of Urban History* 20, 311–39,

Hamper, A. 1785. *The economy of health*. London.

Hannaway, C. 1972. The Société Royale de Médecine and epidemics in the Ancien Régime. *Bulletin of Medical History* 46, 257–73.

——1974. Medicine, public welfare, and the state in eighteenth-century France: the Société de Médecine de Paris (1776–1793). Ph.D. dissertation, Johns Hopkins University.

——1993. Environment and miasmata. In *Companion encyclopedia of the history of medicine*, vol. 1 (eds) R. Porter & W.F. Bynum, 292–308. London: Routledge.

Harley, T. 2003. 'Nice weather for the time of year': the British obsession with the weather. In *Weather, climate, culture* (eds) S. Straus & B. Orlove, 103–20. Oxford: Berg.

Harrison, M. 1999. *Climates and constitutions: health, race, environment and British imperialism in India*. Oxford: University Press.

Hastings, C. 1834. *Illustrations of the natural history of Worcestershire*. London.

Haviland, A. 1853. The iatro-meteorology of Hippocrates. *Association Medical Journal* 1: 44, 961–4.

——1855. *Climate, weather and disease: being a sketch of the opinions of the most celebrated ancient and modern writers*. London: John Churchill.

Hingeston, J.A. 1853. Atmospheric phenomena in relation to the prevalence of Asiatic cholera. *Association Medical Journal* 42, 927–9.

Hippocrates 2004. *Airs, waters and places*. Whitefish, Mont.: Kessinger.

Hoolihan, J. 1989. Health and travel in nineteenth-century Rome. *Journal of the History of Medicine and Allied Sciences* 44, 462–85.

Hooper, G.S. 1837. Observations on the topography, climate and prevalent diseases of the island of Jersey. *The Lancet* 2, 900.

Hunt, L. 1817. Essays. In *The round table: a collection of essays on literature, men and manners* (ed.) W. Hazlitt. Edinburgh: Archibald Constable and Co.

Huxham, J. 1759. *Observations of the air and epidemical diseases, made at Plymouth from 1728–1737* (trans. from Latin). London: J. Hinton.

Jankovic, V. 2000. *Reading the skies: a cultural history of English weather 1650–1820*. Chicago: University Press.

——2002. The politics of sky battles in early Hanoverian Britain. *Journal of British Studies* 41, 429–59.

Jeffreys, J. 1842. On the artificial climates for the restoration and preservation of health. *London Medical Gazette* 1, 27–30.

——1850. *A word on climate and atmospheric influences*. London: Longman, Brown, Green & Longmans.

Johnson, J. 1818. *The influence of the atmosphere*. London: T. & G. Underwood.

——1839. *Change of air or Pursuit of Health*. London: Highley.

Jordanova, L.J. 1979. Earth sciences and environmental medicine. In *Images of the earth: essays in the history of the environmental sciences* (eds) LJ. Jordanova & R. Porter, 119–46. Chalfont St Giles: British Society for the History of Science.

——1981. Policing public health in France, 1780–1815. In *Public health* (ed.) T. Ogawa, 12–32. Tokyo: Taniguchi Foundation.

Kingsley, C. 1909–14 [1889]. Ode to the North-East Wind. In *English Poetry III: from Tennyson to Whitman*, vol. XLII. New York: P.F. Collier & Son.

Knellwolf, C. 2002. The exotic frontier of the imperial imagination. *Eighteenth-Century Life* 26, 10–30.

The Lancet 1886. The east wind. 127, 559.

——1915. Nervous manifestation due to the wind of explosives. 186, 348–9.

——1992. *Ill winds*. 340,171.

Laqueur, T.W. 2002. The places of the dead in modernity. In *The age of cultural revolutions: Britain and France, 1750–1820* (eds) J. Jones & D. Wahrman, 17–32. Berkeley: University of California Press.

Lee, W.R. 1973. The emergence of occupational medicine in Victorian Britain. *British Journal of Industrial Medicine* 30, 118–24.

Levine, G. 2002. *Dying to know: scientific epistemology and narrative in Victorian England*. Chicago: University Press.

Livingstone, D. 1999. Geographical inquiry, rational religion and moral philosophy: Enlightenment discourses on the human condition. In *Geography and Enlightenment* (eds) D.N. Livingstone & C.W.J. Withers, 93–120. Chicago: University Press.

McCombe, A., J. Binnington, D. Donovan & T.S. McCombe 1992. Motorcyclists and wind noise. *The Lancet* 340, 911–12.

MacLeod, R. 1982. The 'bankruptcy of science' debate: The creed of science and its critics, 1885–1900. *Science, Technology, & Human Values* 7, 2–15.

Miller, G. 1962. Airs, waters, and places in history. *Journal of the History of Medicine and Allied Sciences* 17, 129–40.

Moffat, T. 1860. *Medical meteorology and atmospheric ozone*. London: J. Churchill.

Murray, J. 1774. Journal containing daily meteorological and monthly medical observations. Wellcome Trust Library Manuscripts MS.7840.

Pelling, M. 1978. *Cholera, fever and English medicine, 1825–1865*. Oxford: University Press.

Phelps, Mr. 1743. *The human barometer: or the living weather glass. A philosophic poem*. London: M. Cooper.

Pilloud, S. & M. Louis-Courvoisier 2003. The intimate experience of the body in the eighteenth century: between interiority and exteriority. *Medical History* 47, 451–72.

Porter, P. (ed.) 1990. *The medical history of waters and spas, medical history*, supplement no. 10. London: Wellcome Trust.

Poynter, F.N.L. 1973. Sydenham's influence abroad. *Medical History* 17, 223–34.

Proctor, R. 1974. The nose, ambient air and airway mucosa: a pathway in physiology. *Bulletin for the History of Medicine* 48, 352–76.

Reed, A. 1983. *Romantic weather: the climates of Coleridge and Baudelaire*. Hanover, N.H.: University Press of New England.

'Review' 1820. *Bateman's* A succinct account of the contagious fever. *Medico-Chirurgical Journal*, 372.

Riley, J. 1987. *The eighteenth-century campaign to avoid disease*. New York: Knopf.

Robertson, H. 1808. *A general view of the natural history of the atmosphere and of its connection with the sciences of medicine and agriculture*. 2 vols. Edinburgh: Abernethy & Walker.

Rome, A.W. 1996. Coming to terms with pollution: the language of environmental reform, 1865–1915. *Environmental History* 3, 6–28.

Rosen, G. 1957. The fate of the concept of medical police. *Centaurus* 5, 99.

Rosenberg, C. 1992. *Explaining epidemics and other studies in the history of medicine.* Cambridge: University Press.

Rousseau, G.S. 1976. Nerves, spirits and fibres: towards defining the origins of sensibility. *The Blue Guitar* 2, 125–53.

—— & R. Porter (eds) 1990. *Exoticism in the Enlightenment.* Manchester: University Press.

Royston, W. 1809. Hints for a medical topography of Great Britain. *Medical and Physical Journal* 25, 13.

Rupke, N. (ed.) 2000. *Medical geography in historical perspective.* (*Medical History*, Supplement 20). London: Wellcome Trust Centre for the History of Medicine.

Rusnock, A. 2002. Hippocrates, Bacon, and medical meteorology at the Royal Society, 1700–1750. In *Reinventing Hippocrates* (ed.) D. Cantor, 136–53. Aldershot: Ashgate.

Rutty, J. 1770. *A chronological history of the weather and seasons and of the prevailing diseases in Dublin.* London.

Sargent, F., II 1982. *Hippocratic heritage: a history of ideas about weather and human health.* New York: Pergamon Press.

Schaffer, S. 1990. Measuring virtue: eudiometry, enlightenment and pneumatic medicine. In *The medical enlightenment of the eighteenth century* (eds) A. Cunningham & R. French, 281–318. Cambridge: University Press.

Schönbein, C.F. 1845. *Ueber die langsame und rasche Verbrennung der Körper in atmosphärischer Luft.* Basel: Schweighauser'sche Buchhandlung.

Scott, Sir W. 1996 [1817]. *Rob Roy.* London: Everyman.

Scott, W.B. 1854. The wind in the casement. In *Poems.* London.

Short, T. 1749. *A general chronological history of the air, weather, seasons, meteors etc. in sundry places and different times.* London: T. Longman.

Shuttleton, D.E. 1995. 'A modest examination': John Arbuthnot and the Scottish Newtonians. *British Journal for Eighteenth-Century Studies* 18, 47–62.

Smith, H. 1816. A brief sketch of the medical topography of Salisbury. *London Medical, Surgical and Pharmaceutical Repository* 6, 108.

Turner, E.S. 1967. *Taking the cure.* London: Joseph.

Vila, A.C. 1997. Beyond sympathy: vapours, melancholia, and the pathologies of sensibility. *Yale French Studies* 92, 88–101.

Walker, J.K. 1818. Medical topography of Huddersfield. *London Medical Repository* 10, 1–14.

Weindling, P. (ed.) 1985. *The social history of occupational health.* London: Croom Helm.

Willich, F.M. 1800. *Lectures on diet and regimen: being a systematic inquiry into the most rational means of preserving health and prolonging life.* London: Strahan.

Wintringham, C. 1727. *Commentarium nosologicum morbos epidemicos et aeris variationes in urbe Eboracenci locisque vicinis, ab anno 1715, usque ad finem anni* 1725. London: J. Clark.

Wohl, A. 1983. *Endangered lives: public heatlh in Victorian Britain.* London: Dent.

Wrigley, R. 1996. Infectious enthusiasms: influence, contagion and the experience of Rome. In *Transports: travel, pleasure and imaginative geography* (eds) C. Chard & H. Langdon, 75–116. New Haven: Yale University Press.

——2000. Pathological topographies and tourist itineraries: mapping malaria in eighteenth- and nineteenth-century Rome. In *Pathologies of travel* (eds) R. Wrigley & G. Revill, 207–28, Amsterdam: Rodopi.

Yeo, R. 1993. *Defining science: William Whewell, natural knowledge and public debate in early Victorian Britain.* Cambridge: University Press.

Part II

Societal and Environmental Change

Environmental Determinism

7

Nature, Rise, and Spread of Civilization

Friedrich Ratzel

What is then the essential distinction which separates natural and civilized races? Upon this question the evolutionist faces us with alacrity, and declares that it was done with long ago; for who can doubt that the natural or savage races are the oldest strata of mankind now existing? They are survivors from the uncultured ages out of which other portions of mankind, who have in the struggle for existence forced their way to higher endowments and have acquired a richer possession of culture, have long ago emerged. This assumption we meet with the question: Wherein then does this possession of culture consist? Is not reason, the basis, nay, the source of it all, the common property of the human race? To language and religion, as in some measure the noblest forms of expression, we must give the precedence over all others, and connect them closely with reason. In the fine expression of Hamann: "Without speech we could have had no reason, without reason no religion, and without these three essential components of our nature neither intelligence nor the bond of society." It is certain that language has exercised an influence reaching beyond our sight upon the education of the human spirit. As Herder says: "We must regard the organ of speech as the rudder of our reason, and see in talk the heavenly spark which gradually kindled into flame our senses and thoughts." No less certainly does the religion of the less civilized races contain in itself all the germs which are hereafter to form the noble flowery forest of the spiritual life among civilized races. It is at once art and science, theology and philosophy, so that civilized life which strives from however great a distance to reach the ideal contains nothing which is not embraced by it. Of the priests of these races the saying holds good in the truest sense that they are the guardians of the divine mysteries. But the subsequent dissemination of these mysteries among the people, the popularising of them in the largest sense, is the clearest and deepest-reaching indication of progress in culture. Now while no man doubts of the general possession of reason by his fellow-men of every race and degree, while the equally general existence of language is a fact, and it is not, as was formerly believed, the case that the more simply constructed languages belong to the lower races, the richest to those who stand highest; the existence of religion among savage races has been frequently doubted. It will be one of our tasks in the following pages to prove the unfoundedness of this assumption in the

The Anthropology of Climate Change: An Historical Reader, First Edition. Edited by Michael R. Dove.

light of many facts. For the present we will venture to assume the universality of at least some degree of religion.

In matters connected with political and economical institutions we notice among the natural races very great differences in the sum of their civilization. Accordingly we have to look among them not only for the beginnings of civilization, but for a very great part of its evolution, and it is equally certain that these differences are to be referred less to variations in endowment than to great differences in the conditions of their development. Exchange has also played its part, and unprejudiced observers have often been more struck in the presence of facts by agreement than by difference. "It is astonishing," exclaims Chapman, when considering the customs of the Damaras, "what a similarity there is in the manners and practices of the human family throughout the world. Even here, the two different classes of Damaras practise rites in common with the New Zealanders, such as that of chipping out the front teeth and cutting off the little finger." It is less astonishing if, as the same traveller remarks, their agreement with the Bechuanas goes even further. Now since the essence of civilization lies first in the amassing of experiences, then in the fixity with which these are retained, and lastly in the capacity to carry them further or to increase them, our first question must be, how is it possible to realise the first fundamental condition of civilization, namely, the amassing a stock of culture in the form of handiness, knowledge, power, capital? It has long been agreed that the first step thereto is the transition from complete dependence upon what Nature freely offers to a conscious exploitation, through man's own labour, especially in agriculture or cattle-breeding, of such of her fruits as are most important to him. This transition opens at one stroke all the most remote possibilities of Nature, but we must always remember, at the same time, that it is still a long way from the first step to the height which has now been attained.

The intellect of man and also the intellect of whole races shows a wide discrepancy in regard to differences of endowment as well as in regard to the different effects which external circumstances produce upon it. Especially are there variations in the degree of inward coherence and therewith of the fixity or duration of the stock of intellect. The want of coherence, the breaking-up of this stock, characterises the lower stages of civilization no less than its coherence, its inalienability, and its power of growth do the higher. We find in low stages a poverty of tradition which allows these races neither to maintain a consciousness of their earlier fortunes for any appreciable period nor to fortify and increase their stock of intelligence either through the acquisitions of individual prominent minds or through the adoption and fostering of any stimulus. Here, if we are not entirely mistaken, is the basis of the deepest-seated differences between races. The opposition of historic and non-historic races seems to border closely upon it. But are historical facts therefore lost to history when their memory has not been preserved in writing? The essence of history consists in the very fact of happening, not in the recollecting and recording what has happened. We should prefer to carry this distinction back to the opposition between national life in its atoms and national life organised, since the deepest distinction seems to be indicated by internal coherence which occurs in the domain of historical fact, and therefore mainly in the domain of intellect. The intellectual history of mankind no less than the social and political is in the first place a progression from individual to united action. And in truth it is in the first place external nature upon which the intellect of man educates itself, seeing that he strives to put himself towards it in an attitude of recognition, the ultimate aim of which is the construction within himself of an orderly representation of Nature, that is the creation of art, poetry, and science.

Showing as they do every possible variety of racial affinity, the "natural" races cannot be said to form a definite group in the anatomical or anthropological sense. Since in the matter of language and religion they share in the highest good that culture can offer, we must not assign them a place at the root of the human family-tree, nor regard their condition as that of a primitive race, or of childhood. There is a distinction between the quickly ripening immaturity of the child and the limited maturity of the adult who has come to a stop in many respects.

Figure 7.1 *Queensland Aborigines. (From a photograph).*

What we mean by "natural" races is something much more like the latter than the former. We call them races deficient in civilization, because internal and external conditions have hindered them from attaining to such permanent developments in the domain of culture as form the mark of the true civilized races and the guarantees of progress. Yet we should not venture to call any of them cultureless, so long as none of them is devoid of the primitive means by which the ascent to higher stages can be made—language, religion, fire, weapons, implements; while the very possession of these means, and many others such as domestic animals and cultivated plants, testifies to varied and numerous dealings with those races which are completely civilized.

The reasons why they do not make use of these gifts are of many kinds. Lower intellectual endowment is often placed in the first rank. That is a convenient, but not quite fair explanation. Among the savage races of to-day we find great differences in endowments. We need not dispute that in the course of development races of even slightly higher endowments have got possession of more and more means of culture, and gained steadiness and security for their progress, while the less-endowed remained behind. But external conditions, in respect to their furthering or hindering effects, can be more clearly recognised and estimated; and it is juster and more logical to name them first. We can conceive why the habitations of the savage races are principally to be found on the extreme borders of the inhabited world, in the cold and hot regions, in remote islands, in secluded mountains, in deserts. We understand their backward condition in parts of the earth which offer so few facilities for agriculture and cattle-breeding as Australia, the Arctic regions, or the extreme north and south of America. In the insecurity of incompletely developed resources, we can see the chain which hangs heavily on their feet, and confines their movements within a narrow space. As a consequence, their numbers are small, and from this again results the small total amount of intellectual and physical accomplishment, the rarity of eminent men, the absence of the salutary pressure exercised by surrounding masses on the activity and forethought of the individual, which operates in the division of society into classes, and the promotion of a wholesome division of labour. A partial consequence of this insecurity of resources is the instability of natural races. A nomadic strain runs through them all, rendering easier to them the utter incompleteness of their unstable political and economical institutions, even when an indolent agriculture seems to tie them to the soil. Thus it often comes about that in spite of abundantly-provided and well-tended means of culture, their life is desultory, wasteful of power, unfruitful. This life has no inward consistency, no secure growth; it is not the life in which the germs of civilization first grew up to the grandeur in which we frequently find them at the beginnings of what we call history. It is full rather of fallings-away from civilization, and dim memories from civilized spheres which in many cases must have existed long before the commencement of history as we have it. If, in conclusion, we are to indicate concisely how we conceive the position of these races as compared with those to which we belong, we should say, from the point of view of civilization these races form a stratum below us, while in natural parts and dispositions they stand in some respects, so far as can be seen, on a level with us, in others not much lower. But this idea of a stratum must not be understood in the sense of forming the next lower stage of development through which we ourselves had to pass, but as combined and built up of elements which have remained persistent, mingled with others which have been pushed aside or dropped into the rear. There is thus a strong nucleus of positive attributes in the "natural" races; and therein lies the value and advantage of studying them. The negative conception which sees only what they lack in comparison with us is a short-sighted under-estimate.

By the word "civilization" or "culture" we denote usually the sum of all the acquirements at a given time of the human intelligence. When we speak of stages, of higher and lower, of semi-civilization, of civilized and "natural" races, we apply to the various civilizations of the earth a standard which we take from the degree that we have ourselves attained. Civilization means *our* civilization. Let us assume that the highest and richest display of what we conceive by the term is to be found among ourselves, and it must appear of the highest importance for the understanding of the thing itself to trace back the unfolding of this flower to its germ. We shall only attain our aim of getting an insight into the nature and essence of civilisation when we understand the impelling force which has evolved it from its first beginning.

Every people has intellectual gifts, and develops them in its daily life. Each can claim a certain sum of knowledge and power which represents *its* civilization. But the difference between the various "sums of acquirement of the intelligence" resides not only in their magnitude, but in their power of growth. To use an image, a civilized race is like a mighty tree which in the growth of centuries has raised itself to a bulk and permanency far above the lowly and transitory condition of races deficient in civilization. There are plants which die off every year, and others that from herbs become mighty trees. The distinction lies in the power of retaining, piling up, and securing the results of each individual year's growth. So would even this transitory growth of savage races—which have in fact been called the undergrowth of peoples—beget something permanent, draw every new generation higher towards the light, and afford it firmer supports in the achievements of predecessors, if the impulse to retain and secure were operative in it. But this is lacking; and so it befalls that all these plants destined for a larger growth remain on the ground and perish in misery, striving for the air and light which above they might have enjoyed to the full. Civilization is the product of many generations of men.

The confinement, in space as in time, which isolates huts, villages, races, no less than successive generations, involves the negation of culture; in its opposite, the intercourse of contemporaries and the interdependence of ancestors and successors, lies the possibility of development. The union of contemporaries secures the retention of culture, the linking of generations its unfolding. The development of civilization is a process of hoarding. The hoards grow of themselves so soon as a retaining power watches over them. In all domains of human creation and operation we shall see the basis of all higher development in intercourse. Only through co-operation and mutual help, whether between contemporaries, whether from one generation to another, has mankind succeeded in climbing to the stage of civilization on which its highest members now stand. On the nature and extent of this intercourse the growth depends. Thus the numerous small assemblages of equal importance, formed by the family stocks, in which the individual had no freedom, were less favourable to it than the larger communities and states of the modern world, with their encouragement to individual competition.

As the essential feature in the highest development of culture, we note the largest and most intimate interdependence among themselves and with past generations of all fellow-strivers after it; and as a result of it, the largest possible sum of achievement and acquisition. Between this and the opposite extreme lie all the intermediate stages which we comprise under the name "semi-civilization." This notion of a "half-way house" deserves a few words. When we see energetically at work in the highest civilization the forces which retain, as well as those concerned with extending and reshaping, the building, in semi-civilization it is essentially the former which are called into most activity, while the latter remain behind and thereby bring about the inferiority of that state of things. The one-sidedness and incompleteness of semi-civilization lie on the side of intellectual progress, while on the material side development sets in sooner. Two hundred years ago, when Europe and North America had not yet taken the giant's stride which steam, iron, and electricity have rendered possible, China and Japan caused the greatest astonishment to European travellers by their achievements in agriculture, manufactures, and trade, and even by their canals and roads, which have now fallen far towards dilapidation. But Europeans, and the daughter races in America and Australia, have in the last two hundred years not only caught up this start, but gone far ahead. Here we may perceive the solution of the riddle presented by Chinese civilization, both in the height it has reached and its stationary character, and indeed by all semi-civilization. What but the light in free intellectual creation has made the west so far outrun the east? Voltaire hits the point when he says that Nature has given the Chinese the organs for discovering all that is useful to them but not for going any further. They have become great in the useful, in the arts of practical life; while we are indebted to them for no one deeper insight into the connection and causes of phenomena, for no single theory.

Does this lack arise from a deficiency in their endowments, or does it lie in the rigidity of their social and political organisation, which favours mediocrity and suppresses genius? Since it is maintained through all changes of their organisation, we must decide for the defect in their endowments, which also is the sole cause of the rigidity in their social system. No doubt the future alone can give a decisive answer, for it will in the first place have to be shown whether and how far these races will progress on the ways of civilization which Europe and North America vie in pointing out to them; for there has long been no doubt that they will or must set foot on them. But we shall not come to the solution of this question if we approach it from the point of view of complete civilization, which sees in the incompleteness of China and Japan the signs of a thoroughly lower stage of the whole of life, and frequently at the same time signs of an entire absence of hope in all attempts at a higher flight. If they possess in themselves only the capacities for semi-civilization, the need of progress will bring more powerful organs to their head and gradually modify the mass of the people by immigration from Europe and North America. This process may have first raised to its present height many a civilized race of to-day; we may refer to the Russians and Hungarians, and to the fact that millions of German and other immigrants have stimulated in many ways the progress of these semi-Mongols in Europe.

The sum of the acquirements of civilization in every stage and in every race is composed of material and intellectual possessions. It is important to keep them apart, since they are of very different significance for the intrinsic value of the total civilization, and above all for its capacity of development. They are not acquired with like means nor with equal ease, nor simultaneously. The material lies at the base of the intellectual. Intellectual creations come as the luxury after bodily needs are satisfied. Every question, therefore, as to the origin of civilization resolves itself into the question: what favours the development of its material foundations? Now here we must in the first place proclaim that when the way to this development is once opened by the utilisation of natural means for the aims of man, it is not Nature's wealth in material but in force—or rather, to put it better, in stimulus to force—which must be most highly estimated. The gifts of Nature most valuable for man are those through which his latent sources of force are thrown open in permanent activity. Obviously this can least be brought about by that wealth or so-called bounty of Nature which spares him certain labours that under other circumstances would be necessary. The warmth of the tropics makes the task of housing and clothing himself much lighter than in the temperate zone. If we compare the possibilities which Nature can afford with those that dwell in the spirit of man, the distinction is very forcible, and lies mainly in the following directions. The gifts of Nature in themselves are in the long run unchangeable in kind and quantity, but the supply of the most necessary varies from year to year and cannot be reckoned on. They are bound up with certain external circumstances, confined to certain zones, particular elevations, various kinds of soil. Man's power over them is originally limited by narrow barriers which he can widen but never break down by developing the forces of his intellect and will. His own forces, on the contrary, belong entirely to him. He cannot only dispose of their application but can also multiply and strengthen them without any limit that has, at least up to the present, been drawn. Nothing gives a more striking lesson of the way in which the utilisation of Nature depends upon the will of man than the likeness of the conditions in which all savage races live in all parts of the earth, in all climates, in all altitudes.

It is due to no accident that the word "culture" also denotes the tillage of the ground. Here is its etymological root; here, too, the root of all that we understand by it in its widest sense.[1] The storage by means of labour of a sum of force in a clod of earth is the best and most promising beginning of that non-dependence upon Nature which finds its mark in the domination of her by the intellect. It is thus that link is most easily added to link in the chain of development, for in the yearly repetition of labour on the same soil creative force is concentrated and tradition secured; and thus the fundamental conditions of civilization come to birth.

The natural conditions which permit the amassing of wealth from the fertility of the soil and the labour bestowed thereon, are thus undoubtedly of the greatest importance in the development of civilization. But it is unsafe to say with Buckle that there is no example in history of a country that has become civilized by its own exertions without possessing some one of those conditions in a highly favourable form. For the first existence of mankind, warm moist regions blessed with abundance of fruits were unquestionably most desirable, and it is easiest to conceive of the original man as a dweller in the tropics. But, on the other hand, if we are to conceive of civilization as a development of human forces upon Nature and by means of Nature, this can only have come about through some compulsion setting man amid less favourable conditions where he had to look after himself with more care than in the soft cradle of the tropics. This points to the temperate zones, in which we may no less surely see the cradle of civilization than in the tropics that of the race. In the high plateaux of Mexico and Upper Peru we have land less fruitful than the surrounding lowlands, and accordingly in these plateaux we find the highest development in all America. Even now, with cultivation carried to a high pitch, they look as dry and barren as steppes compared with the luxuriant natural beauties of many places in the lowlands, or on the terraces not a day's journey distant. In tropical and subtropical countries the fertility of the soil generally diminishes at high elevations, and in whatever climatic conditions, high plateaux are never so fruitful as lowland, hilly countries, and mountain slopes. Now these civilizations were both situated on high plateaux; of that in Mexico, the centre and capital, Tenochtitlan—the modern city of Mexico—lay at a height of 7,560 feet, while Cuzco, in Peru, is no less than 11,500. In both these regions temperature and rainfall are considerably lower than in the greater part of Central and South America.

This brings us to the recognition of the fact that, though civilization in its first growth is intimately connected with the cultivation of the soil, as it develops farther there is no necessary relation between the two. As a nation grows its civilization sets itself free from the soil, and, in proportion as it develops, creates for itself ever fresh organs which serve for other purposes than enabling it to take root. One might say that in agriculture there resides a natural weakness, which may be explained not only through want of familiarity with weapons, but through the desire of possession and a settled life enfeebling to courage and enterprise. We find, on the contrary, the highest expression of political force among the hunter and shepherd races, who are in many respects the natural antipodes of the agriculturists—the shepherds especially, who unite agility with the faculty of moving in masses, and discipline with force. The very faculties which are a hindrance to the agriculturist in developing that power, can here be turned to advantageous account,—the absence of settled abode, mobility, the exercise of strength, courage, and skill with weapons. And, as we look over the earth, we find that in fact the firmest organisations among the so-called semi-civilized races result from a blend of these elements. The distinctly agricultural Chinese have been ruled first by the Mongols, then by the Mantchus; the Persians by sovereigns from Turkestan; the Egyptians successively by Hyksos, or shepherd kings, Arabs, and Turks—all nomadic races. In Central Africa the nomadic Wahuma founded and maintained the stable states of Uganda and Unyoro, while in the countries that surround the Soudan every single state was founded by invaders from the desert. In Mexico the rougher Aztecs subdued the more refined agricultural Toltecs. In the history of places in the borderland between the steppe and cultivated lands a series of cases will be found establishing this rule, which may be recognised as a historical law. Thus the reason why the less fertile high plateaux and the districts nearest to them have been so favourable to the development of higher civilization and the formation of civilized states, is not because they offered a cooler climate and consequent inducement to agriculture, but because they brought about the union of the conquering and combining powers of the nomads with the industry and labour of the agriculturists who crowded into the oases of cultivation but could not form states. That lakes have played a certain part as *points d'appui* and centres of crystallisation for such states, as seen in the cases of Lake Titicaca in Peru, the lagoons of Tezcoco and Chalco in Mexico, Lakes Ukerewe and Tchad in the interior of Africa, is an interesting but less essential phenomenon.

Beyond the historic operation of climatic peculiarities in favouring or checking civilization, differences of climate interfere most effectually by producing large regions where similar conditions prevail—regions of civilization which are disposed like a belt round the globe. These may be called civilized zones. The real zone of civilization, according to all the experience which history up to the present day puts at the disposal of mankind, is the temperate. More than one group of facts corroborates this. The most important historical developments, most organically connected, most, steadily progressing in and by means of this connection, and externally most exciting, belong to this zone. That it was no accident which made the heart of ancient history beat in this zone on the Mediterranean Sea, we may learn from the persistency of the most effective historical development in the temperate zone even after the circle of history had been widened beyond Europe, ay, even after the transplantation of European culture to those new worlds which sprang up in America, Africa, and Australia. No doubt an infinite number of threads are plaited into this great web; but since all that races do rests ultimately upon the deeds of individuals, the one which has been most fruitful in results is undoubtedly the crowding together in the temperate zone of the greatest possible number of individuals most capable of achievement, and the arrangement in succession and comprehension of the individual civilized districts in one civilized belt, where the conditions were most favourable to intercourse, exchange, the increasing and securing of the store of culture; where, in other words, the maintenance and development of culture could display its activity on the largest geographical foundation.

Old semi-civilizations, whose relics we meet with in tropical countries, belong to a period when civilization did not make such mighty demands upon the labours of individuals, and when for that very reason its blossom sooner faded. A study of the geographical extension of old and new civilization seems to show that as the tastes of civilization grew, the belt comprising it shrank into the regions where the great capacity for achievement co-existed with the temperate climates. This observation is important for the history of the primitive human race and of its extension, and for the interpretation of the relics of civilization in tropical countries. Another mode in which civilization may perish is through the absorption of higher races by lower, who profit by the advantage of better adaptation to conditions of hardship. The despised Skraelings have merged themselves in the Northmen of Greenland. And has not every group of Europeans that has penetrated the Arctic ice-wastes, during the period of its stay in those dreary fields, been obliged to accustom itself to Eskimo habits, and to learn the arts and dexterities of the Arctic people in order successfully to maintain the fight with Nature's powers in the Polar zone? But so has many a bit of colonisation on tropical and polar soil ended in falling to the level of the wants of the natives. The colonising power of the Portuguese in Africa, the Russians in Asia, lies in their ability to do this more effectually than their competitors.

Yet a civilization, self-contained and complete, even with imperfect means, is morally and aesthetically a higher phenomenon than one which is decomposing in the process of upward effort and growth. For this reason the first results of the contact between a higher and a lower civilization are not delightful where the higher is represented by the scum of a world, the lower by people complete in a narrow space and contented with the filling up of their own narrow circle. Think of the first settlements of whalers and runaway sailors in countries rich in art and tradition like New Zealand and Hawaii, and of the effects produced by the first brandy-shop and brothel. In the case of North America, Schoolcraft first pointed out the rapid decay which befell all native industrial activity as a result of the introduction by the white men of more suitable tools, vessels, clothing, and so forth. European trade provided easily everything which hitherto had had to be produced by dint of long-protracted, wearisome labour.[2] and native activity not only fell off in the field where it had achieved important results, but saw itself weakened, and lost the sense of necessity and self-reliance, and so in course of time art itself perished. As we know, the same is going on to-day in Polynesia, in Africa, and among the poorest Eskimo. In Africa it is a declared rule that on the coast you have a region of decomposition, behind that a higher civilization, and the best of all in the untouched far interior. Even the art of Japan, independent as it was, deteriorated after a glimpse of artistically inferior European patterns.

NOTES

1 Of course its employment to denote the cultivation or refinement of the mind and manners (which though found in classical Latin seems comparatively recent in English) is a mere metaphor, without any suggestion of the fact noticed in this paragraph.

2 Cf. Lang, *Myth, Ritual, and Religion*, vol. i. p. 187, "He created the white man to make tools for the poor Indians," said the Winnibagoes to a white inquirer.

8

Environment and Culture in the Amazon Basin

An Appraisal of the Theory of Environmental Determinism

Betty J. Meggers

"Culture and environment" has at least three different meanings in anthropology: 1) that all cultures exist in an environmental setting; 2) that a culture is generally adapted to the characteristics of the environment; and 3) that a culture is molded, or determined by the characteristics of the environment. The first is universally recognized and the second generally accepted, but the third is usually rejected today, although it was a leading theoretical concept in anthropology's earlier years. Its renunciation began with the demonstration that some of the environmental determinists' assertions were exaggerated. Ridicule of this overembellishment brought about the disgrace of the theory, with little serious effort to determine whether or not the core was sound.

In recent years there has been a revival of interest in the relationship between environment and culture; and the biological concept of ecology – that the organism and its environment are interrelated – has been accepted as equally applicable to culture. There has been no serious effort to explore further and to re-evaluate the question of whether or not the environment can exert a deterministic influence on a culture. The misdeeds of the early environmental determinists have been kept fresh in the minds of each new generation of students and the mere use of the word "determinism" can still provoke a vigorous retort (Baerreis 1956: 315). There is, however, the risk that the environmental determinists were on the track of a useful concept, and that we have been uncritical in discarding the whole idea along with the errors and overemphases that poor documentation and an excess of enthusiasm made inevitable. Our knowledge of both culture and environment is now infinitely greater than it was then, and we are in a position to re-examine their relationship from the point of view of determinism. It is appropriate, therefore, that this discussion of environment and culture in the Amazon Basin be directed toward such an end. If the evidence suggests that the Tropical Forest Pattern of culture is a product

The Anthropology of Climate Change: An Historical Reader, First Edition. Edited by Michael R. Dove.

of well defined environmental characteristics, then we can conclude that determinism exists. If no such relationship can be detected, we may decide either that the theory is wrong, or that the evidence is as yet insufficient to permit a final judgment.

To place the discussion on an objective footing, it is necessary to have an impartial definition of "determinism". Our authority here will be Webster's New International Dictionary of the English Language, Unabridged, 1952 Edition, which gives the following meanings for the verb "determine": "To set bounds or limits to; to fix the form or character of beforehand, to foreordain; to establish causally; to bring about as a result; to give a definite direction, impetus, or bias to; to impel; to fix the course or end of." The question to be kept in mind is, therefore: Does the environment limit, establish causally or give definite direction to the culture it supports? Can the culture be shown to respond to characteristics of the environment that fix the course it must take if it is to survive? Finding the answer will require detailed analysis of that portion of the environment most directly affecting human exploitation – the subsistence resources, and particularly the agricultural potential.

Those familiar with some of the recent books extolling the unexploited wealth of the tropics may find it strange to pick the Amazon basin as the test ground for such a theory. The luxuriant vegetation exceeds in variety and exuberance the indigenous flora of any other climate of the world. On seeing it, men have reasoned: if temperate lands with poorer natural flora can be made to yield abundant crops, how much greater must be the harvest where nature unassisted succeeds so well? There is, however, another side to the story. Soil experts, agronomists, plant ecologists and other specialists have come to the conclusion, after detailed and extensive investigations of the properties of the soil, climate and natural vegetation, that the widely held impression of exceptionally fertile soils in the lowland tropics is contrary to the facts, and that to regard such lands as needing but an ax and a plow to transform them into the bread basket of the world is to believe in a myth.

The Standard Climate of the Wet Lowland Tropics

The Amazonian rain forests form part of a belt of similar vegetation that encircles the globe between the tropics of Cancer and Capricorn, or roughly 23° on either side of the equator. Its northern and southern extension is erratic because of the intrusion of mountain ranges, plateaus, and other factors that create climatic conditions beyond the tolerance of rain forest vegetation. The three major regions of the world where this vegetation occurs – tropical America, tropical Africa and tropical Asia – are distinct in plant species but show striking similarities in forest structure, plant succession and adaptation to the standard climate (Richards 1952: 6–7). Consequently, much of what will be said here is also generally applicable to the tropical rain forest regions in Asia and Africa. However, to avoid the pitfalls that come from too wide generalization, this discussion is specifically directed toward the Amazon basin. Although this area supports the largest continuous body of tropical forest in the world, its homogeneous geological origin, its slight variation in altitude and its equatorial location result in a remarkably uniform set of conditions over some million and a half square miles. Most important of these conditions is what may be termed the standard climate[1] – the characteristics of temperature and rainfall that produce tropical rain forest vegetation.

The tropics lack the extremes of both heat and cold characteristic of the temperate zone. The mean annual temperature averages 78°F (26°C), with the mean for the coldest month rarely differing from that of the warmest month by as much as 13° (Richards 1952: 136–7). At Manaos, in the center of the Amazon basin, the data on record show a maximum of 100.9°F (38.3°C) and a minimum of 65.8°F (18.8°C) (Richards 1952: Table 10). These extremes are balanced by more moderate temperatures on other days so that over a series of years the average is 85.1°F (28.2°C) for the hottest month and 79.7°F (26.5°C) for the coldest month (ibid). Lack of marked seasonal variation is related to the small annual variation in the length of the day and the relatively constant angle of the sun throughout the year in contrast with the large fluctuation in both these factors at greater distances from the equator. The marked day-to-day fluctuations in temperature characteristic of temperate regions especially in spring and fall are also rarely met with in the tropics (Richards 152: 136).

Like temperature, rainfall has certain characteristics that contrast with its pattern in temperate regions. Three of the most distinct differences are in quantity, intensity and variability. In typical tropical rain forest localities, the average amount of annual rainfall is at least twice that of normal temperate locations or about 80 inches (200 cm) per year. Manaos, with 66 inches (165.4 cm) is near the estimated minimum requirement for the maintenance of rain forest vegetation. Toward the other extreme is the Mazaruni Station, British Guiana, where the annual average is 98.7 inches (246.9 cm) or a little over 8 feet.

The presentation of rainfall data as averages indicates the large total quantity of precipitation, but obscures the two other important characteristics of the tropical rainfall pattern: intensity and variability. Extended soaking rains of the sort welcomed by farmers in temperate regions are relatively less common in the tropics and cloudbursts are correspondingly more frequent. Technically, a cloudburst is defined as a shower "having an intensity of at least 1 mm per minute for not less than 5 minutes" (Mohr and Van Baren 1954: 41). Converted into inches, this represents a minimum of 1/5 inch in 5 minutes, and tropical rains often exceed this specification in both intensity and duration. Although figures do not appear to be available for comparing the rainfall intensity of temperate and tropical areas using the Amazon as the point of reference, a study has been made using observations taken on the frequency of cloudbursts at 13 stations in Indonesia and at 13 stations in Bavaria (Germany). Analysis shows that cloudbursts furnish 22% of the total rainfall in Indonesia, but only 1.5% in Bavaria. Since tropical regions like Indonesia receive a much larger quantity of rain annually than Bavaria, it has been estimated that "the amount of water precipitated by tropical cloudbursts ... (is) 40 times more than ... in temperate latitudes" (Mohr and Van Baren 1954: 42).

Another important characteristic of tropical rainfall is its high degree of variability. One might say paradoxically that its unpredictability is predictable. There is a wide fluctuation in both monthly and annual totals. A given month may receive 2 inches (5.1 cm) one year and 12 inches (30.4 cm) the next.[2] A wet year receives at least twice as much rain as a dry one and a difference of 400–500% is relatively common (Visher 1923: 4). When one considers that the tropical averages are at least twice as great as the temperate ones, the variability in terms of actual amount of rainfall received becomes very marked.

The characteristics of temperature and rainfall outlined here constitute the dominant factors in the production of the standard climate of tropical lowland areas around the world. They have profound effects on the production and maintenance of soil fertility, and consequently on the food supply available to man. For an understanding of these effects it is necessary to examine the process by which soils become suitable for the growth of plants.

A discussion of soil fertility may properly begin with a description of the flora and fauna that inhabit the soil – earthworms, ants, algae, bacteria, fungi, etc. – because their activity creates the fertility. Of this flora and fauna, the most important and most completely studied are the microscopic plants that abound in all soils and play a major role in the reduction of organic matter to the basic elements from which it was originally derived. Bacteria, the major group of microflora both in quantity and in number of species, are the smallest. They are present by the millions in every gram of soil (Mohr and Van Baren 1954: 277). Larger and less numerous are the fungi or molds, which range from microscopic size to readily visible plants like giant puffballs, and which number from a few thousand to a million or more per gram of soil (Jacks 1954: 93). The activities of these and other micro-organisms result in the complete decomposition of almost every kind of organic compound so that only under special circumstances (such as those leading to the formation of peat or coal) can any organic matter accumulate.

Bacteria and fungi perform partly complementary and partly overlapping functions in the decomposition of organic material. While bacteria are frequently specialized to perform a specific task like nitrogen fixation, cellulose decomposition or ammonia production, fungi are more versatile and act as scavengers, attacking not only wastes that the bacteria are not able to deal with but also some of the same materials that are subject to bacterial decomposition. Lignin, the main constituent of wood, is resistant to bacterial action but falls prey to the voracious fungi. When the micro-organisms themselves die, their bodies are in turn attacked and decomposed. Ultimately, everything that was taken from the soil to build a living plant or animal is returned to the soil in its original form to be reutilized by another generation of living things. However, under certain circumstances some of the organic matter is temporarily diverted from the path toward complete mineralization and converted into an amorphous colloidal substance known as humus.

Although soil scientists and practical farmers agree that humus is a vital component of fertile soils, its actual composition is poorly understood in spite of considerable research. This matters little here, because for our purposes what humus does is more significant than what humus is. There is general agreement that it has two important functions: 1) it increases the capacity of the soil to absorb and hold plant nutrients so that they cannot be dissolved and washed away, and 2) it combines with clay minerals to form the crumb-like texture that is both favorable to the growth of plants and resistant to erosion (Jacks 1954: 110–13). It is also generally agreed that the presence of humus is indispensable to the achievement and maintenance of soil fertility, especially under continuous agricultural exploitation. Since the major producers of humus appear to be fungi, conditions favorable to their activity will result in the maximum degree of humus production (Jacks 1954: 93; Mohr and Van Baren 1954: 279). Contrariwise, in situations where bacteria flourish there will be no humus accumulation, because these micro-organisms reduce organic wastes rapidly and completely to their constituent minerals and return the latter in soluble form to the soil. From this it is evident that climatic factors favoring the activity of fungi will also favor the production and maintenance of soil fertility. At this point, the characteristics of the standard climate of the lowland tropics assume major significance. The ratio of fungi to bacteria is a function of the temperature of the soil, and higher temperatures are more favorable to bacteria. Experiments have shown that temperatures below 77°F (25°C) encourage the development of fungi and that increased warmth above this point increasingly inhibits their activity. The higher the temperature rises above 77°F (25°C), the more bacterial activity will predominate over that of fungi, and the less humus will be formed. Bacterial activity is also promoted by humidity in the neighborhood of 98%, whereas fungi prefer conditions of less moisture (Mohr and Van Baren 1954: 277). Since high temperature and high humidity are both characteristic of the standard climate of the lowland tropics, bacterial activity is favored over that of fungi, and without ameliorating circumstances no humus can accumulate in the soil (Richards 1952: 218; Mohr and Van Baren 1954: 280). In view of the important role of humus in the production and maintenance of soil fertility for agricultural purposes, this situation presents a serious problem.

The abundance and intensity of tropical rainfall compounds the difficulty by thoroughly leaching from the soil all of the soluble minerals deposited by bacterial action, some of which might otherwise remain accessible to growing plants (cf. Richards 1952: 208). The fact that precipitation exceeds evaporation for the greater part of the year means that there is a continuous downward movement of water in the soil. This percolation carries nutrients out of reach of plant roots, and leaves behind little but insoluble aluminum and iron oxides. When the full force of the standard climate is allowed to exert itself, the ultimate result of this process of mineralization and leaching is the production of a compact and infertile compound called laterite, on which only a sparse and hardy grass can make an effort to survive. Much tropical savanna owes its existence to the presence of this type of soil.

The important aspects of the standard climate of the tropics, to be borne in mind as we move on to a discussion of the problem of natural and human exploitation, can now be summarized as:

1. A high and even temperature that favors bacterial activity in the soil, so that the formation of humus essential to continued agricultural utilization is inhibited;
2. Abundant annual precipitation, which exceeds the moisture lost by evaporation and results in the leaching from the soil of soluble plant nutrients;
3. Intensity of rainfall, much of which comes in the form of cloudbursts, creating a great potential for erosion;
4. Variability of the total annual rainfall, expressed in wide differences in the amounts received in wet and dry years, which subjects agricultural crops to uncertain water supplies.

The success of any utilization of a tropical area with this standard climate can be measured by the degree to which this utilization is able to minimize or inhibit the undesirable effects the climate produces.

The Standard Climate and the Natural Vegetation

A visitor from the temperate zone requires only one look at the lush growth of a tropical rain forest to find himself doubting or opposing the suggestion that the soil may be infertile. Statistics are as impressive as the sight itself. Tropical rain forest is the most exuberant, luxuriant and varied of all flora. Trees average 50% taller than those in temperate woodlands, and a comparable area of ground will support many more species. Tropical forests are often so diverse in composition that it is difficult for the expert to locate two trees of the same kind. To the untrained eye, however, the same forest gives an effect of monotony because diversity of species is masked by uniformity of appearance. The trunks are nearly all straight and slender, the base is often expanded into prominent buttresses, the bark is smooth, and the leaves are dark green, leathery, oval and of similar size. Since such a forest is evergreen, not even a change of season does much to vary its appearance.

The dense canopy of foliage formed wherever the sun sheds its light is composed not only of trees but also of quantities of climbing plants, many of which are epiphytic and trail their roots like streamers down from the treetops. An impenetrable-appearing shield of vegetation is presented to the viewer whether he looks from an airplane in the sky or from a dugout on the waterways. This dense and compact surface has led to the mistaken impression that a similar condition exists in the depths of the forest. On the contrary, in a primary forest the canopy of shade is so complete that undergrowth is kept at a minimum and a traveler can frequently pass through on foot with little clearing of the way.

Forests of this sort flourish where the conditions of temperature and rainfall have all the detrimental features described for the standard climate of the lowland tropics. In explaining this apparently contradictory situation, it may be well to begin with the words of P. W. Richards, an outstanding authority on the tropical rain forest: "The most important common characteristic of all rain-forest soils ... is ... their low content of plant nutrients. This being so, it seems paradoxical that rain-forest vegetation should be so luxuriant. The leached and impoverished soils of the wet tropics bear magnificent forest, while the much richer soils of the drier tropical zones bear savanna or much less luxuriant forest. This problem has been considered by Walter (1936) and Milne (1937) for African forests, and by Hardy (1936) for those of the West Indies, and all these authors reach a similar conclusion. In the rain forest the vegetation itself sets up processes tending to counteract soil impoverishment and under undisturbed conditions there is a closed cycle of plant nutrients. The soil beneath its natural cover thus reaches a state of equilibrium in which its impoverishment, if not actually arrested, proceeds extremely slowly" (Richards 1952: 219).

Among the processes favoring the maintenance of soil fertility, which are furthered by the rain forest in spite of obstructing features in the standard climate, is the formation of humus. Although the conditions of high temperature and high humidity work to inhibit its development, the shelter afforded to the soil from rain and sun by the dense mass of foliage mitigates their full expression sufficiently to permit some humus accumulation. The amount is very slight, usually not exceeding 1–2% of the soil content (Richards 1952: 218), but even this small quantity exerts considerable effects on soil development by holding plant nutrients where they can be returned to the vegetation and by retarding leaching. As a result rainforest soils never reach the point of complete laterization that renders savanna soils barren and agriculturally useless.

In addition to creating a microclimate that promotes continuing soil fertility, rain forest vegetation plays an important role in moderating the destructive potential of the precipitation for leaching and erosion. This it does in two ways: 1) by reducing the quantity and intensity of the water that strikes the surface of the ground, and 2) by obstructing the erosional action of the runoff. Reduction in quantity and intensity is accomplished by the foliage. Data collected in Surinam show that when the daily rainfall is .04 inches (1 mm), 80% of the moisture is absorbed before it reaches the ground. As the quantity of precipitation increases, the percentage retained by the vegetation declines, but even when the amount is as high as 1.6 inches (40 mm) per day, about 20% fails to penetrate the vegetation shield. On the basis of these and other similar observations, it has been estimated that tropical forest vegetation allows only about 75% of the annual precipitation to reach the ground surface (Mohr and Van Baren 1954: 96). In view of the fact that such rainfall generally amounts to 80 or more inches annually, a reduction of 25% is a significant mitigating influence on erosion. Furthermore, what does penetrate to the ground comes in a fine spray rather than a heavy shower because of the deflection provided by the leaves, further reducing the erosion potential. A more familiar erosion-controlling mechanism of forests is their root system. This is so effective that forest soils do not develop the kind of erosion-resisting structure characteristic of less protected soils, with the consequence that they are highly subject to erosion once the forest cover is removed (Jacks 1954: 180).

In spite of this considerable role played by the vegetation in shielding the soil from the direct and total effects of the precipitation, a great deal of soil erosion occurs in rain forest areas. Its extent may be illustrated by contrasting the Amazon with the Rhine. Whereas the silt carried by the Rhine amounts to an average lowering of the land surface in the drainage

Efficiency of Soil Protection Under Natural Vegetation and Agricultural Exploitation

Environmental Requirements (Conditions for Permanent Production)	Types of Utilization: Rain Forest Vegetation	Slash-and-Burn Agriculture	Total Clearing
Retardation of Humus Destruction:			
Minimum Exposure of Soil to Sun	++	+	–
Minimum Aeration (Cultivation)	++	++	–
Mitigation of Erosion Potential:			
Diminution of Quantity of Precipitation Reaching Soil	++	+	–
Deflection of Intensity of Precipitation Impact	++	+	–
Reduction of Runoff Erosion	++	++	–
Retardation of Leaching	++	+	–

Figure 8.1 *Relative success of rain forest vegetation, slash-and-burn agriculture and total clearing (temperate zone technique of agriculture) in counteracting the detrimental effects of the standard climate of the wet, lowland tropics. The symbols have the following meaning: ++, approaching complete protection; +, partial protection or brief exposure period; –, no protection.*

area by only 0.002 inches (0.06 mm) per year, that carried by the Amazon amounts to 0.03 inches (0.8 mm) per year, or about 14 times more (Richards 1952: 207). This is still a very minute amount, but it raises a question: if this degree of erosion is possible with the maximum protection provided by forest cover, what would be the consequences of any considerable amount of deforestation, which would expose the soil to the full effects of the standard climate?

The adaptation of the natural vegetation to the requirements of the standard climate of tropical lowland South America can be scored according to the success with which it mitigates or counteracts the unfavorable results of high temperature combined with high humidity. The double plus mark opposite each of the 6 major factors in Figure 8.1 indicates that bacterial breakdown of the humus, leaching, and erosion are all markedly reduced, and in some cases brought to a standstill. The extravagant language often employed in the description of the luxuriant vegetation is thus deserved, not for the more obvious characteristics usually described but because of the marvelous adaptation to stringent climatic conditions that it represents.

The Standard Climate and Agricultural Exploitation

When man turns his hand to making a rain forest like that of the Amazon basin agriculturally productive, he faces the same standard climate with the same problems of soil conservation that nature has faced and solved. To the extent that this standard climate is unalterable and uncontrollable, the bounds within which man makes his adaptation are preordained. Numerous observers have characterized slash-and-burn agriculture as wasteful, inefficient, unproductive and primitive, and have stigmatized groups who practice it as lazy, ignorant, or too stubborn to abandon a practice whose only justification is to be found in tradition. On this basis, the fact that no advanced civilization has developed in the tropical rain forest is attributed to human failure rather than to the existence of a low level of agricultural potential in the environment. We have already seen that visitors from the temperate zone are inclined to evaluate the natural vegetation in terms of their own experience and consequently to pronounce the soil wondrously fertile, although quite the opposite is true. It seems possible that the deprecation of tropical shifting cultivation may spring from a similar misunderstanding of the kind of problems that confront an agriculturalist in the tropics.

First, what is the aboriginal situation? In tropical rain forests around the world, the most common agricultural technique is a variety of shifting cultivation, known in South America as slash-and-burn agriculture. By this method, a patch of forest is cut and the fallen trees are allowed to dry for a period varying from a few months to a year. At the beginning of the rainy season they are burned, a process that frequently consumes only the smaller branches. Some of the unburned wood is removed before planting, but the prepared field remains thickly strewn with trunks and large limbs, which gradually decay where they fell. Immediately after burning, manioc and other crops are planted in the bare spots with a hoe or digging stick. There is no cultivation of the soil and little if any weeding. Such a field will produce rather well the first year, less the second, and by the third is generally so unproductive that the clearing of a new field is begun, although bananas, papayas and other tree crops continue to be exploited until killed out by secondary forest growth. The field may be used again in 25, 50 or 100 years, or never. This method of agriculture requires a high ratio of land per person, with the consequence that only a small fraction of the forest is cleared at any one time. It has been recently estimated that only about 500 square miles out of more than a million square miles in the Amazon Basin are agriculturally productive today (Osborn 1953: 143), when farming is still carried on with aboriginal methods. The main characteristics of slash-and-burn agriculture as practiced in South America are thus small size of clearing, retention of stumps and root mat of the natural vegetation, minimal period between burning and the growth of the crops during which the soil is exposed to rain and sun, return to the soil of some plant nutrients from slashings left to decay, minimal cultivation and aeration, and short interval between the destruction of the forest and the beginning of its regeneration.

Turning now to the effects on the soil of removing the natural vegetation, we find the following sequence of events: "The removal of the forest cover at once changes the illumination at ground-level from a small fraction to full daylight. The temperature range greatly increases.... There is a change from the complicated system of microclimates characteristic of high forest to conditions closely approximating to the standard climate of the locality. Exposure to sun and rain very quickly alters the properties of the soil. Where the slope is sufficient, erosion will begin to remove the surface layers or their finer fractions. The rise in soil temperature leads to a rapid disappearance of humus" (Richards 1952: 401).

The length of time that this situation continues affects the length of time required for recovery. If the growth of the secondary forest is allowed to begin immediately, a vegetation cover will totally shield the soil within a few weeks and there will be a minimum of soil damage to be repaired. It is generally agreed that the longer the onset of secondary succession is postponed, the longer it will take for the soil to return to its original condition of fertility. The length of time required to complete the transition back to primary forest is not well established, but former agricultural land along the Brazilian coast is still occupied by secondary vegetation 150–200 years after clearing (Richards 1952: 399–400).

Under these circumstances, slash-and-burn agriculture seems reasonably well suited to the demands of the standard climate. The small size of the cleared area, and the fact that the stumps are not uprooted, keeps the erosion damage at a minimum. The absence of plowing or other cultivation, and the short time during which the soil surface is exposed to the sun keep humus destruction to a slow rate in comparison with what would result from the kind of extensive clearing and preparation of the soil practiced in temperate regions. Leaving the debris in the cleared area permits the return to the soil of some of the nutrients taken up by the former vegetation during growth, whereas all would be lost if the field were completely cleaned off. Finally, the short period of agricultural exploitation keeps modification of the soil at a minimum and facilitates recovery of the original fertility with the return of primary forest. Marking these adaptive features on the same chart (Figure 8.1) used to score the adaptation of the natural vegetation shows that slash-and-burn agriculture is not as successful as the forest in counteracting the damaging effects of the standard climate. However, it minimizes some and allows others only a brief period of operation before the forest cover returns.

To conclude that slash-and-burn agriculture is relatively well adjusted to the requirements of the standard climate in the Amazon Basin does not end the discussion of this form of agriculture. The disrepute in which it is held does not stem from a conservational effect on the landscape, but rather from the conservative influence it exerts over the local culture. It is these cultural characteristics that concern anthropologists most because they represent evidence of environmental determinism if they can be shown to stem from the character of the agricultural exploitation, which in turn is dictated by the requirements of the standard climate.

It has been noted many times that the centers of development of civilization lie outside the boundaries of the tropical rain forests. Various explanations have been offered, but the one most pertinent to this inquiry is that the limitations of the subsistence resources in tropical rain forest areas prevent cultural development from advancing beyond a relatively simple level. Just as the standard climate has features to which the vegetation must adapt, slash-and-burn agriculture has characteristics to which a culture must adapt. The most significant of these derive from the short period of time that a garden remains productive, which means 1) that a relatively large amount of land per capita must be available for agricultural use, and 2) that the settlement cannot remain permanently in one place. Let us consider the manner in which these two requirements affect cultural development.

The amount of land under cultivation at any time must always be slight in comparison with that held in reserve because of the long period required after use for recovery of fertility before reclearing is profitable. This situation exercises a strong control over population density and also over population concentration (cf. Palerm 1955: 30–1). Since the amount of agricultural land in an accessible radius of the village is limited by the irregularity of the terrain, by and large the smaller the community, the longer it will be able to remain settled in one spot. Differences in local hunting and fishing resources sometimes make local exceptions, and permit larger population concentrations than could be supported were agriculture the only important food resource. However, generally speaking 1000 individuals is a large population for villages in the South American tropical forest, and settlements with less than 300 people are typical.

Such a low level of population concentration allows little room for occupational specialization. If we note in our own culture how occupational specialization increases with the size of the urban population, it is readily understood how its absence is normal and inevitable in the small settlements characteristic of Tropical Forest life. Specialization provides a favorable atmosphere for the acquisition of detailed knowledge and for improvement of technical skill, and the archaeological picture in the centers of civilization around the world shows an advance in all fields of endeavor under its influence. Where specialization is absent, and each generation transmits what it has learned in much the same form as it was received, the normal result is slight variation but not marked increase in the scope or quality of techniques or products. One does not need to search for psychological explanations or to point to geographical isolation from centers of diffusion. The answer seems to be adequately provided by the series of causes and effects just described – without occupational specialization, progress in technology is slight and slow; without population concentration there is little specialization; with shifting cultivation, there can be no large permanent settlements.

In addition to a low population concentration, dependence on slash-and-burn agriculture necessitates periodic moving of the village. While it might be feasible in a temperate climate to exploit fields at a considerable distance from the settlement, this solution is not practical in the tropics. The staple root crops do not mature for seasonal harvest as do temperate crops, and cannot be stored in the humid, warm climate without sprouting or rotting. Harvesting must be daily or several times a week. If the fields were not readily accessible, the time needed to exploit them would be so great that agriculture would provide little more return for the labor than wild food gathering. Hence, shifting of the village periodically is not simply a matter of choice but an adaptation to a subsistence base whose most effective utilization is continuous rather than seasonal.

The inability to settle long in one spot has certain cultural effects that can be most effectively indicated by recalling a saying current in our own culture: "Three moves are as good as a fire." Even with all the opportunities for accumulation of material goods afforded to us, frequent moving inhibits their full realization. It is often more convenient and economical to reacquire an object than to move it. How much less incentive is there for preserving and transporting goods in a culture where everyone can make what he needs from materials freely available in the forest. If equipment is expendable, it also is not normally lavishly made, so that the expertness provided by specialized artisans finds no market. This factor reinforces the small size of population concentrations in inhibiting the growth of occupational specialization. Since property ownership is an effective basis for social differentiation and the concentration of power, the absence of incentive for property accumulation helps to maintain social organization on a simple, unstratified level, largely dominated by kinship.

Environmental Determinism

The picture that emerges from this analysis shows a complex interrelationship between climate, agriculture and culture in the Amazonian rain forests. Agriculture must take account of climatic factors, and the culture must be adjusted to the requirements of the subsistence exploitation. This conclusion is supported not only by theoretical analysis of the processes promoting or inhibiting cultural advance, but by the fact that no advanced culture has developed in the Amazon Basin or has been maintained there by a subsistence pattern dominated by slash-and-burn agriculture. In the sense of

Webster's definition of determinism – namely, to set bounds to, and to give definite direction to – it seems acceptable to conclude that the Tropical Forest Type of culture characteristic of the Amazon Basin shows the effects of environmental determinism.

To the extent that a culture is a functional whole, this determinism may be felt even in aspects remote from the subsistence sphere. There is a difference between this concept and the earlier determinists' tracing of everything to environmental causes. The statement that thatch houses are built because of the abundance of thatching materials is easy to refute by pointing to wattle and daub or even stone as equally available. However, if thatch houses are associated with slash-and-burn agriculture, it can be argued that quickly made structures of short term utility represent the least expenditure of effort for the maximum in effective shelter, and are therefore a secondary reflection of an environmentally determined way of life. This is not to argue that all cultural traits can be traced to environmental causes, even secondarily removed. Certain types of human cultural activity place so little burden on subsistence that they are relatively free from its restraints. However, it is probable that a larger proportion than is generally suspected is the direct or indirect reflection of an environmental characteristic.

It can be shown that the environments of the world do not offer equal potential for human exploitation (Meggers 1954). Reciprocally, all environments are not equally deterministic. In areas where the agricultural potential is absent or limited, the boundaries within which a cultural adaptation must be made are circumscribed. There is little latitude and successful alternative methods of exploitation are few. In areas where agricultural potential is naturally high or can with simple techniques be greatly increased, the boundaries are wide and environment is more permissive than deterministic in its action. In such situations, both greater range of cultural development and greater variety of result might be expected. It would be easy to express this difference by saying that where the environment is strongly deterministic (i.e. offers little latitude) a culture *must* remain simple, but where the environment is permissive it *may* remain simple or it may advance. However, the fact that all the areas of the world with great natural agricultural potential served as cradles of civilization suggests the environment plays an active rather than a passive role even when it is most indulgent.

The strong emphasis currently placed on man's influence on his environment (Heizer 1955; Princeton Symposium 1955) has tended to obscure those situations where man conforms rather than dominates. Changes wrought by hunters, gatherers and primitive slash-and-burn agriculturalists are comparable to those effected by birds, animals and other natural forces: all scatter seeds, selectively kill other creatures and make similar minor alterations in their habitat. Such modifications are inevitable by-products of remaining alive. However, man alone has evolved the capacity to alter his environment purposefully, on a large scale and in a permanent way, and the development of this capacity is a crucial factor in cultural evolution. To say that all human beings modify their environment is to lose sight of the fundamental difference between transporting a seed and flattening a mountain or extinguishing a forest. All cultures have not been equally successful in achieving this mastery, which is another way of saying that all environments are not equally malleable; modifications that some environments reward, others resist. Those that resist may not necessarily be less plastic, but only have their plasticity in directions that cross-cut rather than parallel human needs. In view of the diversity of climate and topography, it would be remarkable if man found all parts of the world equally congenial, equally easy to master, equally unresistant. In the Amazon basin he has conspicuously failed to make more than a fleeting mark, and denial of the environmental obstacles does not make this failure less real. Recognition of the deterministic quality of environment, on the other hand, provides one more tool for the solution of our ultimate problem – the understanding of how and why culture develops when, where and as it does.

Tropical Forest Agriculture in the Future

The recognition that there is a cause and effect relationship between environment and culture, far from being a defeatest doctrine, establishes a realistic basis for the evaluation of the potential of the Amazon Basin for future human exploitation. If any important alteration in culture depends on the eradication of the inhibiting aspects of slash-and-burn agriculture, then the problem is to find a means of retaining fields in more permanent agricultural production so that increased sedentariness and increased population concentration can be achieved. Keeping in mind the requirements of the standard climate, let us examine a few of the solutions that have been proposed.

One suggestion involves the use of alluvial land, whose fertility is renewed by the annual deposition of silt, and which provides the basis of agricultural production in many other parts of the world. A certain amount of native use of the low river banks exists in the Amazon Basin (Smith 1954: 96–7), but the pattern of exploitation is at present no more stable than that associated with slash-and-burn agriculture. Expansion of the flood plain usage has a number of drawbacks, some of which stem from the characteristics of the river. Although the Amazon is vast, its flood plain is not proportionately large. Survey has shown that it is typically narrower than that of the Mississippi below its junction with the Ohio River. The total area occupied by recently deposited alluvium amounts to only about 10% of the entire Amazon Basin (Marbut and Manifold 1925: 622 and 632) and most of this is subject to annual inundation. Furthermore, the flood plain is not continuous, but is subdivided by innumerable lakes and sloughs into isolated and irregularly shaped patches. Engineering attempts to consolidate these into larger units must contend with the river at flood, a mass of water that makes the raging Mississippi seem small and docile by comparison, and whose harnessing seems unlikely (Pendleton 1950: 120–1). Plans for expansion of this type of agriculture must also take into consideration the dietary requirements of the local population. A major factor limiting its aboriginal use is its unsuitability for growing manioc and other staple tropical root crops. Food habits are highly resistant to change and people often prefer to go hungry rather than eat something they do not consider fit for human consumption. It matters little that what they refuse may form a major part of the diet in another part of the world.

A second solution that has been mentioned also raises the question of consumption. This takes cognizance of the fact that tree crops are ideally suited to the requirements of the tropical lowland climate. However, tropical tree crops do not figure prominently in world subsistence, and there seems to be no prospect of making bananas, for instance, a staple food on a large scale. The failure of plantation rubber in the Amazon Valley, furthermore, illustrates one of the undesirable factors encouraged by the standard climate, but not allowed full expression by the scattered distribution of species of plants in the natural vegetation. The continuously high temperature and humidity favors rapid multiplication and dispersal of diseases such as the one that wiped out plantation rubber in Brazil. The absence of a dormant winter period of cold makes such invasions more difficult to combat than in the temperate zone, and the possibility that similar epidemics may defeat other attempts to develop plantation agriculture looms large. That the problem is not physically insuperable is shown by promising results of attempts to produce blight resistant strains of rubber plants. However, this development has taken several decades, and in the interim synthetic rubber has been perfected so that there is little demand for the hard won solution.

A third possibility is the development of a synthetic substitute for humus that will be resistant to breakdown at high temperatures, so that soil fertility can be maintained with complete clearing of the land. This problem seems near solution (Jacks 1954: 113), but there is still the irregularity of the rainfall to contend with and the problem of erosion control. When these are solved, the question remains as to what crops can be raised that will be in sufficient demand on the world market to justify and repay the tremendous expense involved in achieving their production on a large scale. For, when all is said and done, successful agricultural exploitation in the lowland tropics requires more than good will, it must meet the hard test of competition in the world market.

What may we conclude, then, about environment and culture in the Amazon Basin? First, the evidence suggests that environment and culture are not independent variables, but intimately related to one another. The climatic conditions characteristic of this tropical region present well defined problems to agricultural exploitation. To date, no better solution has been found than that developed by the indigenous population, namely, slash-and-burn or shifting cultivation. This type of food production has a conservational effect on soil and soil fertility, which is desirable, but also exercises a conservative influence on the culture, keeping it in a relatively simple level of development. Just as attempts to alter the agricultural pattern that do not recognize its environmental adjustment are doomed to failure, so attempts to maintain a more highly developed type of culture on a subsistence base of slash-and-burn agriculture have not had permanent success. The existence of these interrelationships indicates that efforts to understand how and why culture has developed to different levels of complexity in various parts of the world must literally work from the ground up.

Recognition of determining factors in the environment places the Tropical Forest Area of South America in a new perspective. Instead of being a backward region, whose cultural development was retarded or stagnated by isolation, warfare or psychological barriers, we see it as an area in which the culture is in equilibrium with the environment, having made a remarkably efficient adjustment to extremely difficult conditions for agricultural exploitation. The recognition of this cause and effect relationship helps not only to understand the development of culture in the Amazon Basin in the past, but opens the door to a more realistic attack on the problem of improving the cultural potential of this area in the future. If we accept the premise that the standard climate determines that agricultural exploitation must have certain features, then man's problem is to find a solution that fulfills these requirements and in addition meets the demands of modern civilization. The recognition of the existence of environmental determinants shows us the direction in which we should look, but unfortunately only the future will tell whether or not a satisfactory solution can be found.

NOTES

1 This term is used by P.W. Richards (1952: 158) to label "the climate recorded by instruments about 1 m. above the surface of the ground, exposed, according to the rules adopted by meteorologists, in more or less extensive clearings." It has been retained in this discussion as a convenient means of referring to the combination of rainfall and temperature that combine to create a well-defined and distinctive environmental situation. Humidity, wind velocity, evaporation and other factors are relevant, but less spectacular than temperature and rainfall. They have been omitted in the interest of simplicity because their effects do not conflict with the conclusions derived from an analysis of the temperature and rainfall pattern.

2 Rainfall records on the upper Essequibo River, British Guiana show December totals of 1.63 inches (4.4 cm) in 1954 and 12.03 inches (30.6 cm) in 1955. May totals were 14.53 inches (36.9 cm) in 1954 and 27.97 inches (71.1 cm) in 1955 (Records, Meterological Department, Georgetown, British Guiana).

REFERENCES

Baerreis, David A. 1956. *Review of New Interpretation of Aboriginal American Culture History* (75th Anniversary Volume of the Anthropological Society of Washington). *American Antiquity* 21: 314–316.

Hardy, F. 1936. Some aspects of cacao soil fertility in Trinidad. *Trop. Agriculture, Trin.*, 13: 315–317.

Heizer, Robert F. 1955. Primitive man as an ecologic factor. *Kroeber Anthropological Society Papers*, No. 13, p. 1–31.

Jacks, Graham V. 1954. *Soil.* London (Thomas Nelson & Sons, Ltd.)

Marbut, C. F. and C. B. Manifold. 1925. The topography of the Amazon valley. *Geographical Review* 15: 617–642.

Meggers, Betty J. 1954. Environmental limitation on the development of culture. *American Anthropologist* 56: 801–824.
——. 1955. The coming of age of American archeology. *New Interpretations of Aboriginal American Culture History* (75th Anniversary Volume of the Anthropological Society of Washington), p. 116–129.
Milne, G. 1937. Essays in applied pedology. I. Soil type and soil management in relation to plantation agriculture in east Usambara. *East Afr. Agric. J.* (July).
Mohr, E. C. J. and F. A. Van Baren. 1954. *Tropical soils: a critical study of soil genesis as related to climate, rock and vegetation.* The Hague and Bandung.
Osborn, Fairfield. 1953. *The limits of the earth.* Boston (Little, Brown and Co.)
Palerm, Angel. 1955. The agricultural basis of urban civilization in Mesoamerica. *Irrigation Civilizations: a Comparative Study.* Social Science Monographs 1: 28–42. Pan American Union.
Pendleton, Robert L. 1950. Agricultural and forestry potentialities of the tropics. *Agronomy Journal* 42: 115–123.
Richards, P. W. 1952. *The tropical rain forest.* Cambridge, England (University Press).
Smith, T. Lynn. 1954. *Brazil, people and institutions.* Baton Rouge (Louisiana State University Press).
Stamp, L. Dudley. 1952. *Land for tomorrow.* Bloomington (Indiana University Press).
Visher, Stephen S. 1923. Tropical climates from an ecological viewpoint. *Ecology* 4: 1–10.
Walter, H. 1936. Nährstoffgehalt des Bodens und natürliche Waldbestände. *Forstl. Wschr. Silva* 24: 201–5, 209–213.

Climate Change and Societal Collapse

9

Management for Extinction in Norse Greenland

Thomas H. McGovern

Five hundred years before Columbus arrived in the Caribbean, Scandinavian communities sharing a common language and a common culture stretched from western Norway to eastern North America. Between ca. AD 800 and 1000, Nordic warriors, traders, and settlers colonized the Faroes, Shetlands, Orkneys, northern Scotland, the Isle of Man, parts of Ireland, England, and Frankland, as well as the western Atlantic in Iceland, Greenland, and Vinland (see Bigelow 1991; Jones 1985; McGovern 1990). By the early Middle Ages, thriving European communities existed in arctic Greenland and subarctic Iceland, and the great saga-men of the western North Atlantic were producing medieval Scandinavia's first histories and finest historical novels.

Because the medieval Scandinavian North Atlantic route to North America was ultimately fruitless, Columbus surely deserves credit for the effective connection of the Old World with the New. By the time Columbus sailed, the Norse North Atlantic had long been an economic and social backwater, and its island communities were undergoing contraction and a desperate struggle for survival. In 1492, it is just possible that the very last of the Norse colonists of Greenland were losing that struggle, slipping into social and biological extinction. This westernmost European colony of the Middle Ages probably fell vacant within a few decades of the first Iberian voyages to the Caribbean (ca. 1475–1500).

What caused the failure of the early, northern route to the New World? Why were the early centuries of Scandinavian exploration and relative prosperity followed by isolation, impoverishment, and extinction? Were the Scandinavian colonists of Greenland simply overwhelmed by the challenges of global climate change and contact with Thule Inuit? Did critical failures of human resource management doom the settlements? Does the grim case of Norse Greenland hold any lessons relevant to a modern world faced by climate change, universal culture contact, and the consequences of human management choices on a global scale?

In this paper I briefly explore the 500-year-long interaction of culture and nature in Norse Greenland, explicitly attempting to draw lessons of wider potential relevance. The data employed are archaeological, documentary, and paleoecological. (Recent archaeology is discussed in greater detail in Berglund 1991; Keller 1991; McGovern 1985a, 1985b, 1992; McGovern et al. 1988. For a more complete discussion of the documentary sources, see Arneborg 1991; Gad 1970; Jansen 1972; Keller in press.)

The Anthropology of Climate Change: An Historical Reader, First Edition. Edited by Michael R. Dove.

Documentary sources for Norse Greenland are few (and historically reliable documentary sources even fewer), and they do not provide much information about basic subsistence, settlement pattern, land-use changes, or culture contact (Gad 1970). They do, however, provide a great deal of information about medieval Norse world view and cultural evaluation of "correct" social and environmental relations. Sagas, law codes, myths, and popular stories all provide a rich data set only now undergoing systematic analysis from an anthropological perspective (Durrenburger 1991; Durrenburger and Palsson 1989; Hastrup 1985; Miller 1986).

Thanks to the work of Scandinavian and English-speaking scholars of Greenland over the past three hundred years we now have a rich archaeological data base for many of the aspects of life not deemed "saga worthy" by the Norse themselves. In Atlantic Scandinavia it is thus possible to combine the long, diachronic perspective of history and archaeology with the emic, "inside view" usually restricted to ethnographic cases. Historical archaeology is only beginning to exploit this sort of interface, but a rich potential surely exists for a combination of standard environmental analyses of resource exploitation and land-use patterns with cognitive approaches incorporating concepts of limited knowledge, management hierarchy and heterarchy, and a world view significantly different from our own.

Settlement Process

As described in Landnamabok, The Book of the Greenlanders, and Eirik the Red's Saga (Jones 1985), the settlement of Greenland began with competition between chieftains in Iceland. The loser in this power struggle was Eirik the Red, who departed hastily to investigate reports of new land to the west. This voyage replicated earlier Norse voyages of exploration and settlement, since chiefly competition in the previously settled island colonies encouraged further emigration under the sponsorship of the defeated elites. Many such desperate or hopeful voyagers simply disappeared in the gray waters of the North Atlantic or were forced to accept subordinate positions in already settled colonies. But for a lucky few like Eirik, the special status of Landnamsmann (first settler) rewarded the exploration of a new settlement area.

A first settler was not necessarily the first to discover a new land; he was the one who pioneered a successful settlement, discovering firsthand both the potentials and the hazards of an unknown region. Not all would-be first settlers were successful. Iceland's legendary history describes two failed settlement attempts before the eventual success of Ingolfur Arnarsson around 870 (Jones 1985). The Vinland settlements did not survive the first few years when they were most vulnerable (McGovern 1980–81; Wallace 1991). The risks of pioneering were exceptional since the normal buffering effects of a larger Norse community were absent, the timing and length of the growing season (and its variability) could only be estimated, and the potential hostility of unpacified native humans, elemental in-dwelling land spirits, and nonhuman magical beings (trolls and elves) would be a real threat to a small and restricted settlement. Success could come only to a strong chieftain with both the economic resources (ships, animals, seed, and laborers) and the personal, semimagical "luck" that could overcome great natural and supernatural obstacles.

Successful first settlers were able to claim lasting privileges in return for their luck and their early economic investment. The primary first settlers (like Eirik) could lord it over later-arriving chieftains and could count on playing a leading role in the new colony for the rest of their lives. Many primary and secondary chiefly pioneers set up pagan temples and, later, Christian churches on their home farm, ritually reinforcing their early authority and forcing other settlers to come to them for religious observances (see discussion in Keller, in press). The first settlers often arranged the location of the local *thing* (assembly, attended annually by free farmers who met minimum property requirements) and had a great deal to say about its operation.

The first settlers named broad tracts (often naming whole fjord systems after themselves) and local landscape features, giving meaning and form to a culturally blank wilderness.

Names had lasting power, both magical and legal, and so did this initial division of landscape on the chieftains' terms. Eirik the Red named the longest fjord of the eastern settlement Eiriksfjord and then claimed the best land in Greenland at its head, naming his manor Brattahlid.

First settlers usually centered their main farms on the best agricultural land, then parceled out less-desirable tracts to followers (see Keller, in press). The Icelandic first settler Skalla-grim was well known for skill in integrating coastal, valley floor, and upland resources. He sent slaves and subordinates to harvest seasonally available seals, birds, and fish. He also assigned workers to *saeters* (shielings, summer pastures with shepherd's huts) in higher elevations, while overseeing the working of the main farm in the prime agricultural zone himself (Durrenburger 1991). Skalla-grim's manor thus was said to "stand on many feet": in modern terms, enjoying a wide niche breadth as well as a large foraging territory.

This early "Skalla-grim effect" may have initially dispersed chiefly centers over the whole habitable landscape rather than encouraging villagelike clusters. Certainly in Greenland the archaeological data (backed by a small but growing series of radiocarbon dates) suggests that all of the pockets of potential pasture vegetation in the fjords of West Greenland were occupied by Norse settlers within a generation of first settlement (Berglund 1991; Keller, in press; McGovern 1981). The Norse in Greenland seem to have settled two major regions: the eastern settlement around the modern Qaqortoq and Narssaq districts in the southwest and the smaller western settlement in the more northern Nuuk district.

Initial settlement may have been broad but thin, but as immigration and reproduction continued, the landscape seems to have filled in with small, less-extensive holdings. Many small holdings may initially have been special-purpose seasonal stations later upgraded to year-round farms, or they may have been created by subdividing early chieftain farms among siblings. As the filling in process continued, the niche breadth of individual farms may have narrowed, and access to localized wild resources may have required the cooperation of a local community rather than the coordination of a single chieftain's household labor. As the Viking period ended and local populations increased, slavery was replaced by the use of tenant labor (Durrenburger and Palsson 1989; Hastrup 1985). By ca. 1100, the initial impact of first settlers and the Skalla-grim effect may have been history, but it was to prove a pervasively influential history for the rest of the settlement.

Later Settlement and Architecture

West Greenland's plant communities are similar to those of Iceland and the eastern North Atlantic (Buckland 1988) but are restricted to the lower elevations of the warmer inner fjords of the south and southwest—precisely the location of the Norse eastern and western settlements. Virtually every meadow is marked by a Norse ruin, with close correlation between pasture productivity and settlement density in most regions (Christensen 1991; Keller 1991, in press). Summer pasture and hay for winter fodder were critical determinants of the number and productivity of domestic animals, and North Atlantic farmers were keen observers of pasture potential (see McGovern et al. 1988 for examples and discussion).

The distribution of these critical resources was certainly a key variable politically and economically, as well as environmentally. By the height of the settlement, productive pasture was far from evenly distributed. Figure 9.1 presents the pasture area available to a series of well-surveyed western settlement sites we have reason to believe were contemporary (McGovern 1985a) and illustrates the correlation of pasture area with the size of the farm. The site of W51 (Sandnes) was a chieftain's (probably first settler's) farm with one of the three churches in the western settlement (and the only church farm in this sample).

Table 9.1 summarizes available floor areas of halls (a proxy for human space), byres (cattle sheds), barns (hay storage), storage features, and churches for both settlement regions. The larger farms had a distinctive architectural pattern, with byres and storage buildings having far larger floor areas relative to human dwelling spaces than is found on smaller

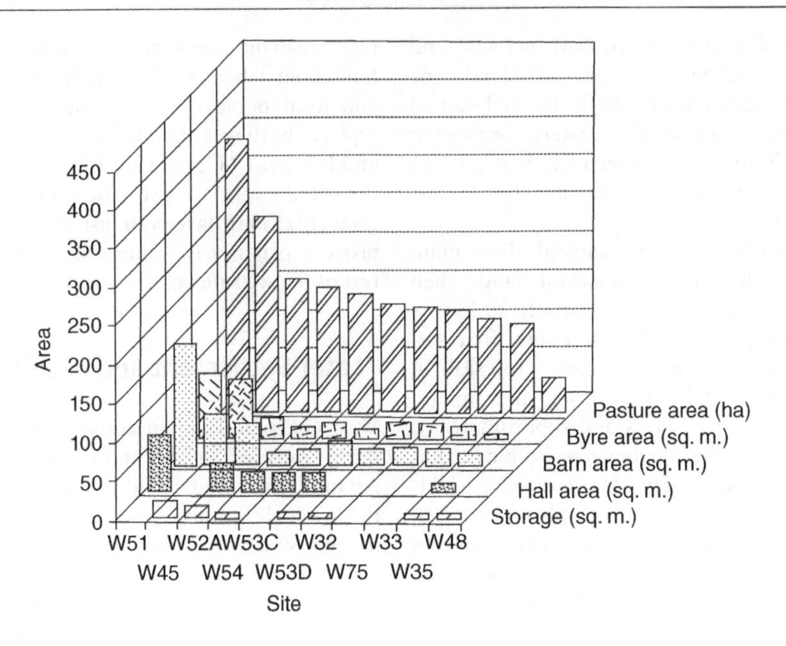

Figure 9.1 *Comparison between pasture area and floor space for various uses at eleven sites in the western settlement.*

holdings (see McGovern 1992 for discussion). The site with the largest total floor area, and also the site showing the greatest surplus of storage capacity and cattle production, is the bishop's manor at Gardar (site E47; modern Igaliko).

The later Norse Greenlanders were indeed Christianized and eventually acquired a small monastery and nunnery as well as a bishop. The first bishop, Arnald, was provided by the king of Norway around AD 1127 in exchange for a live polar bear. According to Einar Sokkason's saga, the bishop became an active player in Greenlandic elite politics, eventually causing the deaths of most of the important men in the eastern settlement (Jones 1985).

The architectural and settlement data certainly indicate that Arnald and his successors were effective land managers as well as serious competitors with the other, nonclerical chieftains. By the later phases of occupation the episcopal manor at Gardar far outstripped the old first-settler's farm at Brattahlid in all our architectural measures. Gardar's massive byres may have held more than 125 cattle (an average farm might have had between three and ten cattle), and an elaborate system of irrigation channels watered the bishop's pastures (Krogh 1982).

It also appears that Gardar was better able to suppress nearby settlement than Brattahlid had been. An excellent locational study by Keller (1991, in press) comparing ruin density and modern pasture productivity in the eastern settlement area reveals that the Gardar area maintained a surplus of pasture relative to site density, while the Brattahlid area did not. In other parts of medieval Europe, churches were often able to maintain and extend landholdings while the secular aristocracy was eventually forced to divide manors among siblings and retainers. A single (rather unreliable) document dating to the mid-fourteenth century and listing church property indicates that the church then owned outright or controlled access to most of the most productive resource spaces in the eastern settlement (Gad 1970; cf. Keller in press).

Whether sponsored by chiefly patrons or ambitious clerics (not mutually exclusive categories in the early Middle Ages), the churches of Greenland clearly played a role in the expression of both status

Table 9.1 Comparison of floor space at sites in the eastern and western settlement of Norse Greenland

		Floor area (sq. m.)				
Site	Site rank (name)	Hall	Byre	Barn	Storage	Church
Eastern settlement						
E47	1 Gardar	131	389	353	361	154
E83	2 Hvalsey	82	53	107		68
E29N	2 Brattahlid	66	127	105	118	59
E111	2 Herjolfsnes	66	48	43	59	86
E105	3		20	20		53
E29R	3	35	77	56	51	
E20	3	32				
E66	3		44	36		
E64C	4	19	18	25	7	
E64A	4	14	10	18	8	
E78A	4				5	
Mean		56	87	85	87	84
Western settlement						
W51	2 Sandnes	72	84	155		40
W7	2 Anavik		50	54	38	58
W45	2		77	64	21	
W52A	3	38	25	52	15	
W54	3	24	15	15	6	
W53C	3	23	20	19		
W53D	3	23	11	30	6	
W8	3	21	12	14		
W16	4	14	14	11	12	
W35	4	11	6	14	6	
W32	4		20	20	6	
W33	4		16	20		
W75	4		18	23		
W44	4		14	13		
W48	4				6	
Mean		28	27	36	13	49

and piety. As has long been noted (Bruun 1918), the stone churches of Greenland are very large by the standards of Atlantic Scandinavia. The episcopal cathedral at Gardar was nearly as large as its counterparts in Iceland (contemporary population estimated at ca. 60–80,000), and it was only one of several large stone churches built during the thirteenth century in Greenland (maximum population ca. 6000). These stone churches were modeled on the latest European fashions and included imported stained glass and English bells as well as costly timber and appropriate vestments. Even if we assume that most of the heavy work of dragging and raising stone was done in the winter, the amount of labor and resources allocated to such ceremonial structures remains impressive for a community whose domestic architecture consisted of a series of low sod huts.

Subsistence Economy

Our evidence for subsistence economy in Norse Greenland is largely zooarchaeological and locational. Faunal data are available for a large number of Norse sites in Greenland (see McGovern 1985a for summary), as is a rich body of modern biogeographical information (Vibe 1967) and a growing number of paleoecological studies (Buckland 1988; Fredskild 1986; Sadler 1991).

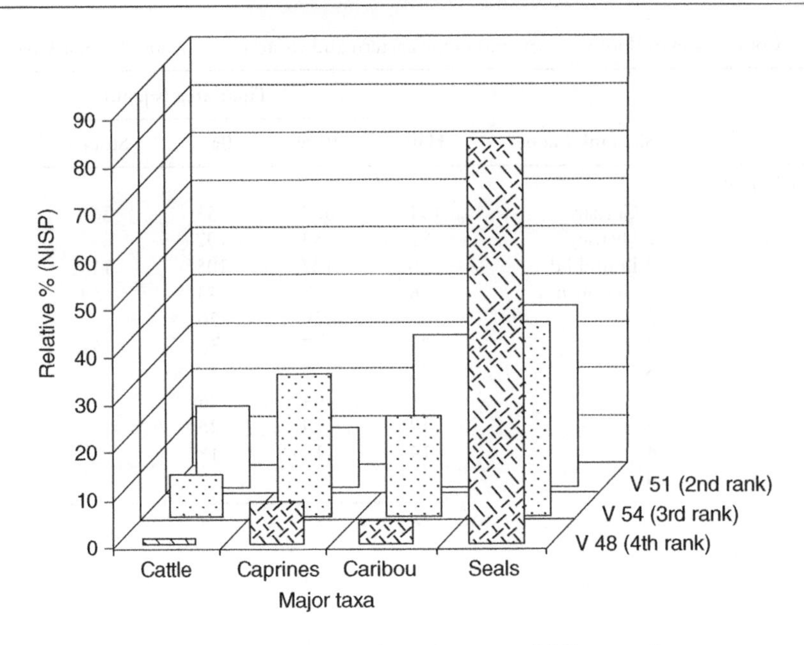

Figure 9.2 *Major taxa in the faunal assemblages from three sites of different ranks.*

From both documentary and climatic studies, we know that cereal agriculture was never possible on an economic scale in Norse Greenland. Herding and hunting supplied the surpluses that fueled clerical and chiefly building projects. Although the location of the Norse farms in the pasture-rich inner fjords suggests the importance of domesticated cattle, sheep, goats, horses, and pigs, the zooarchaeological evidence indicates a major reliance on wild species of the coast and mountains (see McGovern 1981, 1985b for discussion). Seal and caribou bone often make up the majority of the site archaeofauna, followed by caprines (both sheep and goats) and cattle (Figure 9.2).

The majority of the seals in the assemblage were migratory harp *(Pagophilus groenlandicus*) or hooded (*Cystophora cristata)* seals, while harbor or common seals (*Phoca vitulina*) were locally important in the eastern settlement. Ringed seals (*Phoca hispida*) are rare or absent. The migratory seals were probably caught in the outer fjord zone during their spring migration. Two small seasonal hunting stations have been documented in the outer fjord islands of the western settlement (Gulløv 1983), and it appears that nets and communal boat drives were used to catch groups of seals at the same time (probably like modern Faroese pilot whale drives). Harpoons, absent from Norse artifact collections, are post-medieval introductions in most of the Scandinavian North Atlantic.

The seasonal round of the Norse Greenlanders (partly based on unpublished tooth sectioning studies carried out by Bryan Hood) may serve to underline the key role of seals (especially the spring harp seal migration) in the Norse subsistence round (fig. 9.3). By late winter, domestic stock would have long since ceased milk production, and stored food would have run short on many farms. Cattle and some sheep and goats were kept in the warm dark byres most of their lives, standing nearly immobile in a rising tide of their own dung. Many of the small Norse cattle had to be carried out to pasture in the spring because their muscles were too weak after a winter of severe inactivity, and it is clear that Greenland was always near the limit for even the most skillful Scandinavian stock-raising techniques.

These difficult conditions, combined with the very small size of most cattle herds (probably between three and eight animals), suggest the importance of the few large herds (and especially

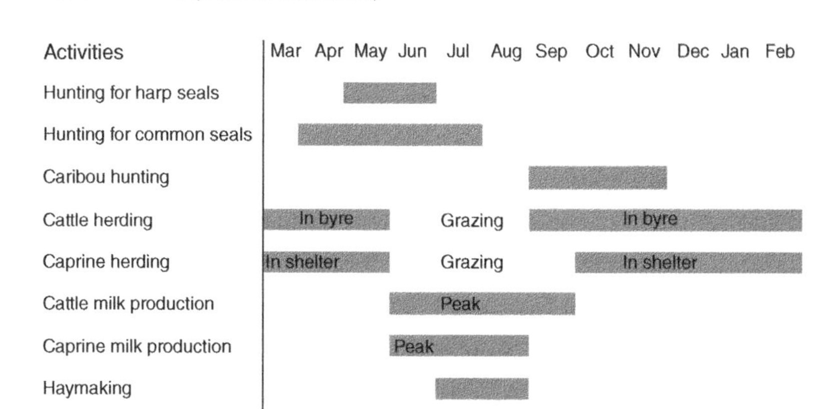

Figure 9.3 *Hypothetical Norse seasonal round (western settlement).*

the 100+ herd at Gardar) in buffering the whole community against catastrophic losses and in maintaining the biological viability of the domestic animal population. If they were organized as in Iceland, local communities (*hreppur*) of 15 to 25 farms centered on a single church operated as minimal subsistence units, pooling labor for critical seasonal tasks (like hay harvests and seal hunts), sharing upland common grazing, and providing fodder, food, and replacement stock to members whose holdings were damaged by fire or other disaster. Several authors have independently employed a range of locational techniques centered on known church locations in an attempt to reconstruct Greenlandic regional divisions in both settlement areas; all of them have modeled basic units in the same size range of 15 to 25 farms (Berglund 1982; Keller 1991, in press; McGovern 1980).

The distribution of animal bones in Norse sites further suggests the communal nature of Norse hunting and resource redistribution. The sites with the highest percentage of harp seal bones are those at the highest elevations, often several hours' walk from the nearest salt water. Seabirds most commonly found in Norse middens (alcids, mainly murres and guillemots) are those best taken by groups of hunters exploiting seasonal concentrations, and these bones too are found many kilometers inland.

At the same time dead seals and sea birds were being moved uphill, dead caribou appear to have been moved downhill. Norse caribou hunters appear to have used crossbows (Roussell 1936) and drive systems (Christensen 1991; McGovern and Jordan 1982) to intercept migrating caribou in the higher elevations. Strangely, the highest percentages of caribou bones are found not on the small holdings nearest the drive systems, but on the larger manors closer to the sea (McGovern 1985b). Patterning so contradictory to the normal assumptions of site territory points to the social dimension of resource exploitation in Norse Greenland.

Social hierarchy is apparent in the bone collections as well as in pasture distribution and site architecture (zooarchaeology, architecture, and locational data show high levels of correlation; see McGovern 1985b). Figure 9.2 compares three archaeofauna typical of western settlement sites of second, third, and fourth rank. The role on the larger farm (W51) of domestic mammals, especially cattle, is clear, as is the inverse role of caribou and seal. While elite farms were able to emphasize cattle production and enjoy a diet of dairy produce and deer meat, lowest-ranking farmers must have been strongly dependent upon seal meat taken in the communal drives.

The Norse subsistence economy in Greenland was thus a balanced exploitation of the inner fjord pastures required by the imported domestic animals and of the outer fjords frequented by the migratory seals. Both communal organization and economic hierarchy seem well marked, with the larger holdings playing a critical role in buffering short-term resource crises. Norse Greenland's productive subsistence economy depended upon careful management of socially structured land and resources in a marginal environment relative to traditional forms of European agriculture. Coordination of labor with seasonally variable resources (especially the migratory seals) was critical for the survival and prosperity of the community as a whole. The skills of Norse community managers at different levels are visible today both in the impressive stone churches built so far from the centers of Christendom and in the piles of seal bones in inland middens so far from the sea. Both communal cooperation and social hierarchy shaped the Norse economy more than simple proximity to resources, either material or social.

Transatlantic Trade and Long-Range Hunting

Communal cooperation, hierarchical control, and willingness to exploit distant resource spaces are all evident in the remarkable trips to the Nordrsetur, or northern hunting grounds. Documentary sources describe long and dangerous journeys far north of the settlement areas by groups of hunters seeking a range of arctic products but concentrating on walrus ivory and hide (which was used to produce exceptionally strong ships' cable). This hunting ground appears to lie in the modern Disko Bay–Holsteinsborg area, still home to one of the largest concentrations of walrus in the Eastern Arctic. The trips north were apparently an annual event, with a few men probably overwintering regularly (see McGovern 1985a for more complete references and discussion of the Nordrsetur hunt).

Despite the documentary and archaeological evidence for the northern hunt, walrus ivory itself is extremely rare on Norse sites in Greenland. Most of the "ivory" artifacts produced for home consumption are in fact fabricated from the peglike post-canines of the walrus. Rich in walrus, Norse Greenland was singularly poor in walrus ivory; most of this valuable arctic treasure was evidently reserved for overseas trade.

Unlike their contemporaries in Norse Shetland, the Orkneys, and Caithness (Batey 1987; Bigelow 1989; Morris 1985), the Norse Greenlanders did not enter into the expanding exchange of fish for grain that enabled survival and some prosperity to the eastern North Atlantic islands in the later Middle Ages. Instead, the Norse Greenlanders seem to have been locked into an early medieval–Viking period pattern of exchange of durable high-value goods suitable for elite consumption (Keller, in press). In exchange for walrus products and skins, we know that the Norse imported stained glass, church bells, elite clothing and church vestments, building timber, wine, and iron. A careful analysis of the documentary sources (Gad 1970) indicates that no great volume of material could have been imported at any time, and that most Greenlanders probably had only occasional access to imported goods. The rarity of imported artifacts in Greenlandic excavations, even in comparison with Iceland and Shetland, and the finds of bone or antler substitutes for items normally made of metal (including a spectacular whalebone battle axe) serve to confirm Gad's analysis.

While overseas trade may have brought only limited benefit to most Norse Greenlanders, the Nordrsetur hunt may have been costly to many. The northern hunt was described as dangerous, and the wild northern heaths were apparently dangerous places for the medieval Norse hunter. Pierced amulets carved from walrus post-canines to represent walrus and polar bear have been found on several sites. Whole walrus and narwhal skulls (which normally would have been broken up at the kill site) were deliberately buried in the consecrated soil of the episcopal churchyard at Gardar (Norlund and Stenberger 1934).

Despite such sacred and magical protection, it seems likely that many hunters failed to return from the north. The vessels used were not the ocean-going ships of the Viking period, but small open "six-oared boats" vulnerable to the ice, rocks, and bad weather of the Greenland coast. Even these small boats must have been extremely valuable to a community lacking most shipbuilding timber, and the loss of such vital tools of production must have been felt almost as keenly as the loss of lives when one of the boats went down. These losses were not the only price of the northern hunt and the transatlantic trade it fueled, since the absent hunters could play little role in the labor-intensive summer activities back on the home farms. The northern hunt was expensive in lost lives and resources, but also in deferred or blocked alternate uses for scarce labor and summer days.

Culture Contact

When the Norse arrived in West Greenland, they found only archaeological traces of the earlier Paleoeskimo inhabitants (Jones 1985). Their first contact with the ancestors of the modern Inuit came ca. 1100–1150, as the Thule folk pushed across from Ellesmere Island and moved down the coast of West Greenland (Schledermann 1990). The *Historia Norvegiae* of ca. 1170 noted that Norse hunters in the Nordrsetur had encountered strange skin-clad *skraelings* (the same term was applied to Native Americans in Vinland one hundred years earlier; see McGhee 1984, Fitzhugh 1985), who did not know metal and used stone and ivory tools. The most remarkable feature of these skraelings was that when merely wounded, they did not bleed, but when finally killed, their blood rushed out dramatically (Gad 1970).

This strange and bloody encounter did not prevent further Inuit migration southwards, and by ca. 1300 large Inuit settlements had been established in the Disko Bay region, and some winter houses had been constructed in the outer fjord zone of the western Norse settlement (Gulløv 1983). The two cultures must have been in increasing contact from ca. 1100–1500, or approximately as long as Native Americans and Europeans have been in contact in the eastern United States following the Jamestown settlement in 1607. Despite ongoing research, we still know frustratingly little about the details of this earlier contact situation (McGhee 1984; McGovern 1979, 1985a). Inuit legends, and scattered references to the skraelings in Scandinavian documents, indicate both peaceful and hostile relationships.

The archaeological evidence for Norse–Inuit contact is strangely one-sided. A growing number of Norse artifacts have been identified in Inuit contexts, including metal objects and woolen cloth, as well as souvenirlike trinkets such as Norse draughtsmen converted into Inuit-style spinning tops. Although no major realignment of Inuit settlement, subsistence, or material culture seems to have occurred, there is abundant evidence on the Inuit side for contact with the Norse settlers.

On the other side, Inuit objects in secure Norse contexts are very rare. A handful of finds, mainly of the nonfunctional souvenir category, are known from the many late phase Norse structures that have been investigated. Notably absent are harpoons or any of the host of refined sea and ice hunting gadgetry so characteristic of Inuit sites. Also absent from Norse collections are the bones of the species of seal most commonly taken with this Inuit technology—the ringed seal (*Phoca hispida)*. In the tens of thousands of seal bones from Norse middens examined by three generations of zooarchaeologists, fewer than a dozen have been identified as ringed seal. Although a mainstay of Thule through modern Inuit subsistence (especially in winter), this common animal was not regularly taken by Norse sealers.

Climate change

As many scholars have observed, the weather turned rotten on the Norse expansion (Lamb 1977[2]). Between ca. 900 and 1200, a period known as the Medieval Climatic Optimum (MCO) produced mean temperatures around 1 to 2°C above the 1930–60 modern baseline. Beginning around 1250, temperatures cooled to around 2 to 3°C below this baseline (Lamb 1977[2]; Ogilvie 1985), producing what many have called the Little Ice Age (LIA). Increasing variation between years and between decades (extreme in the fourteenth century; Dansgaard in Lamb 1977[2]) may have further complicated human economic response. These global climate shifts were of a magnitude to cause significant impact all across the Scandinavian

North Atlantic, but they were probably felt first and most strongly in Greenland.

The growing season for pasturage would decrease, probably irregularly, reducing fodder yields and complicating harvest labor scheduling. Longer winters would increase winter fodder demand, just as the capacity to produce fodder would be reduced. Increased drift ice made navigation more dangerous, requiring a new sailing route from Iceland and probably discouraging transatlantic contacts. All these effects would certainly have presented challenges to the Norse Greenlandic economy as we now understand it, and many writers have cited the case of Norse Greenland as a prime remaining example of simple climatic determinism in human affairs. Many climate impact theories have been proposed (see McGovern 1986 for discussion), but most may be reduced to the simple statement "it got cold and they died." The Norse are seen purely as passive victims, overcome by an unanticipated and unstoppable force of nature.

This view is attractively simple, but it ignores several relevant points:

1. The Inuit of West Greenland did not die out, but instead spread and prospered during the same period that saw the extinction of the Norse Greenlanders. Greenland clearly did not become uninhabitable for humans, even during the depths of the LIA, ca. 1650–1700.
2. The Norse Greenlanders also did not die out the moment the climate changed. Much of the impressive church construction occurred after the onset of LIA conditions, and the colony probably endured in some form until ca. 1475–1500. Medieval Scandinavians had developed a multilevel subsistence system, with extensive social buffering, that had survived many short-term climate shifts. Norse Greenland did not succumb to the first, or the hundredth, bad winter. If climate change did play a major role in Norse Greenland's extinction, it cannot have acted alone.
3. When Norse society in Greenland did perish, it did not die in a resource-depleted environment after using up all possible means of survival. In fact, the Norse Greenlanders had failed to make full use of resources locally available, while continuing to deform the subsistence economy to produce inedible prestige goods for export.

These points, particularly the last, require further discussion, for the Norse Greenlanders did indeed die out, and a resource crisis of some sort remains the most likely proximate cause. If simple climatic determinism fails to explain the collapse adequately, then why did they die when it got cold?

Combining modern pasture productivity data for southwest Greenland and Iceland with archaeological and biological survey data and zooarchaeological evidence in a simple spreadsheet model, we can get an idea of the relative productivity of pasture communities within the holding of farms of different rank, and the adequacy of this pasture productivity to supply the fodder needs of the farm's domestic animal stock (see McGovern et al. 1988). Figure 9.4 illustrates the model output for optimum levels of pasture productivity and for a 30 percent reduction in this level (a realistic worst-case scenario given the results of temperature decline impact models and 1920s–80s weather data).

Note that colder periods will not affect all farms equally. The second-rank farm (Gardar is the only example of the first rank) typifies an old first-settler's farm near sea level with relatively abundant pasture and a major investment in cattle. The modeled effects on third- and fourth-rank farms are based on data from less well sited farms with poorer pasturage, smaller numbers of domestic animals, and a greater emphasis on sheep and goat herding. Although all three types experience fodder shortfalls that would require stock reduction, note the disproportion between good year/bad year outcomes for the three types. In the western settlement, between 60 and 70 percent of sites fall into the fourth rank. Any adverse climate impact would affect these smallholders first and worst. The pattern of extensive use of wild species (particularly seals) by smallholders observed in the zooarchaeological data makes sense in light of this model prediction.

Using floor space as a proxy indicator, we can also model the balance between probable human population and domestic animal production (meat and milk) on a given site.

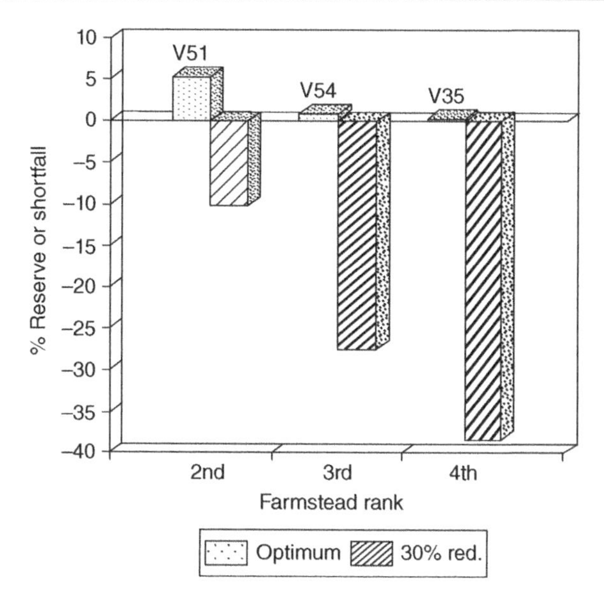

Figure 9.4 *Effect of variation in fodder supply on farms of different ranks, modeled under situations of optimum productivity and 30 percent reduction therein.*

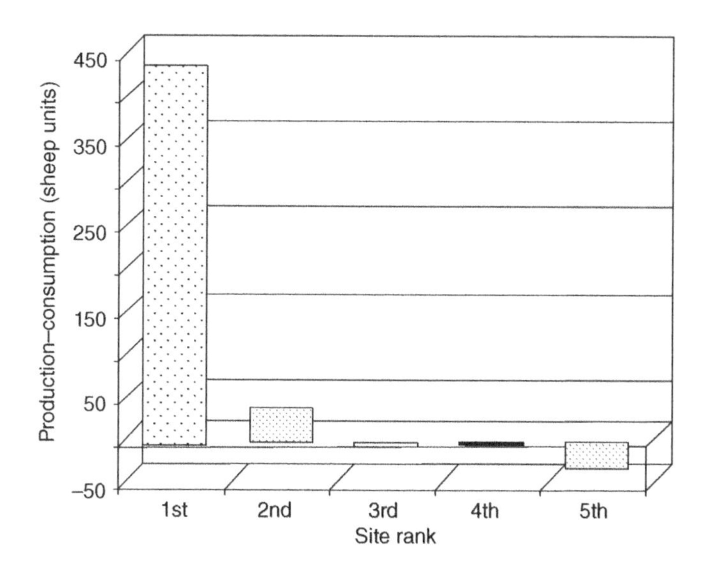

Figure 9.5 *Ability of farms of different ranks to be self-sufficient (based on floor space as an indicator of population).*

Again using excavated sites as examples of a given site class, Figure 9.5 compares the production–consumption balance for all four site ranks. The unique position of the episcopal farm at Gardar is evident. Even if model assumptions are only broadly correct, it is clear that Gardar could potentially produce a surplus above its own consumption needs far in excess of any other holding in Greenland.

These models and our locational and zooarchaeological data indicate that by the later phases of occupation, elite farms had a set of characteristics in common:

1. They were least and last affected by adverse climatic impacts on pasture productivity. Bad seasons will first impoverish lower- and middle-ranking farmers, while having comparatively minor effects on upper-ranking farmers. The immediate impact of LIA conditions would have been to steepen the economic and political gradient in the favor of the heirs of the first settlers.
2. They were most involved in domestic mammal (especially cattle) production. Although wild resources were clearly important to elite farms, they probably did not serve the same critical staple role as they did on smaller farms. The commitment of the elite farmers to cattle and to the best pastures would tend to increase as they became increasingly scarce and valued commodities.
3. They played a buffering role as sources of replacement stock and dairy produce for middle- and lower-ranking farms suffering periodic shortages. The more frequently the buffering role was played, the greater probability of loss of real or effective independence on the part of the recipients. The economic functioning of the local community would increasingly depend upon a few major households, even if the majority of farmers did not become actual tenants as they did in contemporary Iceland (McGovern et al. 1988).

The episcopal manor at Gardar played a critical buffering role for the settlement as a whole during large-scale resource crises. Whether managed by Greenlandic elites or foreign-born bishops, Gardar must have exerted unrivaled economic and political influence by later phases.

Social stratification and economic management

Our current data and the results of our various land-use models suggest that later Norse Greenland was neither a haven for independent-minded individualists nor a home for any form of primitive democracy. Although Norse Greenland lacked all the trappings of high medieval feudal society (Keller 1991), it was clearly no longer the flexible, pioneering society of the first settlement era. Getting through the average year required community cooperation, and that community was increasingly likely to be dominated by a few great landowners. Bad years activated a nested series of community buffering strategies, none of which were cost-free to the recipient.

Studies by Arneborg (1991) and Keller (1991) have raised the possibility of significant doctrinal conflict between Greenlanders and the papacy, as well as probable internal elite competition, rather than unquestioned compliance with royal and papal demands (contra McGovern 1981). As in Iceland, the partly ecclesiastical elite may have had more than changing weather to worry about, and they may have been as concerned over the latest papal edicts as over this year's seal harvest. The disproportionately large and elegant churches of Norse Greenland may reflect growing internal competition and local display as well as orders from the continental core. Certainly the remarkable northern hunt could not have been carried out without the active sponsorship and direction of the local elite, who must have played a leading role in ensuring the remarkably efficient collection of processed walrus tusk for export.

In any case, the Greenlandic elite did not hesitate to employ a full range of positive and negative sanctions to enforce their leading role in this tightly managed society. Our last written record of Norse Greenland (from 1408; see Gad 1970) documents both the proper reading of wedding banns and the burning alive of the unfortunate Kolbein for witchcraft the same year.

Management failure and population extinction

In retrospect, we can see that many of the management decisions made by the Greenlandic elite were disastrous (McGovern 1981). Although the scale and timing of the climate changes that affected Greenlandic managers were beyond human control, the vulnerability of their society to particular impacts and the response to perceived challenges and opportunities were certainly subject to culturally mediated political choice. As argued elsewhere (McGovern 1981), it would appear that Norse managers simply managed badly, and their society died as a result.

As Vibe (1967) demonstrates, ocean current dynamics make the west coast of Greenland unstable on the scale of centuries, with any given point on the coast passing through repeated cycles of resource scarcity and plenty as marine resource concentrations move up and down the coast and terrestrial resources undergo boom and bust cycles. The Inuit of West Greenland have survived and prospered by maintaining locational flexibility, moving up and down the long coastline as climate changed and seasonal resource zones shifted.

The Norse were far less mobile. The domestic animal component of their subsistence economy (especially cattle) was closely tied to a few restricted pockets of low-arctic vegetation in the fjords of the southwest. The elite farmers with the richest pastures and heaviest investment in cattle were least likely to consider any adaptive strategy that would devalue these resources. While Norse sealing parties might extend their hunting ranges, the inhabitants would still be tethered to the home farms (and the stone churches) deep in the southern fjords. The politically dominant (but climatically threatened) herding component of the Norse subsistence economy thus limited the possibilities for effective development of the marine hunting component.

The elite connections to the markets and political centers of continental Europe must have proved increasingly difficult to maintain. Documentary sources describe increased East Greenland drift ice as a growing hazard to transatlantic navigation by the mid-fourteenth century. About the same time, changing fashions in Europe were replacing walrus and elephant ivory with Limoges enamels in luxury goods and religious artifacts. Demand for the main product of the costly Nordrsetur hunt was dropping just as the climate-related dangers of both the hunt and the trip across the Atlantic were increasing. We have no evidence for an attempted restructuring of trade goods production, and a late court case describes Greenlanders forcing unwanted trade goods on chance visitors looking only for provisions (Gad 1970). While other North Atlantic communities successfully shifted to commercial fishing, the Greenlanders apparently continued to offer increasingly devalued, old-fashioned goods to the few traders willing to risk the drift ice.

Since the lower-ranking Norse farmers were effectively dependent upon sea mammal hunting anyway, and seventeenth- to nineteenth-century West Greenland was to become an international center for whale, seal, and walrus exploitation, we might imagine that the prospects for enhanced Norse maritime adaptation would be considerable. Even if the Norse farms remained nailed down to the inner fjords, expansion of marine hunting efforts in the local area should have proved productive.

This strategy would almost certainly require acquisition of more Inuit technology and expertise than the Norse Greenlanders apparently were willing to absorb. As we have seen, the Norse lacked both harpoons and the ringed seals normally taken with them, despite nearly three centuries of contact with the Inuit. The complex of whaling gear (large harpoons, float valves, etc.) so successfully employed by the Thule Inuit in West Greenland for five hundred years is also absent. There is also no evidence for the acquisition of Inuit skin boats, despite the shortage of timber suitable for the keels of traditional Scandinavian clinker-built ships.

Modern catch data from southwestern Greenland indicate that the failure of the Norse to exploit ringed seals effectively may have been increasingly costly during climatic fluctuations. Figure 9.6 presents a 21-year record of major seal species taken in one part of the former Norse eastern settlement (for a discussion of these data, see Vibe 1967; McGovern 1986). In the first third of the chart (approximately 1954–60) the total seal catch is dominated by migratory harp seals. The well-documented cooling of the early 1960s (just under 1°C) altered harp seal migrations, causing a crash in catches all along the southwest coast. Harp seal catches did not begin to recover until the 1980s. Note, however, that after a few hard years, the total seal catch reported at Qaqortoq recovered to nearly pre-crash levels—propelled by a steady increase in ringed seal hunting in the period 1960–75. Similar patterns of crash and rapid recovery through shift in target

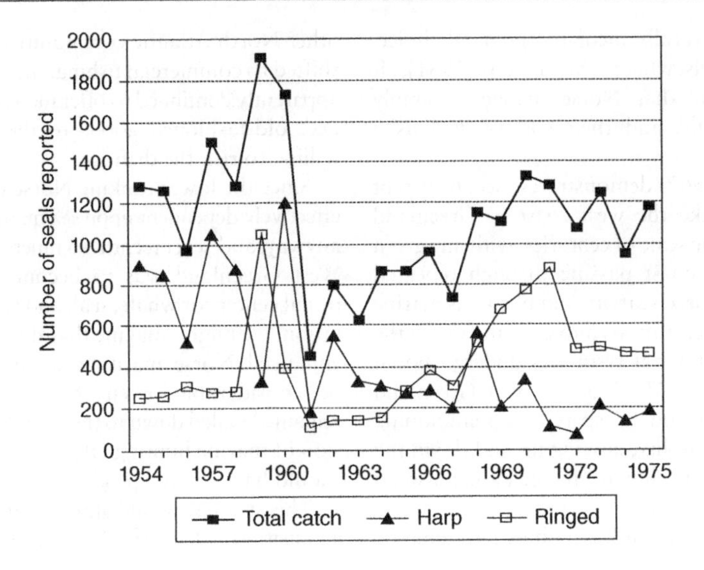

Figure 9.6 *Reported seal catch, 1954–1975, Qaqortoq, Greenland.*

species are evident in catch records for the Narssaq and Nuuk districts in the eastern and western settlements. Because we have no reason to suspect a radical change in seal distribution since Norse times, it would appear that the Norse Greenlanders would have been able to manage a similar shift in critical resources, if they had acquired the necessary technology and skills.

More speculatively, we might also wonder what might have developed from an integration of Norse and Inuit populations and adaptive strategies. The Norse could have provided milk, wool, metal, and access to European markets in exchange for sea mammals, boats, and arctic products. Had the skraelings been properly Christianized and suitably socialized they could even have provided a welcome source of additional profit to their Norse patrons, much as the Sami of arctic Scandinavia provided income to elites in northern Norway. A variation of this pattern was actually developed by the Danish colonial administration in the eighteenth century and provided the basis for a culturally and technologically diverse society that persists to the present day in Greenland.

Speculative scenarios could be further elaborated, but it is clear that Norse Greenland did not perish in a barren wasteland devastated by the Little Ice Age. Instead, they starved in the midst of unexploited resources, with a working model for maritime-adapted northern survival camped on their doorsteps. The death of Norse Greenland was not caused by nature, but by culture.

World View and Mental Templates

If the leadership of this tight little society did indeed manage so badly that the whole population failed to react appropriately to climate change, cultural competition, and market peripheralization, we might well ask why. After all, there is no lasting advantage to managing your own society so you have the privilege of starving last. We may assume that the managers of Norse Greenland did not intend the outcome that resulted from their self-serving, short-term choices. Why did they choose so badly, dooming themselves and their culture?

As many have observed, humans make decisions based upon culturally mediated perception and interpretation of reality and may or may not consistently observe environmental variables that other humans would find important. Some managers may be more interested in closely tracking church politics and the latest building styles than in accurately recording variability in seal catches. A variety of cognitive maladies common to all managers may have afflicted the Norse elite (discussed in McGovern et al. 1988):

1. False Analogy. The managers' cognitive model of ecosystem characteristics (potential productivity, resilience, stress signals) may be based on the characteristics of another ecosystem (Norway, Iceland) whose surface similarities mask critical threshold differences from the actual local ecosystem.
2. Insufficient Detail. The managers' cognitive model is overgeneralized and fails to allow for the actual range of spatial variability in an ecosystem whose patchiness is better measured in resilience than initial abundance (Moran 1984).
3. Short Observational Series. The managers lack a sufficiently long memory of events to track or predict variation in key environmental factors over a multigenerational period and are subject to chronic inability to separate short-term and long-term processes.
4. Managerial Detachment. The managers are socially and spatially distant from agricultural producers who carry out managerial decisions at the lowest level and who are normally in closest contact with local-scale environmental feedbacks.
5. Reactions out of Phase. Partly as a result of the last two factors, the managers' attempts to avert unfavorable impacts are too little and too late, or apply the wrong remedy.
6. SEP (Someone Else's Problem) (Buckland 1988: 7). Managers at many levels may perceive a potential environmental problem but do not feel obligated to take action because their own particular short-term interests are not immediately threatened. Some adverse impacts may actually enhance the managers' position by differentially impoverishing unruly subordinates and potential rivals.

The Norse Greenlanders were probably also subject to a series of obstacles to perception and management that were far more culture-specific. The world view of a particular culture at a particular moment may be a product of both current social and environmental conditions as well as traditional knowledge of variable time depth.

The world view of the medieval Scandinavians of the North Atlantic is set out in considerable detail in law codes and sagas, and it has been the subject of a series of scholarly analyses (Andersson and Miller 1989; Byock 1988; Durrenburger and Palsson 1989; Hastrup 1985; Miller 1986; Palsson 1991). In her extended analysis of the twelfth-century Grågås law code, Hastrup describes a world view that rigidly partitions land and society into a series of dual oppositions. Law, society, home, and order lay on one side, and dangerous, lawless chaos lay on the other (Hastrup 1985: 136–51).

These categories had a strong locational aspect. The farmyard and home field of the family farm were bounded by a low earthen dike (still visible archaeologically), which provided legal and symbolic security. An assault within a man's home field was legally more serious than one farther from home, and a variety of malign nonhuman beings were magically confined outside the dike. In Iceland, the center of human order and society lay at Thingvellir (the site of the major annual assembly), and the opposing center of natural chaos and evil lay in the desolate "lava field of misdeeds" in the unpopulated arctic interior. Persons sentenced to permanent exile were both literally and figuratively set outside the bounds of human society. They could be killed on sight without the need to pay for spilling their blood (blood money), and they had to live in the wilderness, where they would be exposed to weather, hunger, and the inhuman creatures that haunted such places (several mountains in Iceland are still named "Trolls-Church"). A heroic outlaw could win readmission to human society by killing other outlaws and struggling successfully against the evil creatures of the wilderness.

Clearly, this was not a nineteenth- or twentieth-century romantic world view that sees pristine nature in positive opposition to corrupting culture, but a more traditional European dichotomy of good, safe (but fragile) human culture and dangerous, potentially evil nature. Both pagan and Christian cosmology saw the natural world as potentially hostile, and both gloomily predicted a steady decline

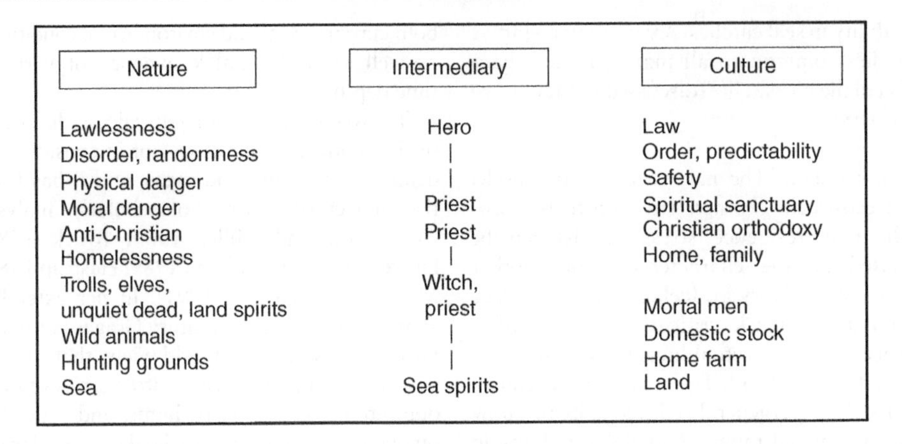

Figure 9.7 *Summary of medieval Greenland/Icelandic world view.*

of humanity and the human world, ending in Ragnarok/Armageddon/Götterdämmerung.

Hastrup (1989) has argued that the identification of wilderness with threat and evil was so extreme in late medieval Iceland that it inhibited effective exploitation of distant saeters (shielings). Although changes in vertical zonation during the LIA probably also played a part in settlement contraction, there seems little debate over the character of this medieval nature–culture division among scholars most familiar with the Norse documentary and literary evidence.

What effect did this cognitive division have on the fate of Norse Greenland? Figure 9.7 presents the world view probably shared by medieval Greenlanders and Icelanders (categories based upon Hastrup 1985, Palsson 1991). Although some individuals (heroes, witches, and Christian clerics) could safely act as intermediaries, the passage between realms was dangerous and uncertain. In this light, the elaborate magical buffering associated with the trips to the Nordrsetur hunting grounds becomes immediately comprehensible.

This world view also places the Inuit in an ambiguous position. Are they normal humans of a different culture (like the English, Franks, Irish, etc.) to fight, trade with, marry, and learn from? Are they instead nonhuman creatures of nature to be shunned? Like trolls, they did not bleed when wounded. Like many water-beings, they did not use iron weapons. Had any Norseman overcome the barriers of language he would have encountered an Inuit world view and ideology full of humanlike animals and powerful sorcery not based upon a nature–culture dichotomy like his own. It is all too easy to imagine Norsemen culturally preprogrammed to reject all innovations from the Inuit, fatally ignoring tainted technology and alien expertise and keeping closer and closer to home, hearth, and church (Hastrup 1989).

This structuralist analysis leaves many questions unanswered. As Palsson argues:

> Some scholars concentrate on the search for abstract semiotic systems, emphasizing cultural codes rather than social action – language rather than speaking. . . . The saga-people become trapped in the superorganic, as prisoners of medieval culture. . . . We cannot observe or participate in the praxis of the natives, but the "world out there" presented in our second-hand field notes is a world of active persons—not of rules and things. (Palsson 1991: 17–18)

Both traditional structuralist and traditional ecosystemic cultural ecological analyses have been criticized as overly static and unreflective of the active role of individual humans in reshaping meaning and realigning system linkages. If the Norse Greenlanders did maintain the ultimately maladaptive world view documented in their laws and literature, maintaining cultural purity at the cost of biological extinction, we still must ask *why*. It is as much of a dead end to blame their extinction on rigid cultural programming as to resort to simple climatic determinism.

In responding to social situations, as well as to environmental changes, the individual Norse colonist in fact had a range of potential options open. The relative rigidity of the Norse world view and its enforcement were variables that could be altered from one generation to the next. The early pioneer settlers like Skalla-grim possessed the same cultural baggage yet showed far greater adaptive flexibility than their fourteenth- and fifteenth-century descendants. The Norse world view was subject to conscious manipulation, reinterpretation, and alteration (as the battles over Christian doctrine indicate). If it solidified into a rigid barrier to successful adaptation, fatally limiting options in a changing world, then this rigidity was a *result* of such manipulation. The choice *not* to innovate is a real choice made by particular people in a particular political and economic context.

During the long contact period between Inuit and Scandinavian in Greenland, many on both sides must have discovered mutual humanity, and it is hard to imagine that Norse seal hunters never admired the sealing skills of the kayaker. Had the Norse been a simple band-level egalitarian society, we might easily imagine that households adopting Inuit technology would gradually replace households that did not, as selection took its toll on the overly conservative. It is hardly an accident that these life-saving skills were so systematically rejected—it took a great deal of effort on somebody's part.

We have seen that Norse Greenland was far from egalitarian, and selection did not operate entirely at the level of the individual household or kin group. Instead, a multifarm community, ritually and economically dependent on a few large manors, was the minimal unit of survival. If alteration of the seasonal round, labor allocation, or social contract was proposed, its adoption or rejection would be a community decision highly influenced by the opinions of the wealthier farmers. According to our data and models, short-term climatic stresses would tend to enhance the authority of the elites and increase the dependency of the majority. Propelled by short-term economic necessity, the Norse social units may have emphasized cohesion and communal solidarity at the expense of innovation and cross-cultural experimentation.

If increased Inuit technology transfer and a declining emphasis on inner-fjord pastures were optimal for the society as a whole, they were by no means optimal for the elites. Some Nordrsetur hunters or seal-dependent smallholders may have seen greater benefits in a more mobile existence and more merit in Inuit lifeways than did the partly clerical Norse elite, proud of their stained glass and church bells. In the contest for community support, however, there is no question who was better placed to manipulate both world view and social sanctions. Increasing the rigidity of the existing northwest European nature–culture split through sermons, thing meetings, and the occasional execution at the stake would be only one means of bolstering social control in a threatened society.

The economic hierarchy that fostered this deadly control was itself a product of pasture distribution and the initial division of land and resources. The Norse pathway to extinction was directed by the distribution of natural resources, by the pattern of climatic impact, by culture-specific cognitive patterns, as well as by divergent class interests. No single decision, but a cascade of decisions closing options and forestalling innovation made Norse Greenland one of the most conservative of the Scandinavian North Atlantic colonies. Culture, ecology, and history together provided the backdrop for the last fatal decision not to choose.

Conclusion

Whatever combination of unenlightened self-interest, class conflict, imperfect knowledge, and maladaptively rigid world view produced the disastrous management choices of the Greenlandic elite, the result was fatal to the whole society. Many important questions remain unresolved in Greenlandic archaeology, and more research is urgently needed to go beyond facile generalization. It seems clear, however, that like characters in a proper classical tragedy, the Norse Greenlanders contributed significantly to their own grim fate.

The case of Norse Greenland may have some disquieting parallels in the modern world. Like the Norse Greenlanders, many humans today are pursuing limited, but intensive strategies of exploitation requiring precarious balancing of distant resource zones and markets. Like the Norse, many economies have developed high levels of vulnerability to sudden change in a changing earth. Like the Norse cathedral in the arctic, there are today many monuments to peoples living beyond the means of local resources. Like the Norse elites, we are today very certain of the complete adequacy of a particular world view, and we are often willfully ignorant of alternate sources of expertise.

If modern managers of global resources respond to change and challenge no better than their predecessors in Norse Greenland, Ragnarok may yet prove more than myth. Like the Norse Greenlanders, however, we are not inevitably the prisoners of history and culture. Like them, we have many potential options. We can choose to broaden, rather than restrict, the management subculture and actively seek alternate courses. If the case of Norse Greenland can spur such efforts, then perhaps the Norse Greenlanders themselves will not have suffered and struggled and finally died in vain.

REFERENCES

Andersson, Theodore M. and William Ian Miller. 1989. *Law and Literature in Medieval Iceland*. Stanford, CA: Stanford University Press.

Arneborg, Jette. 1991. "The Roman Church in Norse Greenland." In: *The Norse of the North Atlantic*. Edited by G.F. Bigelow, pp. 142–50. *Acta Archaeologica* (Copenhagen) 61 (special issue).

Batey, Coleen. 1987. "Freswick Links, Caithness: A Reappraisal of the Late Norse Site in Its Context." *British Archaeological Reports* 179. Oxford: BAR.

Berglund, Joel. 1982. Kirke, hal, og Status. *Grønland* 8–9:310–42.

Berglund, Joel. 1991. "Displacements in the Building-over of the Eastern Settlement: A Sketch." In: *The Norse of the North Atlantic*. Edited by G.F. Bigelow, pp. 151–7. *Acta Archaeologica* (Copenhagen) 61 (special issue).

Bigelow, G.F. 1989. "Life in Medieval Shetland: An Archaeological Perspective. *Hikuin* 15:183–92.

Bigelow, G.F., editor. 1991. *The Norse of the North Atlantic. Acta Archaeologica* (Copenhagen) 61 (special issue).

Bruun, Daniel. 1918. "The Icelandic Colonization of Greenland and the Finding of Vineland." *Meddelelser om Grønland* 57(3).

Buckland, Paul C. 1988. "North Atlantic Faunal Connections: Introduction or Endemics?" *Entomologica Scandinavica*, 1988 supplement: 8–29.

Byock, Jesse. 1988. *Medieval Iceland: Sagas, Society, and Power*. Berkeley and Los Angeles: University of California Press.

Christensen, K.M.B. 1991. "Aspects of the Norse Economy in the Western Settlement of Greenland." In: *The Norse of the North Atlantic*. Edited by G.F. Bigelow, pp. 158–65. *Acta Archaeologica* (Copenhagen) 61 (special issue).

Durrenburger, E. Paul. 1991. "Production in Medieval Iceland." In: *The Norse of the North Atlantic*. Edited by G.F. Bigelow, pp. 14–21. *Acta Archaeologica* (Copenhagen) 61 (special issue).

Durrenburger, E. Paul and Gisli Palsson, editors. 1989. *The Anthropology of Iceland*. Iowa City: University of Iowa Press.

Fitzhugh, W.W. 1985. "Early Contacts North of Newfoundland Before AD 1600: A Review." In: *Cultures in Contact: The Impact of European Contacts on Native American Cultural Institutions, AD 1000–1800*, pp. 19–22. Washington, D.C.: Smithsonian Institution.

Fredskild, Bent. 1986. "Agriculture in a Marginal Area: South Greenland, AD 985–1985." In: *The Cultural Landscape, Past, Present, and Future*. Edited by H.J.B. Birks, pp. 28–41. Bergen: Botaniske Institut.

Gad, Finn. 1970. *A History of Greenland, vol. 1*. London: D. Hurst.

Gulløv, H.C. 1983. Nuup kommuneani qangarnitsanik eqqaassutit inuit-kulturip nunaqarfii. Kalaallit Nunaata Katersugaasivia and Nationalmuseet, Nuuk, Greenland.

Hastrup, Kirsten. 1981. "Cosmology and Society in Medieval Iceland: A Social Anthropological Perspective on World-View." *Ethnologia Scandinavica*: 63–78.

Hastrup, Kirsten. 1985. *Culture and History in Medieval Iceland: An Anthropological Analysis of Structure and Change*. Oxford: Clarendon Press.

Hastrup, Kirsten. 1989. "Saeters in Iceland 900–1600: An Anthropological Analysis of Economy and Cosmology." *Acta Borealia* 6(l):72–85.

Jansen, H.J. 1972. "A Critical Account of the Written and Archaeological Sources' Evidence Concerning the Norse Settlements on Greenland." *Meddelelser om Grønland* 182(4).

Jones, Gwyn. 1985. *The Norse Atlantic Saga*. 2nd Ed. Oxford: Oxford University Press.

Keller, Christian. 1991. "Vikings in the West Atlantic: A Model for Norse Greenland Medieval Society." In: *The Norse of the North Atlantic*. Edited by G. F. Bigelow, pp. 126–41. *Acta Archaeologica* (Copenhagen) 61 (special issue).

Keller, Christian. In Press. *The Eastern Settlement Reconsidered: Some Analyses of Norse Medieval Greenland*. Glasgow: University of Glasgow Press.

Krogh, K.J. 1982. *Eirik den Rodes Grønland*. Copenhagen: Nationalmuseet.

Lamb, H.H. 1977. *Climate: Past, Present, and Future*. 2 vols. London: Methuen.

McGhee, Robert. 1984. "Contact Between Native North Americans and the Medieval Norse: A Review of the Evidence. *American Antiquity* 49:12–29.

McGovern, T.H. 1979. "Thule-Norse Interaction in Southwest Greenland: A Speculative Model." In: *The Thule Eskimo Culture: An Anthropological Retrospective*, pp. 171–89. National Museum of Man Mercury Series no. 88. Ottawa: National Museum of Man.

McGovern, T.H. 1980. "Site Catchment and Maritime Adaptation in Norse Greenland." In: *Site Catchment Analysis: Essays on Prehistoric Resource Space*. Edited by F. Finlow and J. Eriksson, pp. 193–209. Los Angeles: University of California at Los Angeles.

McGovern, T.H. 1980–81. "The Vinland Adventure: A North Atlantic Perspective. *North American Archaeologist* 2: 285–308.

McGovern, T.H. 1981. "The Economics of Extinction in Norse Greenland." In: *Climate and History*. Edited by T.M.L. Wigley et al., pp. 404–34. Cambridge: Cambridge University Press.

McGovern, T.H. 1985a. "The Arctic Frontier of Norse Greenland." In: *The Archaeology of Frontiers and Boundaries*. Edited by S. Green and S. Perman, pp. 275–323. New York: Academic Press.

McGovern, T.H. 1985b. "Contributions to the Paleoeconomy of Norse Greenland." *Acta Archaeologica* 54:73–122.

McGovern, T.H. 1986. "Climate, Correlation, and Causation in Norse Greenland." Paper presented at the meeting of the Society for American Anthropology, Philadelphia.

McGovern, T.H. 1988. "Bones, Buildings, and Boundaries: Patterns in Greenlandic Paleoeconomy." Paper presented at the Bowdoin Conference on Norse Archaeology of the North Atlantic, April.

McGovern, T.H. 1990. "The Archaeology of the Norse North Atlantic." *Annual Review of Anthropology* 19:331–51.

McGovern, T.H. 1992. "Bones, Buildings, and Boundaries: Paleoeconomic Approaches to Norse Greenland." In: *Norse and Later Settlement and Subsistence in the North Atlantic*. Edited by C.D. Morris and D.J. Rackham, pp. 193–230. Glasgow: University of Glasgow Press, Archetype Publications.

McGovern, T.H. and R.H. Jordan. 1982. "Settlement and Land Use in the Inner Fjords of Godthaab District, West Greenland." *Arctic Anthropology*. 19(1):63–80.

McGovern, T.H., G.F. Bigelow, T. Amorosi, and D. Russell. 1988. "Northern Islands, Human Error, and Environmental Degradation: A View of Social and Ecological Change in the Medieval North Atlantic." *Human Ecology* 16(3):176–221.

Miller, W.I. 1986. "Dreams Prophecy, and Sorcery: Blaming the Secret Offender in Medieval Iceland." *Scandinavian Studies* 58:101–23.

Moran, E.F., Editor. 1984. *The Ecosystem Concept in Anthropology*, Boulder, CO: Westview Press.

Morris, Chris. 1985. "Viking Orkney: A Survey." In: *The Prehistory of Orkney*. Edited by Colin Renfrew, pp. 210–42. Edinburgh: Edinburgh University Press.

Nørlund, P. and Martin Stenberger. 1934. "Brattahlid." *Meddelelser om Grønland* 88(1).

Ogilvie, Astrid. 1985. "The Past Climate and Sea-Ice Record from Iceland, Part I: Data to AD 1780." *Climatic Change* 6:131–52.

Palsson, Gisli. 1991. "The Name of the Witch: Sagas, Sorcery, and Social Context." In: *Oromo Studies and Other essays in Honor of Paul Baxter*. Edited by David Brokenshaw. Syracuse and New York: FACS Publications.

Roussell, Aage. 1936. "Sandnes and the Neighboring Farms." *Meddelelser om Grønland* 88(3).

Sadler, Jon. 1991. "Beetles, Boats, and Biogeography: Insect Invaders of the North Atlantic." In: *The Norse of the North Atlantic*. Edited by G.F. Bigelow, pp. 199–211. *Acta Archaeologica* (Copenhagen) 61 (special issue).

Schledermann, Peter. 1990. "Crossroads to Greenland: 3000 Years of Prehistory in the High Arctic." In: *Komatik* no.2. Calgary: Arctic Institute of North America.

Vibe, Chr. 1967. "Arctic Animals in Relation to Climatic Fluctuations. *Meddelelser om Grønland* 170(5).

Wallace, Birgitta. 1991. "L'Anse aux Meadows: Gateway to Vinland." In: *The Norse of the North Atlantic*. Edited by G.F. Bigelow, pp. 166–98. *Acta Archaeologica* (Copenhagen) 61 (special issue).

10

What Drives Societal Collapse?

Harvey Weiss and Raymond Bradley

The archaeological and historical record is replete with evidence for prehistoric, ancient, and premodern societal collapse. These collapses occurred quite suddenly and frequently involved regional abandonment, replacement of one subsistence base by another (such as agriculture by pastoralism), or conversion to a lower energy sociopolitical organization (such as local state from interregional empire). Each of these collapse episodes has been discussed intensively within the archaeological community, commonly leading to the conclusion that combinations of social, political, and economic factors were their root causes.

That perspective is now changing with the accumulation of high-resolution paleoclimatic data that provide an independent measure of the timing, amplitude, and duration of past climate events. These climatic events were abrupt, involved new conditions that were unfamiliar to the inhabitants of the time, and persisted for decades to centuries. They were therefore highly disruptive, leading to societal collapse—an adaptive response to otherwise insurmountable stresses.[1]

In the Old World, the earliest well-documented example of societal collapse is that of the hunting and gathering Natufian communities in southwest Asia. About 12,000 years ago, the Natufians abandoned seasonally nomadic hunting and gathering activities that required relatively low inputs of labor to sustain low population densities and replaced these with new labor-intensive subsistence strategies of plant cultivation and animal husbandry. The consequences of this agricultural revolution, which was key to the emergence of civilization, included orders of magnitude increases in population growth and full-time craft specialization and class formation, each the result of the ability to generate and deploy agricultural surpluses.

What made the Natufians change their lifestyle so drastically? Thanks to better dating control and improved paleoclimatic interpretations, it is now clear that this transition coincided with the Younger Dryas climate episode about 12,900 to 11,600 years ago. Following the end of the last glacial period, when southwest Asia was dominated by arid steppe vegetation, a shift to increased seasonality (warm, wet winters and hot, dry summers) led to the development of an open oak-terebinth parkland of woods and wild cereals across the interior Levant and northern Mesopotamia. This was the environment exploited initially by the hunting and gathering Natufian communities. When cooler and drier conditions abruptly returned during the Younger Dryas, the harvests of wild resources dwindled, and foraging for these resources could not sustain Natufian subsistence. They were forced to transfer settlement and

The Anthropology of Climate Change: An Historical Reader, First Edition. Edited by Michael R. Dove.

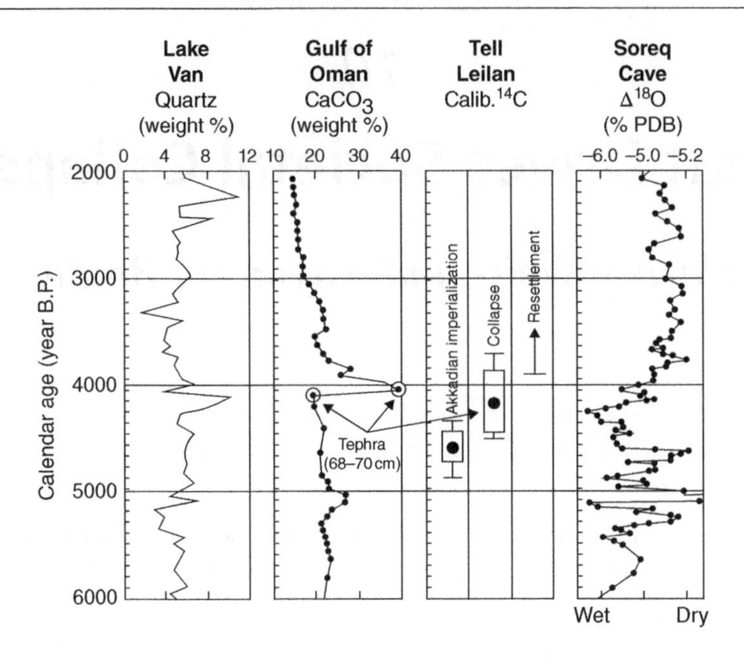

Figure 10.1 *Climatic effects. High-resolution lake, marine, and speleothem cores and tephrochronostratigraphy document abrupt aridification and linkage with Akkadian empire collapse at Tell Lielan, Syria*[9–11].

wild cereals to adjacent new locales where intentional cultivation was possible.[2]

The population and socioeconomic complexity of these early agricultural settlements increased until about 6400 B.C., when a second postglacial climatic shock altered their developmental trajectory. Paleoclimatic evidence documents abrupt climatic change at this time,[3] the last major climatic event related to the melting continental ice sheets that flooded the North Atlantic.[4] In the Middle East, a ~200-year drought forced the abandonment of agricultural settlements in the Levant and northern Mesopotamia.[5,6] The subsequent return to moister conditions in Mesopotamia promoted settlement of the Tigris Euphrates alluvial plain and delta, where breachable river levees and seasonal basins may have encouraged early southern Mesopotamian irrigation agriculture.[7]

By 3500 B.C., urban Late Uruk society flourished in southern Mesopotamia, sustained by a system of high-yield cereal irrigation agriculture with efficient canal transport. Late Uruk "colony" settlements were founded across the dry-farming portions of the Near East.[8] But these colonies and the expansion of Late Uruk society collapsed suddenly at about 3200–3000 B.C. Archaeologists have puzzled over this collapse for the past 30 years. Now there are hints in the paleoclimatic record that it may also be related to a short (less than 200 year) but severe drought.[9,10,11]

Following the return to wetter conditions, politically centralized and class-based urban societies emerged and expanded across the riverine and dry-farming landscapes of the Mediterranean, Egypt, and West Asia. The Akkadian empire of Mesopotamia, the pyramid-constructing Old Kingdom civilization of Egypt, the Harappan C3 civilization of the Indus valley, and the Early Bronze III civilizations of Palestine, Greece, and Crete all reached their economic peak at about 2300 B.C. This period was abruptly terminated before 2200 B.C. by catastrophic drought and cooling that generated regional abandonment, collapse, and habitat-tracking. Paleoclimatic data from numerous sites document changes in the Mediterranean westerlies and monsoon rainfall during this event (see Figure 10.1), with precipitation reductions of up to 30% that diminished agricultural production from the Aegean to the Indus.[9,10,11]

These examples from the Old World illustrate that prehistoric and early historic societies—from villages to states or empires—were highly vulnerable to climatic disturbances. Many lines of evidence now point to climate forcing as the primary agent in repeated social collapse.

High-resolution archaeological records from the New World also point to abrupt climatic change as the proximal cause of repeated social collapse. In northern coastal Peru, the Moche civilization suffered a ~30-year drought in the late 6th century A.D, accompanied by severe flooding. The capital city was destroyed, fields and irrigation systems were swept away, and widespread famines ensued. The capital city was subsequently moved northward, and new adaptive agricultural and architectural technologies were implemented.[12] Four hundred years later, the agricultural base of the Tiwanaku civilization of the central Andes collapsed as a result of a prolonged drought documented in ice and in lake sediment cores.[13] In Mesoamerica, lake sediment cores show that the Classic Maya collapse of the 9th century A.D coincided with the most severe and prolonged drought of that millennium.[14] In North America, Anasazi agriculture could not sustain three decades of exceptional drought and reduced temperatures in the 13th century A.D, resulting in forced regional abandonment.[15]

Climate during the past 11,000 years was long believed to have been uneventful, but paleoclimatic records increasingly demonstrate climatic instability. Multidecadal- to multicentury-length droughts started abruptly, were unprecedented in the experience of the existing societies, and were highly disruptive to their agricultural foundations because social and technological innovations were not available to counter the rapidity, amplitude, and duration of changing climatic conditions.

These past climatic changes were unrelated to human activities. In contrast, future climatic change will involve both natural and anthropogenic forces and will be increasingly dominated by the latter; current estimates show that we can expect them to be large and rapid.[16] Global temperature will rise and atmospheric circulation will change, leading to a redistribution of rainfall that is difficult to predict. It is likely, however, that the rainfall patterns that societies have come to expect will change, and the magnitude of expected temperature changes[17] gives a sense of the prospective disruption. These changes will affect a world population expected to increase from about 6 billion people today to about 9 to 10 billion by 2050. In spite of technological changes, most of the world's people will continue to be subsistence or small-scale market agriculturalists, who are similarly vulnerable to climatic fluctuations as the late prehistoric/early historic societies. Furthermore, in an increasingly crowded world, habitat-tracking as an adaptive response will not be an option.

We do, however, have distinct advantages over societies in the past because we can anticipate the future. Although far from perfect and perhaps subject to unexpected nonlinearities, general circulation models provide a road map for how the climate system is likely to evolve in the future. We also know where population growth will be greatest. We must use this information to design strategies that minimize the impact of climate change on societies that are at greatest risk. This will require substantial international cooperation, without which the 21st century will likely witness unprecedented social disruptions.

REFERENCES AND NOTES

1 Weiss, H. 2000. in *Confronting Natural Disaster: Engaging the Past to Understand the Future*, G. Bawden and R. Reycraft, eds., pp. 75–98. Albuquerque: University of New Mexico Press.

2 Bar-Yosef, O. 2000. *Radiocarbon* 42, 23.

3 Casse, F. 2000. *Quat. Sci. Rev.* 19, 189.

4 This flooding may have altered thermohaline circulation (THC), although there is as yet no direct paleochemical data demonstrating a shutdown or reduction in THC at this time.

5 Goring-Morris, A. N., and A. Belfer-Cohen 1997. *Paléorient* 23, 71.

6 Kozlowski, S. K. 1999. *The Eastern Wing of the Fertile Crescent.* BAR Intl. Series 760, Oxford.

7 Adams, R. M. 1981. *Heartland of Cities.* Chicago: University of Chicago Press.

8 www.science.widener.edu/ssci/mesopotamia.
9 Cullen, H. M. *et al. 2000. Geology* 28, 379.
10 Bar-Matthews, M. *et al.* 1999. *Earth and Planetary Science Letters* 166, 85.
11 Lemcke, G. and M. Sturm, 1997. In *Third Millennium BC Climate Change and Old World Collapse*, H. N. Dalfes, G. Kukla, H. Welss, eds., pp. 653–678. NATO ASI 49. Berlin: Springer.
12 Shimada, I. *et al.* 1991. *World Archaeol.* 22, 247.
13 Kolata, A. *et al.* 2000. *Antiquity* 74, 424.
14 Brenner, M. *et al.* 2001. In *Interhemispheric Climate Linkages*, V. Markgraf, ed., pp. 87–103. New York: Academic Press.
15 Dean, J. S. *et al.* 1993. In *Themes in Southwest Prehistory*, G. J. Gumerman, ed., pp. 53–86. Santa Fe: School of American Research Press.
16 www.grida.no/climate/ipcc/regional
17 The leaked Summary for Policy Makers of the upcoming Third Assessment Report by the IPCC gives estimates of 1.5° to 6.0°C.

Climatic Events as Social Crucibles

11

Natural Disaster and Political Crisis in a Polynesian Society

An Exploration of Operational Research

James Spillius

Some Principles of Operational Research

In recent years there has been much discussion of what the role of the anthropologist is and should be.[1] Departures from 'pure' anthropology are variously described as applied anthropology, action anthropology, technical advising, participant intervention, and operational research. Although there is little agreement on what each of these terms means, there are several issues that keep recurring in the discussion of them. Should anthropologists confine themselves to 'pure' research? Why do administrators make so little use of anthropological writings? Is it possible to do research that is simultaneously of theoretical and practical importance? Can social change be studied in process, and should anthropologists try to affect the course of such change? Many questions concern ethics and values, among them: can anthropologists, as scientists, take a moral stand on anything connected with their work? Is it morally right to supply administrators with information that they may use in a way the anthropologist disapproves of? Should anthropologists take issue with government policy or accept it as given? What are the anthropologist's responsibilities to the people he studies and works with? It is not my purpose in this paper to answer each of these questions specifically, but I hope to give some partial answers by drawing out the broader implications of my experiences on Tikopia in what I have chosen to call operational research.[2]

Operational research is here defined as the attempt to produce an effect on processes of social change while they are being studied.[3]

Throughout this paper I have tended to discuss the social anthropologist in operational research in terms of a role. But the fact that the situation is a changing one means that operational research itself is a process and cannot be rigidly defined as a set of role prescriptions. Actions that may be appropriate at one phase, such as my assuming command on the day of

The Anthropology of Climate Change: An Historical Reader, First Edition. Edited by Michael R. Dove.

the 'evil *fono* [gathering of the people]', described in Spillius (1957), may be inappropriate at a later phase of the process. The precise techniques the operational research worker uses depend on the problem he must deal with and on the developing situation he is part of. But it is possible to state in general terms some of the principles of operational research that emerged on Tikopia and that might be useful in other field situations.

1. *The unit of study in operational research is Government and indigenous community*, and especially those members of each group who interact with one another.

In social anthropology Schapera (1938) and Fortes (1938) clearly express the importance of treating European and native groups as one community when studying a process of culture change or contact. Gluckman further underlines and gives vivid illustration to this point (1940). Although the theoretical importance of treating European and indigenous community as one social system has long been recognized, it is not always put to practical use in applied anthropology. Colonial situations are usually analysed either as the administration of an indigenous community, or as the indigenous community's reaction to colonial administration; the assumption is that there are two distinct social systems instead of one, and the essential process, the *interaction* of colonial administration and indigenous community is not fully analysed.

In a colonial situation there are several specific relationships such as those between master and servant, bishop and local priest, plantation manager and labourers, Government and chief or headman, doctor and medical assistant, trader and customer, and so on. It is with such specific contexts of interaction that the operational research worker must deal. Schapera, in particular, stresses that certain offices like District Commissioner and Bishop are of such a nature that the indigenous group cannot avoid taking them seriously in the conduct of their affairs (Schapera 1938). In most places these offices are occupied at the moment by Europeans. But it is not inconceivable that they will be occupied by members of the indigenous people, as has indeed happened in several territories.

In Tikopia the problem was primarily to improve communication between the Government officers and chiefs, overseas labourers, and the Mission, and between chiefs and labour recruiters. These were the relationships that needed to be clarified and for which a *modus operandi* had to be agreed on. I did not try to study the Government hierarchy in itself, nor did I try to improve communication between its various levels. My isolation prevented this, and in any case Government felt no need for such study. With regard to the Tikopia, I did not succeed in improving communication between the factions or in helping them resolve their own conflicts—nor did I even try to do so. This does not mean that operational research cannot be carried out within a community that is part of a larger society. In fact it was begun in the study of the operations of such a community, the R.A.F. (Dobbs 1949–50), and has been further developed in studies of factories (Jaques 1951; Rice 1953). But on Tikopia the most pressing and immediate problem was the relation between Government and chiefs; the partial solution to this problem had repercussions throughout the Tikopia social structure, solving and creating problems in turn.

2. Since the unit of study is Government and indigenous people, *the anthropologist must occupy a position outside both systems, especially outside the Government hierarchy.*

This is defined by Dobbs as the *sine qua non* of operational research (Dobbs 1949–50: 29–30). If I had been a member of Government, I should not have been in the advantageous position of being able to interpret both sides to one another as freely as I did;[4] I could not have worked as quickly towards the goal of placing responsibility for dealing with outside problems upon the Tikopia. Similarly, if I had been completely identified with the Tikopia, I should have had difficulty in seeing Government's point of view.

I was exceedingly fortunate in having a good working-relationship with Government. Several factors contributed to this. Basically I agreed with Government on policy. Government shared my feeling of sympathy for the Tikopia and my desire to help them. The officials were men of high calibre and they were fully prepared to make decisions quickly and realistically. I had easy access to all levels of the Government hierarchy. Without this kind, or at least some degree, of co-operation with Government, operational research would be impossible.

According to this principle, the anthropologist is studying and has relations with both Government and indigenous people but is not fully identified with either. This raises the question of the possibility of a community, whether Government or indigenous people, being able to assimilate a role for which they have no precedent in their social structure. The anthropologist may have a perfect conception of his role and play it perfectly, but if it is not known or comprehended by the society there may be little or no communication. The fact must be faced that the anthropologist, wittingly or not, is assigned a number of roles. He has to begin with those that are understood; then, as the study proceeds, he can develop the operational research role, which he has preconceived. Awareness of the potentialities of the operational research role rather than its precise definition is the important first stage.

Operational research, then, cannot be carried out unless both Government and indigenous people are willing to give positive sanction to the anthropologist. For the primitive society, particularly if it is isolated and unsophisticated, this is likely to be a rather academic question. They may object with passive resistance but they are usually not in a position to keep the anthropologist out. With Government the situation is different, for unless they give their active assent, this kind of research becomes impossible. I should hazard the guess that Government officials are most likely to welcome the idea of operational research if they feel there are administrative problems in the area but that these problems are not so serious as to appear insoluble; or if conditions are basically in good order but more and speedier communication is required for adequate formation and execution of policy.

The anthropologist's position outside the hierarchy should also be maintained in the financing of operational research. In the development of such work it seems advisable that financing should take place in phases. In the initial phase the anthropologist would most probably be paid by a private foundation. In a second phase when the nature of the work is better understood, he would be paid by a distant branch of Government; financing by the local government would identify the anthropologist too closely with its interests. In a third phase operational research would be financed by the local unit of study, the Government and the indigenous people. This would imply considerable sophistication and monetary development of the indigenous community. At the present time many indigenous peoples can pay only in food, housing, and the enormous prestige which even now anthropologists are frequently accorded by the groups they study. The status and attention given to an anthropologist by a primitive community should be recognized as payment for services that the anthropologist provides with little difficulty to himself, services which the indigenous people may find it impossible to procure for themselves.

3. *Both parties involved must recognize that a problem exists and there must be some agreement in principle about its solution.*

Both Government and Tikopia agreed that people should not die of hunger, that violence should be prevented, and that labour recruiting should be more strictly controlled. Violent antagonism between Government and Tikopia would have impeded or made impossible a satisfactory solution to these problems. It is doubtful whether a government would want a social anthropologist to do operational research if relations between Government and governed were openly hostile or if complete tranquility were the order of the day.

But in some field situations the recognized problem is not the essential one. One of the tasks of the operational research worker is to point out that the vital problems are not always the obvious ones, as in the case of the *inaki* [upkeep of ancestral temples], when Government was unaware of the political implications of the rite. Mair also suggests that anthropologists can make themselves most useful if they '... do not seek to recommend action so much as to lay a finger on maladjustments, to indicate trends, and to check the accuracy of current assumptions' (Mair 1950: 187). This view is also expressed by Schapera (1949, 1951) and is exemplified by Firth's discussion of the social aspects of the Colombo Plan (1951). While I agree in general with this point of view, I should add that although the anthropologist may not seek to recommend action, nevertheless when he indicates trends, which amounts to making interpretations of the indigenous people to Government, he is in effect implying a course of action.

4. *The main task of operational research is the continuous interpretation of events to both parties of the social system*, enabling them to take one another into account in their decisions and formation of policy.

Operational research is concerned with a situation that is constantly developing and changing, and the most helpful interpretations are those rendered while the process of change is going on.[5] Problems are dealt with as they crop up instead of being saved up, as so often happens in applied anthropology, for recommendations in a report written some time after leaving the field, by which time the problems and the government personnel will probably have changed. As Evans-Pritchard has pointed out (1951), reports written after leaving the field are most useful to the administrator as general background information. If administrators have time and interest—which in many areas they have not—such reports can deepen their sociological understanding. But anthropological monographs are written primarily for other anthropologists, and as the discipline becomes more advanced technically, the monographs become more and more difficult for the administrator to understand and apply. (Forde (1953) and Schapera (1951) both make this point.) Administrators require special reports, to meet their needs. But in my view it is even more important that *specific* problems should be dealt with on the spot, preferably in person, instead of only in a written report.

5. *The operational research worker should assume professional as distinct from personal responsibility for his actions.*

Professional responsibility is not easy to define. Its precise meaning is known only for those arts or professions such as medicine and law that have had a long history of public practice and service. In the full professions a code of ethics has been established governing standards of work and relations with the public. The doctor or lawyer, by virtue of his training, is expected to assimilate the conscience he will share with other members of his profession. He is handed a formal reminder of it on graduation day. In the case of social anthropologists, no such standardized code governs their relations to the people they study, for anthropologists do not provide a regular, continuous service to clients. Anthropologists do of course adhere to an academic code of ethics, but this relates to university affairs rather than to the provision of practical services to clients. For these reasons I view social anthropology as an academic discipline and not as a profession. I am not suggesting here that anthropologists may achieve full professional status only through carrying out operational research. I am stating that professional responsibility is a necessary component of operational research.

As a first attempt to define professional responsibility in an anthropological field situation, I should say that it means accepting the consequences of one's presence in the area and attempting to understand how it affects the life of the community, not only in the present but in the future as well. It means working towards the goal of leaving both Government and indigenous community—but particularly the Tikopia in the case reported here—better able to cope with such problems as may arise after the anthropologist has gone.

I am sometimes asked if what I did on Tikopia was any different from what any European would have done. No anthropologist, however set he was on sticking to the role of observer, would have done nothing in this situation. A Government officer, a trader, even an entomologist would most certainly have tried to prevent violence and to help the Tikopia get food. But professional responsibility implies more than coping with an immediate crisis. On Tikopia it meant working towards the long-term goal of leaving Government and Tikopia better able to communicate with each other and to make realistic decisions in terms of each other.

The idea of professional responsibility has many implications. First, the field worker's position outside the hierarchy implies an attempt at impartiality towards both sides, which in all probability can be maintained. But this does not mean that the field worker has no effect on the situation, or that he is a detached observer. Claims that one is only an observer, a recorder of facts, a disinterested spectator, are phrases of cant, or they indicate self-deception. When Government and informants ask the anthropologist a question, his answer is in fact an interpretation. On Tikopia several difficult incidents might have been avoided if I had recognized from the outset that what the Tikopia really sought was clarification of Government's attitudes and intentions and not simply information. To vexing questions, silence itself is treated as an answer.

Second, having accepted the fact that his mere presence has some effect on the situation, the anthropologist must constantly make judgments about the course of action most likely to achieve the long-term goal of leaving the people studied better able to cope with their problems. On Tikopia I decided that at this stage of their contact with the outside world, it would be best to try to get the chiefs and *maru* to cope with the political crisis. It would have been easier to have presented Government with one spokesman and representative of the Tikopia, Robinson Vaka. Nevertheless I chose the more laborious and difficult course of getting all the chiefs to approach Government individually or as a group through an interpreter of their own choosing. While Robinson Vaka would have been speedier and more efficient, using him exclusively would have only postponed the day when Government would have had to discuss and clarify basic issues with each of the chiefs. Supporting Robinson exclusively would have undermined an already weakened traditional authority, which in my judgment would have reduced the society to chaos. My judgment may have been wrong. But at this stage it seemed best to keep Robinson in his traditional role as a 'strong *maru*' and to bolster the chiefs' prerogative of dealing with external affairs.

In retrospect, it appears that I made several errors of judgment, partly because I was learning as I went along, and partly because I was not fully aware of the effects of my own sympathies and prejudices at the time. My direct intervention to prevent violence, for example, although necessary at the time, could probably have been avoided if I had not been trying previously to withdraw into the observer role.

I think that some of the 'objective' information I gave to various officials was heavily slanted so that they would come to the policy decision I considered best. I am still convinced that this bias in what I then considered factual information resulted in a just solution, but it was also motivated by my anxiety to get Government to take back the task of command so that I could recapture my former role as an impartial observer of the Tikopia instead of being their commander. It occurs to me that at the time of [the colonial officer] Mr. Davidson's visit, when I was so busy shedding my 'command' role in Tikopia, by suggesting policy to him in his dealings with the Tikopia I was actually once again issuing commands to them in an indirect way. I did not realize this at the time, but I see now how interpretations can be so distorted unintentionally as to make only one decision possible. It is a rare man whose supposed objective statements are not coloured by hidden subjective elements in his personality. The only antidote to this all too human foible is the awareness by the anthropologist that he is capable of committing it.

Another error of judgment, both naïve and emotionally based, was my expectation that the Melanesian priest on the island would have the same code of ethics as a European priest. He had the Western status but not the Western norms of the position. In retrospect, it is evident that my naïveté blocked my understanding of the fact that he had to act in a way personally objectionable to me not only to satisfy the Mission's expectations of him, but also to survive in the famine. If I had been more aware of my biases at the time I might have reduced somewhat the increasing tensions between the Mission and the Christian chief on the one hand and the pagan chiefs and myself on the other.

Endemic factionalism is a common characteristic of many small communities. Working through or closely with a faction may stabilize the situation while the operational research worker is on the spot. It does not remove the conflicts or guarantee that the people will be able to cope with new conflicts without fitting them into old factional patterns. There is no doubt that I was 'captured' by the pagan faction. I think it more than possible that one can avoid the pitfall of acting spontaneously according to one's personal feelings of sympathy for one faction and antagonism for the other. But I do not think that an anthropologist living in a small community can avoid emotional identification with the people in it. He must have recreation and he must work in the company of other people if he is living in isolation with them. For the period of his field trip he is bound to have special friendships with persons of the community in which he lives. It is difficult indeed to have an ascetic morality guiding what has to be gregarious behaviour.

The anthropologist's *personal responsibility* for his everyday behaviour depends partly on the mores of the people he is living with and partly on his own character. But the split from his professional responsibility cannot be complete. The important questions are whether his behaviour jeopardizes his professional responsibilities and how it affects his relations to the groups with which he is working. In brief, the anthropologist cannot avoid being fitted into the local social system through being assigned pseudo-kinship status and roles, adopting some of the local mores, and forming special sympathies and prejudices; but he can avoid acting uncritically according to his sympathies without being aware of what he is doing. The object is not to eliminate these prejudices and sympathies, which is impossible, but to be aware that they do colour judgment and interpretation. It is just as important to recognize that one cannot clarify everything and that clarification itself may be biased as it is to be in a position to make an interpretation. The point is not to fulfil a role for the sake of fulfilling a role. The point is to know what one's aims are, how best to fulfil them, and when and why one is departing from a preferred mode of conduct.

In work of this kind special problems arise over publication and the communication of confidences. It seems to me the principle to be followed here is that if repeating a confidence is likely to have repercussions or to cause embarrassment to any of the people involved, it must be treated as a secret. But practices that can only be termed criminal by the European administration, such as infanticide and sadistic punishments and abortion, do occur in primitive societies, either through ignorance of European law or through the heavy pressure of tradition. Should the research worker report to Government the individual cases? If it is an undesirable practice, the important issue is not that it does occur, but why it occurs. Such explanation will be far more valuable to Government for long-range understanding and control than any immediate punishment it may choose to inflict. If a specific case is reported to Government, it has no option but to act immediately on the information. But I think such practices can be discussed with Government in such a way as to indicate that they are probably occurring but can be understood and controlled by indirect means.

Publication of matters concerning the Government with whom one has enjoyed a confidential and cooperative relation, without discussion of the material beforehand, amounts to a betrayal of confidence. It would make cooperation in future work extremely difficult for oneself and for other anthropologists. As primitive peoples become more westernized and more sophisticated, the same rules of confidence will apply to them.

6. *Operational research has a double aim: (a) the conceptual aim of contributing to the analysis of social processes and social change, and (b) the practical aim of effecting a change in the situation under study.*

(a) The *conceptual contribution* of operational research rests on the recognition by social anthropologists that all primitive societies change continually, not only because of external influences, but also because of internal developments. No anthropologist can seriously entertain the idea of a community immune from all change, or, as Nadel put it, no anthropologist would now want 'to preserve them (primitive peoples) in a sort of living museum or anthropological zoo, presumably in order to study them undisturbed' (Nadel 1953: 11). Any attempt by Government or anthropologist to prevent change would be defeated not only by missionaries and traders, but also by such people as the Tikopia themselves, who desire, welcome, and are willing to pay the price of social change.

In anthropology social change is often studied by examining a primitive society at two separate time periods, but this approach is limited because it is frequently difficult to tell what has gone in the interim. Or an attempt is made to identify long-range trends, a method that is much easier to use when there are historical records of past developments. In operational research, change is studied while it actually occurs. The operational research worker makes predictions and sees whether they are valid. The advantage of this method is that one can study the exact mechanisms of social change. The disadvantage is that many social changes take so long to run their course that they are not completed while the anthropologist is doing his field work. Field trips of two to three years would allow one to deal with short-term changes, but the analysis of general trends would require historical records or return field trips.

In operational research the ideal is an experimental situation in which variable factors can be controlled by the experimenter. This ideal is rarely approached in most groups, but certain types of social change have been very effectively planned and studied by operational research, especially in industrial situations. Thus Rice was able to study the socio-technical system of weaving in a cotton mill, to predict what form of work organization would be most appropriate, and to show that when this form of organization was adopted production went up (Rice 1953).

On Tikopia I could not make this precise, controlled sort of experiment. I did make predictions, but, with the exception of the migration scheme and the effect of the sophisticates [Tikopians with more knowledge of the world beyond Tikopia], they were not concerned with general social change. They dealt with such problems as crops, time of recovery, and the likelihood of violence. Most of these forecasts turned out to be inaccurate. Fortunately Government put me in the position to correct these mistakes. Allowed this luxury of correction, my subsequent interpretations and predictions in these matters proved reasonably accurate. In the case of migration, my prediction was correct, but it was of little use to Government because, in terms of resources and organization, the recommendation could not be implemented in a hurry.

It is sometimes asked if this sort of activity is research or just inquiry. If the criteria of research are a systematic formulation of the problem, collection of data, and testing of hypotheses, then operational research can be classified as a scientific endeavour. It differs from other disciplines in that the standards of work are more exacting, since hypotheses are tested immediately and there is little room for speculation about their validity. Operational research is most effective in analysing short-term processes and changes. Their interpretation by the anthropologist and the practical application of his interpretations affect the process of change. But, what is more important, he is able to follow up the results of his efforts and to correct his errors of interpretation or the method of applying them. But such a concentration on social process and change need not preclude the collection of conventional anthropological data. Indeed, such data are essential for proper understanding of social processes, and on Tikopia I should have been seriously handicapped if I had not been able to build on Firth's previous work.

Several aspects of Tikopia social organization were clarified because I watched them changing in response to crisis. It became apparent that the chiefs could deal with external crises and people (hurricanes and Government), but that they could not do much to cope with internal violence. This was not their province. The *maru* coped with the internal problems of the island and applied the rigorous discipline so alien to the general permissive behaviour of the chiefs towards their people.

Through the activity of the *maru*, especially the charismatic leader, Robinson Vaka, there was a sudden development of the institution of the *fono*, and a decline in the *fono* as the crisis lessened. This indicated that the *fono* was an emergency institution that lay dormant in normal times, being used only in situations of crisis. The function of such an institution in a society without written records can be learnt only by observation. It also indicates that the development of an institution is not necessarily an irreversible process of social change. One needs to observe the whole process of rise and decline to be sure of this point. This raises questions of some interest: in future crises and changes will the *fono* be used again by the young sophisticates to achieve power and to bring European notions of authority and legal sanctions to the island? Will the young sophisticates again shift from a revolutionary position to one of support of traditional authority? Or will they actively campaign to weaken further the power of the chiefs and older *maru*? The nature of the kinship and land-holding systems and the present policy of Government preclude the abolition or overthrow of chiefs, but do not preclude a loosening of traditional mechanisms of social control.

Another set of processes may also be indicated briefly. On Tikopia a series of methods was employed by the people in coping with the crisis. At first they made an attempt at a realistic solution in proposing migration of labour to the New Hebrides, but this was frustrated. The Tikopia propensities for idealizing individuals and spawning rumours were then intensified and given impetus. This was followed by a struggle for power between factions and an attempt at a violent solution of their problems. After this had been averted, tension remained but was controlled by the reinforcement of the dominant institutions, the chiefs and the *fono*, and the lead was given in control of punishments and theft by a charismatic leader. Plans for migration were begun, but interest in them declined as the situation improved. These phases of development may be compared with reactions to similar crises in other societies. Although such a comparison is beyond the scope of this paper, it might yield results of predictive value.

(b) The *practical aim* of operational research is twofold: (i) to effect changes in the situation while the anthropologist is in the field, and (ii) to leave the social unit, that is, Government and indigenous people, better able to cope with their problems in terms of one another.

Certain short-term practical results can be assessed for Tikopia. Rice was fairly distributed. Violence was kept under control. The traditional authority structure was strengthened. All these results, as I said earlier, with the possible exception of the last, might have been achieved by any European taking personal initiative. But an understanding of such features as factionalism, the roles of chiefs and *maru*, the function of the *fono*, and the *inaki*, requires specialist training if interpretations of some lasting use are to be made.

Other results were gained by my attempts to get the Tikopia to take responsibility for their own decisions. Communication with Government and with traders was greatly improved. Their interpretations of Government decisions were more realistic in terms of Government policy. The Tikopia realized their responsibilities in economic relations with the outside world, although I do not think they were necessarily better able to cope with these themselves.

Judging from recent reports, it would seem that the Tikopia understood what I was trying to do while I was there, but have not yet reached the point where they feel they can dispense with someone who will live with them and interpret European governmental aims to them and their own aims to Government. Mr. Pat De Burgh Thomas got to know the Tikopia well in his last three years as manager of the Lever Brothers plantation on which Tikopia labourers worked. He was extremely fond of them and received in turn rather hectic Polynesian affection. Before he left the Solomons, he was asked by the Tikopia to stay on the island, and they made fantastic promises for this eventuality. They explained, 'We don't need another chief; we have our chiefs; but we need a man like you who will explain our mind to Government.' In brief, it seems that the Tikopia wanted a man outside the hierarchy who could interpret the attitudes of each side to the other. It is also clear that if the job on Tikopia had been officially defined as operational research, I left before the task at hand was fulfilled. In any attempt to control social change, it is only too true that for every problem solved, a dozen more are created.

The issue of paganism versus Christianity on Tikopia was evidently resolved after a disastrous influenza epidemic in July–August 1955. Many people died, including two pagan chiefs. According to recent reports, all the remaining pagans, chiefs included, were baptized into the Christian faith. There are no details on the factions, the *fono*, or the powers of the chiefs and *maru*.

Another practical point from the Tikopia experience concerns the phases of the food crisis—from hurricane to famine recovery. Comparison with similar disasters in places with like conditions might be of predictive value, especially in the timing of relief supplies. It would seem that it takes three to four months for acute food shortage to develop after a severe hurricane on an isolated tropical island like Tikopia. This is the period of grace for organizing and distributing relief supplies. On Tikopia ten tons of rice, enough for two weeks, were sent four months after the hurricane, but this was not enough to halt the increasing shortage of food. The food shortage became a famine in the sixth month. Seven months after the hurricane twenty tons of rice were sent, enough for four weeks. It would seem that once the death-rate starts to increase because of malnutrition and lowered resistance to disease, relief supplies cannot be counted on to arrest it immediately; on Tikopia the death-rate began to decline four to six weeks after the second shipment of rice. The Tikopia experience also suggests that relief should be timed to arrive after the fallen crops have been eaten but before people are driven to tear up unripe crops from the ground. If the population can be persuaded to subsist almost exclusively on relief supplies for a four- to six-week period so that growing crops remain almost untouched, the speed of recovery is hastened by proportionately more than four to six weeks.

The practical aims of operational research can only be achieved if the social anthropologist shows some sensitivity to the social structure of the government he is dealing with. Too frequently anthropologists are so obsessed with the exotic elements of their field work that they forget by whose courtesy they are able to conduct their investigations. This does not mean that an anthropologist must agree with government policy; but it does mean that he demonstrates his ineptness in his subject if he offends officials by disregarding courtesies or proper acknowledgement of those occupying the offices of government. It is curious that sometimes anthropologists embarked on 'objective' studies display compassion and understanding for the indigenous community but find it much more difficult to take a dispassionate view of Government as a social system. If an anthropologist is engaged in operational research, he cannot identify himself exclusively with the interests of the indigenous people; he must also think in terms of government policy for a wide area. Such policy he may or may not find repugnant. If he does not find broad government policy too disturbing, then his only real concern is with government policy on specific problems. My expectation would be that in most cases policy connected with a *specific* problem and its solution would not involve to any great extent the moral or political beliefs of either anthropologist or Government. If the anthropologist disagrees fundamentally with *general* government policy, such disagreement would soon become obvious, and would pose the choice between compromise of beliefs and the expediency of solving an immediate problem on the one hand, and adherence to general moral or political beliefs on the other. Thus operational research would not be possible in all situations. But all too frequently anthropologists who disagree with general colonial policy seek to take issue with it at the local level, where it is only executed, not formed.

But these are issues for which no general rule can be stated; they are decisions that only individuals can make for themselves, without reference to a directive. But these decisions must be made primarily in terms of the government one is dealing with, as it is they who hold the power of implementing directed social change.

Thus, in a practical sense, the chief value of operational research (and here I am paraphrasing Dobbs) is that it couples the capacity of a technical analysis of a given situation from someone outside the organization with the recognition of what is possible in terms of resources. The social anthropologist presents the administrator and the indigenous people with an analysis that is not only relatively impartial but also *immediate*, an analysis that can be realized in action.

Summary

Operational research is here defined as the attempt to produce an effect on processes of change while they are being studied. The principles of operational research that emerged from the Tikopia experience are described as follows:

- (i) The unit of study is Government *and* indigenous community.
- (ii) The anthropologist must occupy a position *outside* this unit. He must not be completely identified either with Government or with the indigenous community.
- (iii) Both parties involved must recognize that a problem exists, and there must be some agreement in principle about its solution.
- (iv) The main task of operational research is the *continuous* interpretation of events to both parties of the social system.
- (v) The operational research worker should assume *professional responsibility* for his actions. This means accepting the consequences of his presence in the field, and attempting to work towards the goal of leaving the social unit better able to cope with its problems after he has gone. It also requires awareness of how his own prejudices, which cannot be eliminated, affect his sociological analysis.
- (vi) Operational research has a double aim: (*a*) the *conceptual* aim of contributing to the analysis of social processes and social change, and (*b*) the *practical* aim of providing an analysis of immediate problems that can affect the situation under study.

NOTES

1 Much of this discussion is contained in *Human Organization*, the journal of the Society for Applied Anthropology. See especially editorials in Vol. 9, No. 3; Vol. 10, No. 2; Vol. 11, No. 2; Vol. 13, No. 1; and Vol. 14, No. 2. For a recent definitive paper representing the views of many British social anthropologists on the relation between anthropological research and social needs, see L. P. Mair (1956).

2 A discussion of anthropologists' views on these issues, particularly on the relation of Government and anthropologist (omitted here for reasons of space), will appear at a later date.

3 This definition is very close to Holmberg's definition of 'participant intervention'. 'His (the participant observer's) job is to assist the community to develop itself and to study the process while it is taking place' (Holmberg 1955: 26.)

4 For a brief discussion of the difficulty of anthropologists becoming administrators, see Tax *et al.* (1953: 181). For a discussion of anthropologists as permanent technical advisers to administrators, see Fosbrooke (1952) and Moffett (1952).

5 For a similar point of view see Kennard and Macgregor (1953: 839–40).

REFERENCES

Dobbs, H. A. C. 1949–1950. 'Administrator and Specialist.' Four articles in *Corona* 1(10): 7–10; 1(11): 18–22; 2(1): 27–31; 2(2): 74–6.

Evans-Pritchard, E. E. 1951. *Social Anthropology*. London: Cohen and West.

Firth, R. 1951. 'Some Social Aspects of the Colombo Plan.' *Westminster Bank Review*, pp. 1–7, May.

Forde, C. D. 1953. 'Applied Anthropology in Government: British Africa.' In *Anthropology Today*, A. L. Kroeber, ed., pp. 841–65. Chicago: University of Chicago Press.

Fortes, M. 1938. 'Culture Contact as a Dynamic Process.' In *Methods of Study of Culture Contact*, L. P. Mair, ed., pp. 60–91. Monograph 15 of the International Institute of African Languages and Cultures.

Fosbrooke, H. A. 1952. 'Government Sociologists in Tanganyika: II—A Sociological View,' *J. Afr. Admin.* 4(3): 103–8.

Gluckman, M. 1940. 'Analysis of a Social Situation in Modern Zululand.' *Bantu Studies* 14: 1–30, 147–74.

Holmberg, A. R. 1955. 'Participant Intervention in the Field.' *Human Organization* 14(1): 23–6.

Jaques, E. 1951. *The Changing Culture of a Factory.* London: Tavistock Publications.

Kennard, E. A., and Macgregor, G. 1953. 'Applied Anthropology in Government: United States.' In *Anthropology Today*, A. L. Kroeber, ed., pp. 832–40. Chicago: University of Chicago Press.

Mair, L. P. 1950. 'The Role of the Anthropologist in Non-Autonomous Territories.' In *Principles and Methods of Colonial Administration*, pp. 178–88. (Colston Papers). London: Butterworth's Scientific Publications.

Mair, L. P. 1956. 'Applied Anthropology and Development Problems.' *Brit. J. Sociol.* 7(2): 120–33.

Moffett, J. P. 1952. 'Government Sociologists in Tanganyika: I—A Government View.' *J. Afr. Admin.* 4(3): 100–3.

Nadel, S. F. 1953. 'Anthropology and Modern Life.' Inaugural Lecture at the Australian National University, Canberra.

Rice, A. K. 1953. 'Productivity and Social Organization in an Indian Weaving Shed.' *Hum. Relat.* 6(4): 297–329.

Schapera, I. 1938. 'Contact Between European and Native in Bechuanaland.' In *Methods of Study of Culture Contact*, L. P. Mair, ed., pp. 25–37. Monograph 15 of the International Institute of African Languages and Cultures.

Schapera, I. 1949. 'Some Problems of Anthropological Research in Kenya Colony.' *Memorandum No. 23 of the International African Inst.*

Schapera, I. 1951. 'Anthropology and the Administrator.' *J. Afr. Admin.* 3(3): 128–35.

Spillius, James. 1957. Natural Disaster and Political Crisis in a Polynesian Society: An Exploration of Operational Research I. *Human Organization* X: 3–27.

Tax, S., *et al.* 1953. *An Appraisal of Anthropology Today*. Chicago: University of Chicago Press.

12

Drought as a "Revelatory Crisis"

An Exploration of Shifting Entitlements and Hierarchies in the Kalahari, Botswana

Jacqueline S. Solway

In his analysis of the domestic mode of production Sahlins (1972: 124, 143) refers to a revelatory crisis as one of the few occasions outside of an 'act of ethnographic will' that exposes to the observer the central contradiction in that mode of production.[1] The drought which occurred in Botswana in varying waves of intensity between 1979 and 1987 constituted such a revelatory crisis. It precipitated the interruption of socio-economic patterns to a degree sufficient to lay bare contradictions in the existing order that had been latent or contained prior to the drought. Structural contradictions such as those between household and kindred, between individualized and communal property claims, between production for market and production for subsistence, between the state's vision of 'rational' peasant production and the realities of daily economic life in the rural areas, and those between classes, were revealed to varying degrees and in different ways both to the outside observer and to the participants themselves. Moreover, the drought also exposed the extent to which conditions in the rural areas had deteriorated. In doing so it revealed a 'crisis of social reproduction' (Watts 1987: 207), a condition related to the above contradictions, but one that points to the fact that the crisis at hand was not simply one of external origin but was, in fact, systemic and indicative of long term change. Things were not going to 'get back to normal' when the rains resumed. And indeed, they have not.

Southern Africa and Botswana, in particular, are drought prone regions. Droughts occur with alarming frequency in Botswana, on average every seven years. One can therefore assume that a cyclical pattern must exist. However, I argue that this drought broke the cyclical pattern and enabled fundamental as opposed to superficial changes to become entrenched. The drought occurred at a critical historical juncture in Botswana; commercialization and class formation had been proceeding at a rapid pace[2] and Botswana's economy,

The Anthropology of Climate Change: An Historical Reader, First Edition. Edited by Michael R. Dove.

newly rich from diamond revenues, was growing dramatically. The drought of the 1980s was thus a 'watershed' event. Changes in meaning and practice reached a point where pre-drought production patterns, in all likelihood, would not return. In addition, the state initiated actions to ameliorate the effects of drought which, in many respects, hastened the trajectory of change and, at the same time, laid the basis for its own expanded presence.[3] The drought was thus an opportunity and provided a point of entry for the state to insert itself in the lives of citizens in new and expanded ways (cf. Ferguson 1990); this was an 'instrument effect'[4] of the drought relief measures.

However, in a paradoxical fashion, while a crisis such as a drought reveals and exposes contradictions and deteriorating conditions, it also allows them to be concealed and mystified. It is precisely because drought is a crisis and therefore believed to be temporary, unusual, externally wrought, and severe in nature that it allows systemic socio-economic problems to be concealed. Drought is a perfect scapegoat; all social and economic dislocation and suffering can be attributed to the drought and underlying problems can be left unacknowledged and, therefore, unconfronted. This is what happened in Botswana when state officials and many citizens alike argued repeatedly that the problems surrounding the drought would go away with the return of the rains. This reluctance to admit to a fundamental crisis of social reproduction contributed to the formulation of drought relief measures which, in some respects, were palliative in the short term but, in the long term, have served to exacerbate socio-economic inequalities.

The third point I wish to make with regard to drought is that it offers a critical occasion when conventional routine is sufficiently disrupted that actors are given licence to innovate with social and moral ideological and behavioural codes. It is a time of experimentation; when taboos can be violated, moral codes flaunted, and tradition invented (Hobsbawm and Ranger 1983). Something that may have been unacceptable before is possible during a drought. Thus drought is a perfect venue for agency and change and therefore offers a perfect lens for viewing the dialectic between structure and agency. The opportunities actors take when certain structural fault lines are exposed are not random nor are they entirely predictable. The process of how such change occurs and what aspects of a system are emphasized, elaborated, or altered are analysed here. Thus, change is not revealed simply as the unfolding of the logic of a pre-ordained system (such as capitalism) nor as simply the result of individual action.

This paper draws upon extended case material from central Botswana based on fieldwork covering a period of seventeen years. It is argued that a decline in the capacity to produce amongst Botswana farming households is related to commercialization, changing conceptions of property, and state intervention. Sen's method of entitlement analysis is utilized to examine social process in the realm of property relations in which narrow exchange rights have increasingly been granted legitimacy over wider use rights. The incomplete shift in entitlements from locally based sources to state based is also examined. In the final section dependency relations are considered and questions relating to various forms of patronage are posed.

Drought and Agriculture in Botswana: The National Context

In post-independence Africa, drought[5] has reputedly been a leading cause of declining production and great human suffering.[6] Yet the manifestations of drought and people's experience of drought conditions are mediated at all levels by non-meteorological factors.[7]

Moreover, within rural communities and even within households the effects of drought are not evenly borne. When the statistics are disaggregated it often comes to light that declining production and/or food shortage is not generalized between households in a community, nor is suffering equally distributed within households; certain members, particularly women, children and retainers, are often the first to suffer.[8]

Although for most of the years between 1979 and 1987[9] Botswana's rainfall levels were sufficiently below average to rank her amongst the five most drought-stricken countries in Africa (Holm and Morgan 1985) the government managed to keep famine at bay. Massive drought relief measures including food relief, labour-based relief, water development, and agro-pastoral subsidies insured that no one starved.[10] Every rural resident received, in one form or several, the benefits of drought relief. Botswana's capacity to generate relief derived from her own relative wealth,[11] from the receipt of foreign aid, and from the fact that the delivery of relief was handled efficiently and with remarkably little corruption.

This was a stunning achievement and has been cited as an exemplary case of famine prevention (Drèze 1990: 151–8; Hay 1988; Valentine 1993). However, although starvation was averted, the drought relief measures have served to accentuate socio-economic inequalities and have quite possibly undermined the capacity of the majority of rural peoples to produce their own subsistence requirements and to support themselves.

On a national level almost 40 per cent more hectarage (408,000 ha.) was under cultivation by the end of the drought in 1988–9, than was under cultivation in 1977–8 (260,000 ha.) (Republic of Botswana 1991, vol. 2: 10). Given that farmers were literally paid to plough by the government from 1985 to 1990 the increase in hectarage is not surprising. One must question seriously whether such figures will be sustainable once the subsidies are withdrawn and farmers must fund their own agricultural initiatives. In addition, the figure for the increase in hectarage must be tempered by information regarding who is doing the farming. Mean farm sizes grew in the 1980s but the increased hectarage under cultivation was not equally distributed (Republic of Botswana 1991, vol 2: 4). Drought relief agricultural subsidies enabled some farmers, particularly the larger and wealthier ones, to dramatically enlarge their farms while at the same time, many smaller, particularly resource-poor farmers, were unable to take full advantage of the programmes (discussed in more detail below). Moreover, many poorer farmers found that the arrangements they had relied upon before the drought to gain access to agricultural fields and resources were no longer as readily available.

The fact that for years after the end of the drought the government was still at pains to withdraw relief again indicates that the crisis exposed by the drought had much deeper roots than the recent rainfall shortages. After the drought ended in 1987 certain relief measures were simply renamed drought recovery; some have been incorporated into Botswana's expanding and ongoing social welfare programme;[12] and some were withdrawn in 1990.[13]

In keeping with a time honoured anthropological tradition of illuminating the general with the particular, I will employ a specific example to illustrate the general processes outlined above.

The Local Context

My specific case material derives from three periods of fieldwork in central Botswana.[14] I worked in the Kalahari region of the western Kweneng district, a semi-arid area receiving approximately 300–350 mm. of rainfall a year. The predominant local 'ethnic' group, both politically and numerically, are the Kgalagadi (cf. Kuper 1966, 1970a, 1970b). The Sarwa (San or Bushmen) pre-date the Kgalagadi in the area. They have a history of hunting and gathering but now largely constitute a casual labour force for the Kgalagadi. The Sarwa are the largest per capita recipients of food relief and have increasingly come to rely on this as their basic food source.

The local economy is based on pastoralism (cattle are the most valuable and important animal), supplemented by rain-fed agriculture (beans, maize, melons, and sweet-reed are grown but sorghum is the principal crop). Livestock are valued for the meat, milk, dung, and draught power they provide; they are valued for purposes of local exchange, both as gifts and as payments; they are valued for the social and spiritual sense of well-being they afford their owner; for the role they play in mediating relations between people, and between people and the *Badimo* (Gods) (cf. Bonte 1975, 1981: 40–41). Cattle are used in curing ceremonies and are sacrificed at rituals; as such they offer a direct link between the material–social domain of people and things and the spiritual domain of gods and power. And, of course, cattle are valued for the price they obtain at the market. Botswana maintains favourable trade arrangements with the EC, which imports her beef at inflated prices. The use of cattle for subsistence purposes undermines their market value, so that the two types of production are not easily compatible. Since all cattle owners are simultaneously involved to varying degrees in both types of production, they are constantly weighing the relative costs and benefits of pursuing the different production strategies.[15]

In addition, virtually all Kalahari residents are dependent, directly and indirectly, upon income derived from wage labour. Until the late 1970s most men migrated at several points in their adult life to the South African mines (in 1978, 84 per cent of Kgalagadi men I interviewed had been to the mines). By 1980 South Africa had cut back its recruitment of foreign miners, in effect withdrawing an employment opportunity upon which rural residents had come to rely. However distasteful working below ground in the mines may have been for local men, it remained one of the few cash earning opportunities. Extremely few wage opportunities existed for women at that time. New formal sector employment opportunities have become available in Botswana since 1980:[16] however, the majority of new jobs have been for skilled and educated individuals which has tended to give an advantage to better-off households as they were the ones who had previously invested in their children's education.[17] Thus the 'trickle down' effect of transfers from formal sector employment are not as equally distributed within rural communities (cf. Good 1992: 92) as South African mine wages were in previous decades. Examples such as this point to the ironies of development.

The Kalahari is rich cattle country, covered with good pasturage and few, if any, poisonous plants for cattle. The limiting factor on pastoral production in the region had been water but since the 1960s boreholes have continued to produce new sources. In the drought of the 1980s – for the first time, according to my informants – pasturage became a limiting factor in this part of the Kalahari. Agriculture is risky because of erratic rainfall but twenty years of data indicate that for farmers well prepared to plough, good harvests outnumber bad ones.[18]

Social Differentiation

In order to understand the role and consequences of drought relief measures in the region it is necessary to summarize recent changes in patterns of social differentiation in the Kalahari (see Solway 1979, 1987, nd. for a fuller discussion of these issues). Tswana–Kgalagadi structures have long been ranked and disparities in political, social, and economic status between households have existed for generations. However, in recent decades it has become apparent that the nature of social differentiation has been transforming; since the 1960s, in particular, opportunities have become available for significant capital investment in productive resources, and this along with explicit state sanction for commercial development has resulted in the formation of a distinct group of accumulators (Peters 1984).

Although the process is at best uneven, capitalist development and class formation are occurring. Amongst the rural élite there has been a shift of emphasis from wealth in people to wealth in things. There has been an increasing trend towards privatization and consolidation of economic resources which is only partly attributable to increasing capital investment in production (see discussion of draught power and property relations below for an example). This changing emphasis has contributed to a productive strategy on the part of the emerging commercial élite which has led them to withdraw, to a certain extent, their economic resources from the pool available for communal use, a process which is affecting the pattern of social reproduction. Thus poorer members of the community experience greater difficulties in gaining access to productive resources both for purposes of use and for accumulation. In this manner discrepancies in productive potential are growing; accumulation is creating a parallel process of dispossession.

The key resource and marker of wealth in the Kalahari has been cattle; herd size in the region ranges from zero to approximately 1000 with most cattle owning households owning less than 40 head. The process of commercialization leads many wealthier households to emphasize the market value of their livestock over local use and exchange value. Thus commercially oriented herders are less likely to milk their cows, to plough with them, to exchange them as gifts, sacrifice them at rituals, or to let others use them for these purposes. What is significant is that such decisions by the commercial élite have ramifications far beyond their households since the majority of villagers have depended upon access to the resources of the wealthier (this will be illustrated below with regard to agriculture).

The current pattern of accumulation reverses previous patterns. For the first two-thirds of this [twentieth] century accumulation had a dual dynamic of both concentration and dispersion of property. For instance, at the turn of the century the Kgalagadi were emerging from a period of political subjugation in which their capacity to accumulate property was extremely circumscribed by the powerful Tswana chiefdom to their east. Initially, in the early part of the century, the political élite accumulated livestock. By 1977, through local means of redistribution and through migrant labourers' wages, only 8.6 per cent of Kgalagadi households in the community where I conducted research owned no cattle; in 1986 that rate had doubled to just over 17 per cent. National figures are consistent with my own. According to the 1991 National Development Plan, farming households (thus excluding urban based and poor rural households who do not farm) with no cattle increased from 28 per cent to 38 per cent between 1980 and 1988 (Republic of Botswana 1991, vol. 2: 4).

Transformations are occurring at other social and ideological levels as well. Ideologies based on western Christian individualism compete with those grounded in the ancestors which affirm principles of mutual responsibility in which one's own well-being is inseparable from that of one's kin. There has been experimentation on the part of some wealthy households with 'new' non-syncretic Christian sects; these illustrate attempts by the élite to find cosmologies more consistent with their emerging world views. Notions of the person, of the body as individually or mutually conditioned, and concepts of morality and kin are also contested terrains (Comaroff 1985). For sake of space I will not elaborate these points further: but it is clear that changes in the productive process are part of a larger complex of related changes in which the entire basis of social reproduction is implicated.[19]

The local élite do harbour some ambivalence towards their declining role as social and economic benefactors. Business is easier to manage without attending to the diffuse needs of others, but, at the same time, benefactors or patrons lose labour, social and political support, and become subject to retribution, the 'weapons of the weak' (Scott 1985). Nonetheless, a diminishing of diffuse organic links between rich and poor in the region is a discernible trend.

Agriculture

At this point I wish to focus upon changes in agricultural production between the late 1970s and late 1980s in order to illustrate the growing discrepancies in productive potential that were thrown into relief by the drought and the manner in which agricultural subsidies, most of which were part of the drought relief measures, provided differential benefits to the rural populace depending on their position within the local political economy. In addition, it will be shown that the agricultural subsidies were based on certain ideological notions which in themselves contributed to reformulating rural relations and realities with the effect of further reinforcing growing disparities. I will present this data by comparing participation in agricultural production during the two seasons for which I collected survey data, 1977–8 and 1985–6, and for 1990, a year for which I have less complete data.

In 1977–8 approximately 90 per cent of Kgalagadi households (n = 70) in the village where I based my fieldwork ploughed.[20] In 1985–6 the number of households ploughing was reduced to about half that figure.

This statistic is rendered all the more striking by the fact that in 1985–6 there were three government sponsored programmes offering substantial subsidies for agricultural production and in the late 1970s there were none. The rains which herald the ploughing season came about two weeks later in 1985 than they had in 1977 and people claimed to be demoralized because of a series of years of poor rainfall. However, neither of these factors can entirely account for the discrepancy especially since 1978 followed a year of poor harvests and the agricultural subsidies of the mid-1980s should have compensated for some of the demoralization. The later rains of 1985 were not insignificant in terms of limiting numbers of people ploughing but they cannot begin to account for the discrepancy between the years. Late (but not too late) rains usually account for fewer hectares under cultivation rather than for fewer households ploughing. In the pre-agricultural subsidy years households often reduced risk by ploughing less when the rains were late; rates of field sharing also increased in such a year. Many poorer households, often in a queue for oxen or wishing to minimize risk in a questionable year, reduced their input into field preparation and, instead, attempted to obtain a small plot in the fields of another household that had better access to draught power and fenced and cleared fields.

Participation in agricultural production declined almost steadily in the first half of the 1980s. It rose again in the second half of the decade (although not to 1978 levels) but virtually all the households ploughing were doing so at government expense. With the withdrawal of some subsidies in the early 1990s I suspect that participation in agricultural production has dropped considerably from recent levels.

'Explanations' of declining participation in production

Before discussing explanations provided for declining production, it is important to distinguish declining levels of production in the form of output or overall yield from declining levels of production in terms of the capacity to produce amongst individual farmers. These two aspects of production may vary independently of each other but are often conflated in discourse on the effects of drought. In a situation of rain-fed arable production the impact of drought on yields is, to a great degree, a technical matter. In dryland farming measures can be taken to offset the effects of drought but insufficient or poorly timed rain can still reduce or eliminate a crop. However, the capacity to produce refers to series of relationships between producers which regulate their access to productive resources. If fewer farmers have the means to produce then aggregate harvest may be diminished. Of course, this may not necessarily be the case if the means of agricultural production become concentrated in fewer hands – fewer large farmers could produce more food than many small farmers.

In Botswana it is clear that both the aggregate harvest and the numbers of people who could produce were diminished at the height of the drought. The two are related but not in a simple linear or consistent fashion. For the purposes of this article, I am primarily concerned with changing patterns of agricultural production in terms of the capacity to produce.

In the latter half of the 1980s, drought was cited in almost every instance by local people as the explanation for declining participation in agricultural production and for declining yields. Official national discourse also implicated the drought as the cause for both. However, the more I investigated the situation the more this answer seemed problematic and limited in explanatory value. National statistics showing lower than average rainfall levels for several years could not be contested but they only tell a partial story; local and yearly variation was significant. In the village where I worked most households claimed they had not reaped a harvest in years because of the drought, yet some households reported years of bumper harvests. One even proudly showed me the German (battery operated) 'sound system' which was purchased from the profits from selling one year's surplus harvest. In addition, the discrepancy in harvest sizes had grown significantly. In 1978 the largest harvests of the principal crop, sorghum, amounted to approximately 40 bags; by the mid-1980s the largest harvests consisted of approximately 140 bags. The vast majority of households continued to harvest well under 10 bags. Such patterns of 'feast during famine' resonate with Sen's observation that: 'in many famines complaints have been heard that, while famine was raging, food was being *exported* from the famine-stricken country or region' (Sen 1982: 161, emphasis in original).

The cause of these stark discrepancies are complex and cannot be explained by any single factor. Insufficient rainfall played a role but that role has been inconsistent; some households increased production while for most production declined. To understand changes in productive capacity it is necessary to move from technical factors and to examine socioeconomic relations.

It is useful at this point to evoke Sen's method of entitlement analysis. In his book, *Poverty and Famine* (1982), Sen locates the source of scarcity, food in this instance, in the realm of social relations which govern people's socially recognized access (entitlement) to a desired good, rather than in the absolute supply of the good. Entitlement refers to a person's legitimately acknowledged right to a certain thing in a specific society.[21] Entitlement relations are linked, in Sen's words, into chains in which one set of entitlements legitimizes another set. Following Sen's typology, what occurred in Botswana was that state based social security entitlements expanded and averted potential famine. Farmers whose harvests failed or those who were unable to plough could command food which was provided through government programmes. In addition, programmes which offered employment and those which supplied agricultural subsidies provided individuals and households with certain entitlements which formed links in an entitlement chain which could eventually lead to entitlement to food. Sen (1982: 7) notes that in countries such as Britain and the United States, where unemployment is high, there would be widespread starvation 'but for the social security arrangements' which grant people entitlement to a number of basic provisions. In Botswana the expansion of social security entitlements compensated for and, at the same time, masked the countervailing process in the realm of locally based entitlements in which there was a contraction. This will be illustrated in the more detailed discussions of agricultural arrangements below.

In the 1970s the entitlements that operated which permitted such high levels of agricultural production included those which granted access to the three most important resources (outside of rainfall) in Kalahari agricultural production – draught power, land, and labour. Of these three, draught power has always been the most critical for Kalahari residents.

Draught power and property relations

Apart from the relations which existed around reciprocal access to water sources, particularly those for watering livestock, the relations which existed around reciprocal access to draught power were the most significant in binding the community in an ongoing network of rights and obligations during the 1970s. It was access to these resources that people often gave as the reason for moving from one village to another and it was the relations surrounding access to these resources that often formed the basis upon which other sets of relations were built.

Prior to the 1979–80 agricultural season all draught power in the region where I conducted field work was animal, predominantly oxen. Of the households ploughing (90 per cent of Kgalagadi households) in 1978 approximately half owned sufficient oxen to form their own spans of six; the other half utilized one or more of the many formal and informal relations which facilitated access to draught power.[22] At the time none of these relations were commodified in the sense that money changed hands. However, they frequently entailed labour exchange as well as more diffuse exchanges that were not directly tied to the oxen exchange but were part of the wider relations between the parties.

Local conceptions of property facilitate the reciprocal use of resources. Rights to property are simultaneously individuated and dispersed (cf. Galaty 1981: 10): virtually all items of productive property can be identified with an individual but at the same time a larger group maintains rights to or, in other words, has entitlement to the 'family estate' (Gray and Gulliver 1964; cf. Carstens 1983: 60). Communal property claims, to a great extent, rest upon the notion that most property derives ultimately from the ancestors and was created and sustained through the labour of past family members; their descendants can claim the fruits of this labour and the patrimony of their forefathers.[23] In most Kgalagadi communities Kgalagadi residents are linked through dense agnatic and affinal ties so that virtually everyone shares the same ancestors. However, inheritance patterns are such that each item of property, for instance each calf, is designated as belonging to one individual although the wider claims are implicit. The rights to dispose of or to exchange an item of property such as a cow typically rest with that individual (if he or she is not a minor) *and* a small circle of kin.[24] Such decisions are conventionally made through consultation amongst a group of kin which is smaller than the group which can exercise claims to the use value of the property. Thus claims to the use value of a particular item tend to be wider than claims to the exchange value.

The manner in which claims can be articulated and the network of relations they can implicate is illustrated by the following example. In 1986 a relatively poor widow told me she had recently ploughed her fields with her 'father's oxen'. This surprised me, as not only had her father been dead for eight years but, being extremely old (he was over 100 when he died), he had divided his cattle and the authority and responsibility over them amongst his children at least a decade before his death – close to twenty years before the woman ploughed with her 'father's oxen'. Upon further questioning it became clear that the cattle she ploughed with were from her brother's kraal, had the brother's brand, had been his responsibility for decades, and were largely considered the brother's cattle. Yet, at the same time, the deceased father's association with the cattle remained alive, justifying the woman's claim to them. In this case the woman's phrasing (that the cattle belonged to her father) already took her well into the process of exercising her kin-based entitlement to the cattle.

The process of commodification leads to an emphasis, particularly amongst the more commercially oriented households, on the private end of the property continuum. However, such an emphasis does not necessarily or easily result in the relinquishing of the communal end. Similarly, the expansion of commodification also grants greater, but not complete, legitimacy to the rights of the narrow group which can command exchange value over the wider group that can command use value. While one person's claim may rule out the viability of another's, it does not preclude the other claim from being made nor does it negate its legitimacy. Such a system invites redefinition, negotiation, and a constant series of claims and counter-claims amongst kin (see Peters 1992 for a discussion of similar processes operating in the redefinition of rights to boreholes and grazing land).

Depending upon their location in the local political–economic structure, farmers have different and, at times, competing interests and interpretations as to the extent of legitimate claims over property. Whatever their interests, local property conceptions still render a variety of often contradictory claims legitimate and denying them requires a rationalization that is rarely universally accepted.

A locally recognized crisis such as a drought offers an opportunity for a redefinition of the range and priority of property relations and claims. Although people suffer differentially during a drought, they all experience and acknowledge its negative effects. It thus provides one of the most easily accepted rationales and, indeed, to some extent, a licence and moral pretext for the act of denying communal claims, acts which may have been considered anti-social prior to the drought. In some instances communal claims may be difficult to honour because of the constraints imposed by late or limited rains, as in the fairly common example of the man who usually ploughs for his widowed mother's household or unit and his unmarried sister's, in addition to his own. When the rains arrive late, and the ploughing season is truncated, he can only plough for one or two. A decision must be made as to which unit's entitlements are the most attenuated – most likely it would be the unmarried sister, and she and her children would suffer first. However, not all denied claims are as directly justified by climatic conditions: for instance, the example later in this article where certain farmers withdraw their oxen for ploughing altogether and turn to tractors.

In the least formal sorts of draught power sharing arrangements, oxen are shared. Between closely related households exchanges occur frequently; a draught power exchange would simply be embedded within an ongoing exchange cycle and not specifically linked to any direct reciprocal exchange. More formal arrangements for gaining access to draught power include *mafisa*, a system of loan cattle whereby a herder takes one or several of another's cattle, cares for and uses the beasts for an unspecified number of years, receives a female calf as compensation, and then eventually returns the animals. Households short of oxen but not of male labour, often ask a wealthier household to loan them one or more oxen for the ploughing season. Such an arrangement almost always involves the loaning of untrained oxen so that the recipient is responsible for the training, an extremely time consuming and difficult process. After one or two ploughing seasons the trained oxen are returned to the owner. Another common strategy, particularly for households short of male labour, is to help another plough and then take the oxen (and occasionally the plough and labour) to their own fields. This option is the least likely to be utilized in a season of late rains. All these methods are most rational when subsistence livestock production is emphasized over commercial factors.

In a very few years, tractors have replaced oxen as the preferred form of draught power. By 1990, in the village where I worked three households owned tractors and several tractors from outside the community were available for hire. Tractors use little labour and can cover large areas quickly. Since the mid-1980s the government has subsidized tractor rental and, according to a senior government official with whom I spoke in 1990, the net effect of this policy has been that many tractor owners have become very wealthy. One local tractor owner, for instance, ploughed for twenty-three people in addition to himself in 1990, all on government subsidies. However, the use of tractors has its consequences. The long term ecological consequences of tractor use are doubtless significant, but I do not have the data available to make such an assessment. However, the socio-economic consequences of tractor use, particularly in a situation where the vast majority of farmers cannot, on their own, afford tractors or their rental are many.

Agriculture is risky in the Kalahari; ploughing is no guarantee of harvest and crops can scorch at any point in the season. The greater the investment, such as that involved in tractor use, the greater the risk. In the last several years the government has assumed the risks in tractor agriculture by bearing its monetary costs but this is unlikely to be sustainable.

Farmers choose tractors over oxen because ploughing with oxen diminishes their market value. Prime young 'unused' oxen command high prices at the abattoir. By the early 1980s, before government tractor rental subsidies became established, commercially oriented households were already attempting to limit their oxen's use. Some were beginning to rent tractors and many were keeping fewer mature oxen so that when others asked to use their oxen they could truthfully say that they had insufficient to loan. This is an example where wide claims to use value can be precluded by narrower claims to exchange value. Herd composition figures reveal the change. For example, in a sample of moderately large herds – 50 to 150 head (n = 10) – herd composition figures from 1978 indicate a percentage of work oxen in the herds of approximately 13 per cent: in 1985 the percentage for the *same* herds was 6 per cent. Similarly, in 1978 approximately 50 per cent of farming households had sufficient oxen to compose a ploughing span, as opposed to only about 10 per cent by 1985. Because of the drought some cattle had been sold to meet food shortfalls, and pasture and water shortages had weakened the animals. But herd composition figures did not alter in similar fashion in the 1960s when an equally serious and long drought occurred. Moreover, it is important to note that the trend in tractor use and withdrawal of oxen from ploughing *preceded* the drought; the drought hastened the process and provided a perfect rationale for farmers to legitimately dispose of their oxen. The fact that in 1990, three years after the drought officially ended, there were proportionately fewer oxen than at the height of the drought is testimony to the fact that factors other than the drought have contributed to the decline in numbers of draught animals. In 1990 I was told (although I was unable to conduct a systematic investigation) that only three households still kept plough oxen.[25] As I have emphasized, such a reduction in oxen affects the ploughing potential of the community at large and not just the individual households who choose to sell their oxen. If the draught power subsidies are lifted for a number of consecutive years the crisis of reproduction, set in motion before the drought and concealed by the relief measures, may well be revealed in alarming proportions.

Tractors have not been entirely assimilated into the property concepts which surround other productive resources such as livestock. The link between tractors and the patrimonial estate is not absent but is less clear, direct and, to many, less compelling. It can be argued, and some make this argument, that the patrimony was used to obtain the tractor: patrimonial wealth was simply converted into cash to purchase the tractor. Cash also has not been entirely assimilated into existing property relations; its association with individuals is much stronger than other forms of property; and cash, unlike livestock, is not visible. This is again a point where the legitimacy of narrow rights to exchange value take precedence over wider communal rights in an increasingly commodified system. In addition, tractors, for a variety of reasons, do not lend themselves to the same sorts of reciprocal arrangements as cattle do. Factors such as their high initial cost, maintenance and running costs, and the fact they do not reproduce themselves, mean that they have not been integrated into existing entitlement patterns.

The majority of oxen-sharing relations involve some sort of labour exchange and many are premised upon asymmetrical relations.[26] Thus they are not great equalizers. But they are levelling mechanisms which function to keep the gap between rich and poor narrower than otherwise might be the case. They are relations of entitlement[27] which permit semi-independent production on the part of the poorer majority. As such they are consistent with the Kgalagadi cultural logic of self-construction (cf. Alverson 1978: 135; Comaroff 1982: 109; Comaroff and Comaroff 1987), the ethic of doing for oneself. The Kgalagadi contrast this with other forms of economic activity, particularly wage labour, which do little to contribute to an individual's 'valourization' – claims to legitimacy and social standing (Ortner 1989).[28] Such entitlements, unlike state based social security entitlements, contribute to the dignity of the individual as defined within Kgalagadi cultural models; they hold out the possibility that any individual will eventually achieve a position in life from which they can grant entitlement to others, with the political and economic benefits that entails.

Land, labour, and agricultural subsidies

Land and labour, the other critical resources for agricultural production, are closely connected in the Kalahari context, and will therefore be considered together. Kalahari land is not ideal farming land but it is usable once sufficient labour has been invested in it; it is non-commodified; and it is readily available.[29] Any adult should be able to obtain a plot. If, after approximately five years, the plot remains unused, the claim can be challenged. On the other hand, the resources, largely labour, to clear, fence[30] and plough land are in short supply. Although 20 ha. plots are the standard land board allocation, very few farmers actually use the full extent of their fields.

It is necessary here to disaggregate the variable of labour along gender lines because what is in abundance is the labour associated with women – weeding, bird scaring, harvesting, crop preparation. It is the labour associated with men – clearing and fencing fields, training and supervising of oxen – that is in short supply. Thus Batswana have evolved a number of strategies which allow extra household access to prepared fields. Many of these arrangements include access to draught power and the associated labour, and most entail a labour but not cash exchange. In the 1970s most households with well-prepared fields allocated subplots regularly; in a year of late rains the number of people requesting subplots rose dramatically. For example, in 1979 the rains came late; one prominent farmer allocated nine plots within his fields to other people, mostly single mothers.

As with oxen-sharing arrangements these arrangements can be placed on a continuum of egalitarian to hierarchical, with some amounting to simple field sharing (not share-cropping) and others resembling but falling short of wage labour arrangements. Amongst the least egalitarian is a system referred to as *majako* or 'putting in hands'. In the case of *majako*, a subplot is rarely granted. Typically, *majako* involves helping in the fields, particularly with the weeding and harvesting (seldom with the ploughing and planting) and receiving as payment a specified portion of the harvest.

Systems of field sharing have come under pressure. Tractors make it easier for people to plough the full extent of their cleared fields; increasing commercialization has influenced some farmers to sell a portion of their harvest. However, commercialization of the arable sector has not followed at the same speed as that of livestock; for a variety of reasons, arable agriculture is less profitable. In addition, well-endowed farmers are less inclined to dispense with the primary benefit provided to them by field sharing – labour. Agriculture, as it is practised in the Kalahari, is more labour intensive than pastoralism. Thus while the greatest blow to oxen sharing arrangements has come from commercialization, with the state hastening the process, the greatest blow for field sharing arrangements has come directly from the state. A brief description of the most radical of the agricultural subsidies will illustrate this point.

ARAP (Accelerated Rain-fed Arable Programme) was announced in late 1985. It was formulated by high level government fiat and was not widely discussed or evaluated prior to being instituted. In fact, most members of the Ministry of Agriculture first heard about the programme on the radio at the same time as everyone else in the country. This was atypical for Botswana[31] and demonstrated that the government, too, was utilizing the drought as a licence for policy-making methods that would have been unacceptable in a 'non-crisis' situation. ARAP was meant to be a one time, one year, affair but instead lasted for six years, with some farmers employing it repeatedly. Agricultural Demonstrators came to the villages and held meetings to explain ARAP. It sounded too good to be true: farmers would be paid the equivalent of P50 (approximately US$20) per hectare for every hectare they ploughed up to ten hectares, and they would receive additional funds for properly destumping and weeding their fields, for using fertilizer, a planter, and so on. Yet ARAP, like the other schemes, was underutilized.[32] Why?

The agricultural subsidies required users to have fields registered in their own names, thereby excluding all people employing field sharing methods. This attempt to rationalize agricultural organization was, perhaps, well-intentioned but it put the subsidies beyond the grasp of a significant proportion of rural residents. Many of the poorer peasants, particularly female-headed households (which comprise over one-third of rural households [Izzard 1979]) who were most in need of help, were defined as ineligible. There is no Botswana law that prohibits women from having land but customary land tenure arrangements are such that areas of agricultural land are associated with groups of kin related through the male line; the normative pattern is for women to gain access to land as wives. While the reality of so high a number of unmarried mothers renders this ideal unworkable, custom is powerful and it is not always easy for women to apply for land. Furthermore, as I have emphasized, even if women have their own fields, as some do, their capacity to mobilize the resources necessary to use the land, largely male labour, is circumscribed.[33] Thus the various arrangements formulated by rural peoples have permitted women to produce subsistence for their families while at the same time minimizing the costs and difficulties involved – including the ones entailed in too overtly violating social norms.

It is ironic to note that ARAP subsidies actually appear to have promoted the least egalitarian sort of field sharing arrangements, *majako*, while limiting field sharing opportunities which grant greater autonomy to extra-household field users.

'Liberation' or New Patrons for Old

In 1986, at a talk I attended in Botswana's capital, a Ministry of Agriculture official was queried about the fact that his ministry's schemes undermined informal rural based land and draught power sharing arrangements. He responded in a dismissive manner – 'We want to liberate people from all those sorts of relations' – implying that they were nothing more than patron–client relations characteristic of a backward nation. What was apparent in his view, and in both the discourse and policy of development planners, was an attempt to project upon Botswana's rural populace an ideal of individualized nuclear family production units which functioned independently of one another but in conjunction with the state.[34] Horizontal linkage (in terms of inter-household, not necessarily in terms of equality) should be minimized but vertical linkage to the state encouraged, initially in the form of aid but eventually through greater involvement in credit schemes, marketing, taxes, etc. To some extent such methods of attempting to restructure and exert control over the rural areas according to an imposed and idealized western model simply follow in a long line of practices such as census taking, tax collecting, land registration, and aid schemes which have emerged from similar ideologies and have had similar effects.

In instituting the government subsidy schemes (and, indeed, all the drought relief measures) greater infrastructure was established – new agricultural demonstrators were hired, new offices created, communication and transport networks were enhanced, more farmers were registered and fields measured, and so on. Such an increase in the presence of the state and in state knowledge of the economic situation and activities of rural people is surely a double-edged sword. It enables the delivery of expanded services but, at the same time, it facilitates greater control and wielding of power through the state's growing bureaucracy; it is an instrument effect of 'the development apparatus' (cf. Ferguson 1990).

Government policy makers want to 'liberate' rural households from relations which keep them beholden and dependent on others. While in the minds of the planners this might have been a fine sentiment, it failed to acknowledge the realities of poverty, interdependence and debt among the majority of rural dwellers.[35] And it failed to recognize the good sense which many of the existing arrangements make in the context of a largely resource-poor rural populace attempting to minimize the risks inherent in producing in a marginal environment.

It is true that many of the rural based entitlement relations promote dependence, are premised on inequality, and serve to perpetuate that inequality. I do not mean to celebrate or romanticize the situation, only to note that the relations do allow wide scale production. As I have already noted, the majority of them also follow a certain cultural logic whereby the poorer member's productive potential is enhanced, to some extent, by the relation. They are not simply the sort of wage labour arrangements which Batswana claim use up a person and do nothing to contribute to his or her own self development (Comaroff and Comaroff 1987).

What appears to be happening in the present situation is not the ending of dependence, rather it is a shift; dependence on the wealthy is being replaced with dependence on the state, one sort of paternalism for another. Dependence is being bureaucratized and 'modernized'. What is more, the degree of dependence being engendered is deeper than it appears. In spite of Botswana's economic expansion, the alternative at the moment is *not* one where the majority of the population have the opportunity to gain satisfactory employment which could offset rural under-production. As the massive subsidies are withdrawn, what will happen? How will people plough if there are too few oxen? Will they mortgage their harvests to tractor owners? Will the rural areas be more vulnerable to future drought as a result of recent state interventions? Questions such as these are endless and all revolve around the fact that the set of entitlements which allowed people to survive previous droughts and which allowed the majority of people to produce in good years have been undermined to the point that they no longer provide basic social security.

Locally based entitlements have been undermined on two fronts: the state, with its image of 'rationally based individualized peasant production', formulated development policy which refused to acknowledge for subsidy the locally based entitlements that facilitated shared use rights to land. More fundamentally the vitality of reciprocal use entitlements has been undermined as a result of long-term structural changes tied to commodification, privatization, and class formation.

Thus far the state has done well by its relief efforts; it has prevented starvation, received international acclaim, and substantiated its own power base. Indeed, drought relief has been established as the single most important reason people give for supporting the current government (Molutsi 1989: 128). The state has fulfilled some of its patronage duties and its clients have reciprocated with votes. But in averting famine has the state simply delayed the full effects of the 'crisis of social reproduction'? Is the state's position sustainable?

Conclusion

It is not drought that produces a crisis of social reproduction; the drought only hastens it and renders visible what had been, up to a certain point, largely latent processes. Yet at the same time, because of the severity of the 'externally wrought' conditions, the crisis can be blamed on the drought, thus obviating the necessity to examine or acknowledge underlying causes. In this manner drought functions as a means of concealing such systemic realities, both for local farmers and for government bureaucrats. In addition, it is the perfect opportunity for breaking taboos, norms, and moral standards; wealthy cattle owners could deny kin based claims to access to their property with licence and the state could institute policy in an unprecedented manner. Both of these processes were already in motion prior to the drought but, as this analysis has demonstrated, the perceived crisis provided the moral pretext for actors to extend them to previously unacceptable levels.

The observation of social activity around a drought is also revelatory for social scientists both for what it reveals of social process and for the questions it poses for comparative purposes. The Botswana case is instructive. A relatively wealthy benevolent state set about to avert famine, a task at which it succeeded when others around it failed. However, in shifting entitlement to itself the state contributed to a parallel process of disentitlement. Is there necessarily a contradiction between state based and local based entitlements? How can they be compatible?

Similarly, such questions lead one to wonder about the relative advantages and disadvantages of different forms of patronage relations in varying contexts. What is entailed for a poor person in being a client either of the state or of a rich neighbour in terms of getting access to productive resources and other needs of daily life? Wolf, following Pitt-Rivers (1954: 140), speaks of patron–client relations as 'lop-sided friendships' (Wolf 1966: 86) which involve some degree of trust and affect; Scott (1985) emphasizes the importance to the poor of the élite being within the same moral community and thus within moral reach. Is the state within moral reach? Do its agents have the inclination or the capacity to respond to the myriad needs which may emerge in a relationship as multifaceted as locally based patronage relations tend to be? Has sufficient state infrastructure – medical, educational, social security, transport, etc. – been established to replace the need for diffuse patron–client relations for the rural poor? Where and when is one form of patronage preferable to the other and what options do peasants have in choosing patrons? What is involved when types and sources of entitlements transform and how is the liberatory quality of any to be assessed? My analysis does not answer these questions but points to the need to consider them in any analysis of rural change and development.

NOTES

1 Sahlins draws upon the work of R. Firth on Tikopia for this analysis. The specific revelatory crisis was a famine.
2 This is particularly true in the Kalahari region where wealthy cattle ranchers began pursuing commercial production on a greatly expanded scale after a borehole loan scheme in the 1960s and the opening up of livestock marketing co-operatives in the 1970s.
3 Of course, it does not automatically follow that a state would direct its revenues to expanded services, but Botswana did. Botswana is noted for its democratic nature (it has held free and openly contested elections every five years since independence in 1966) and its lack of corruption (see Holm and Molutsi 1989; Molutsi and Holm 1989, 1990).
4 I borrow the concept of 'instrument effects' from J. Ferguson's analysis of development efforts in Lesotho (Ferguson 1990). Drawing on Foucault's analysis of prison systems, Ferguson argues that development projects, in spite of their repeated failures in terms of their intended goals, nonetheless have regular effects that were not necessarily part of their articulated goals. One effect Ferguson identifies for Lesotho is the expansion of bureaucratic state power. The state's presence increases through the administration of the development project, producing new and greater possibilities of exercising power through its offices and infrastructure.
5 For the purposes of this article drought is defined simply as significantly lower than average rainfall levels.
6 Much has been written on this theme; see, for example, Bush (1988); Drèze and Sen (1990); Franke and Chasin (1980); Glantz (1987); Lofchie and Commins (1982); Shipton (1990); Swift (1993).
7 Sen (1982) puts this point most forcefully and eloquently in his analysis of poverty and famine when he argues that food availability decline alone tells one very little about the cause of starvation. Food supply statements describe a thing while those of starvation or its opposite imply a relationship. It is that relationship which must be examined.
8 On these points see, for example, Berry (1984); Vaughan (1987); Watts (1987); Wylie (1989).
9 The drought actually came in two waves, late 1978–79, and 1981 through mid-1987, with two years of better rains in between. However, in the minds of many rural residents, the entire period is thought of as 'the drought'.
10 Relief measures escalated during the latter period of the drought; they included the development of an early warning system for the detection of malnutrition, agro-pastoral subsidies, and an emphasis on labour-based employment generation relief (Drèze 1990: 153–5). In 1985 the equivalent of 26,413 full time jobs were created by the drought relief programme. In the same year the formal sector (excluding drought relief) accounted for 105,000 full time job equivalents (Valentine 1993: 120). According to the 1991 Development Plan approximately US$200 m. was spent on drought relief between 1982 and 1990 (Republic of Botswana 1991). In the mid-1980s 380,000 people (over a third of the population) were *direct* recipients of food relief (Republic of Botswana 1989, emphasis added).
11 By most standards, Botswana is an African economic miracle. The World Bank Development Report (1992: 219) reveals that Botswana had the fastest growing economy in the world between 1965 and 1990; its growth rate averaged 8.4 per cent per annum during this period (South Korea, the second fastest growing economy, averaged a growth rate of 7.1 per cent). As a result, Botswana has gone, according to the United Nations, from being amongst the twenty-five poorest nations in the world at independence in 1966, with a per capita income of approximately US$80 (Colclough and McCarthy 1980), to being one of Africa's wealthiest in the 1990s: in 1991/2 per capita income was approx US$2850 (Republic of Botswana 1993: 5). However, these stunning figures should not disguise the existence of marked discrepancies in wealth and deepening inequalities within the country (Good 1992).
12 For instance, between 1984 and 1988 the number of registered destitutes receiving a monthly government allowance doubled. By 1990 the number of 'official destitutes' equalled the number of employees in the mining sector. In 1990 the mining sector

accounted for 80 per cent of Botswana's export earnings (Good 1992: 90).

13 The rains failed again in 1992 and some were reinstated. Good rains returned in 1993.

14 Twenty months in 1977–9, eight months in 1985–6, and very briefly in 1990.

15 Subsistence oriented and commercial or market oriented production rationales are best viewed as 'ideal types'; neither exist in pure form on the ground but represent poles in a continuum upon which existing situations can be placed.

16 Formal sector employment has grown at a rate of 9 per cent per annum since 1966 while the population has grown at a rate of 3.4 per cent in the same period (Republic of Botswana 1991, vol. 2: 19). Still, unemployment is a serious problem.

17 Only in the 1980s did public education become free.

18 The notion of 'good yield' versus 'poor' is a subjective one and is derived from my informants. A 'good yield' does not imply self-sufficiency in grain; rural Batswana have long been incorporated into a money economy and prefer to vary their diets with purchased carbohydrates. A 'good yield' ideally gives a household enough sorghum for its own defined nutritional needs and a sufficient surplus to sell some, give some as gifts, pay agricultural labourers, brew beer for ritual events, and display hospitality. The largest harvest of a village household in 1978 (the last year before tractors were used and yields began to vary to a much greater degree) consisted of approximately 2940 kg of sorghum, 665 kg of maize, and 105 kg of beans. The household sold approximately 25 per cent of the harvest, used about the same in payment for labourers, gave less away as gifts, and reserved the rest for household consumption.

19 One current domain of struggle concerns the appropriateness of wage labour relations between kin. Kgalagadi have long employed each other but these labour relations were conventionally couched in terms of mutual 'help' in which the employer received labour and the employee received access to productive resources such as land or livestock or was paid in kind, often livestock or agricultural produce. Kgalagadi have also participated in wage labour for a century but employers have almost exclusively been external, faceless businesses. In recent years the commercial élite have attempted to expand local wage labour relations and have met with resistance culminating in a dramatic witchcraft case in the late 1980s (see Solway nd.). Eliminating the veneer of mutual help with internal wage labour relations violates the moral basis of kin relations in which kin are meant to contribute to each other's capacity to 'develop' themselves. A cash wage implies short-term and limited association and discharges and denies the larger and long-term links between the parties. Bohannan's (1959) discussion of spheres of exchange and the varying moral assumptions associated with different mediums of exchange is relevant here.

20 Sarwa households, which comprise less than 20 per cent of households, rarely pursue independent agro-pastoral production.

21 De Waal (1990) has offered a compelling critique of Sen's theory of entitlements and has demonstrated some of its limitations in explaining famine, particularly in instances of violence and where farmers choose to suffer in the short-term in order to preserve assets for the long-term. This critique, while insightful, is not directly relevant to the Botswana case. In spite of de Waal's critique of the theory of entitlements, as a method for elucidating social process during crises such as droughts, entitlement analysis remains an excellent tool (cf. Gasper 1993: 689–90).

22 This region lagged behind the rest of the country in utilizing tractors – although tractors were still relatively uncommon in most of the country (with some important exceptions) at this time. According to 1976–1981 National Development Plan, 90 per cent of Botswana farmers still ploughed with oxen; only 38 per cent of them were able to plough exclusively with their own oxen (Republic of Botswana 1977: 159). This figure contrasts to the 50 per cent figure from the Western Kweneng region and attests to the more equitable distribution of cattle amongst Kgalagadi in this part of the Kalahari.

23 This system, of course, denies the very significant contribution that Sarwa labour has had in building Kgalagadi wealth. A discussion of the ways in which ethnicity operates in

Kgalagadi property ideologies to exclude the Sarwa from claims to productive resources is beyond the scope of this paper. See Hitchcock (1987), Solway (1979, 1987) and Solway and Lee (1990) for further elaboration of this issue.

24 I have deliberately left the categories of small versus wide circles of kin vague in order to emphasize the relational quality of rights and obligations within a field of relations. This also facilitates comparison with other cases along the lines outlined in the paper. Amongst all peoples there are cultural assumptions regarding the differences between kin and the types of access they have to certain resources. Amongst the Kgalagadi these differences are influenced by considerations of the past, the present, and the future with agnatic descent being the most important constant variable. Property devolves agnatically to all descendants and the closer the relation along agnatic lines, generally the stronger the claim. Sisters' claims are compromised by the fact that ideally their children belong to another agnatic unit and the locus of their rights and responsibilities lies elsewhere. However, given the fact that discrete descent groups do not form amongst the Kgalagadi, it is not always practical to observe ideal rules based on agnatic ideology and descent.

25 Milking practices have followed parallel patterns. In the late 1970s all cattle owners milked; milk was widely shared; and fresh and soured milk were dietary staples for all members of the community. Powdered and UHT milk were unavailable locally for purchase. By the mid-1980s the majority of cattle owners had ceased milking, and many that continued milked only a few of their cows. Cans and cartons of powdered and UHT milk lined the shop shelves. The lack of moisture and shortage of pasture reduced the cow's lactation but in previous droughts such dramatic changes in milking patterns had not occurred. Most farmers maintained they had stopped milking because of the drought but few returned to their former milking patterns after the rains and pasture returned. Reduced milking enhances the herds' cash value because calves left to suck uninterrupted reach maturity more quickly and the cows calve more frequently.

26 These inequalities which exist between households or between household units involve varying degrees of patronage and paternalism. In some instances the economic differences are slight and short-term (tied to the household's developmental cycle) but in other instances, the differences are more systemic in nature and households become involved in cycles of dependency.

27 Following Gasper's (1993) refinement of entitlement analysis, one could say that draught power sharing relations constitute both entitlement and endowment. In instances where oxen have been sold, thus precluding their sharing amongst close kin, endowment loss prevails. In the case of more formal relations, where an exchange is acknowledged, the limiting of oxen exchange results in entitlement loss.

28 Attitudes surrounding wage labour or, more accurately, formal sector employment, were undergoing rapid change throughout the 1980s. Previously, most residents' experience with formal sector employment was the South African mines; an individual took a contract with the idea of investing in agro-pastoralism (despite the actual circumstances which dictated how they disposed of their income). Wage labour was a means to an end, the end being more cattle and crops. With the opening up of formal sector opportunities in Botswana, particularly skilled and professional salaried employment, this attitude has undergone some transition, and the acquisition of such employment by a privileged minority brings dignity in its own right. Investment in local agro-pastoralism is no longer the only reason for taking employment. Gulbrandsen's evidence (personal communication) supports this view.

29 The fact that agricultural land is not commodified in Botswana, with the exception of a few freehold blocks, is significant and has had a limiting effect on the degree of socio-economic inequality amongst the Kgalagadi. It has contained, within certain limits, the distinctions between locally based patrons and clients, and has prevented the growth of a landless agricultural labour class (it can be argued that the Sarwa constitute such a class but that is a complex issue beyond the scope of this paper; cf. Wilmsen 1989). It has also contributed to flexibility and diffuseness in

patron–client bonds in that clients are less likely to be bound, through land, to any particular patron. Thus, although the commercialization of agro-pastoralism and its influence on social differentiation in the Kalahari bears strong resemblances to patterns that have existed elsewhere, there are important differences which mute certain common effects (cf. Scott 1972). Inter-ethnic and extra-local patterns of patronage follow different models and have their roots in nineteenth century patterns of settlement, warfare, and trade.

30 Animal (wild and domestic) damage to fields is a serious problem. Building a good thornbush fence is labour intensive and difficult especially when good supplies of thornbush are not available nearby. Wire fences are both labour and capital intensive.

31 For example, ALDEP (Arable Lands Development Programme), while not specifically tied to drought relief, is an agricultural subsidy programme which was introduced in 1981. Its introduction followed years of planning and discussion and a two year pilot phase.

32 Over the years they were more widely used, but still disproportionately by the wealthier farmers. By the end of the scheme the terms had been changed somewhat: for example, instead of P50 for ploughing each hectare up to 10ha., P70 was granted for each hectare up to 7ha. Interestingly, in spite of ARAP's underutilization by poorer households, 'ARAP's share of total nonrecurrent drought expenditures increased from 25 per cent in 1985–86 to roughly 48 per cent in 1988–89' (Valentine 1993: 117).

33 The labour problem is exacerbated with tractor use because the fields must be more thoroughly cleared. Whereas oxen can move around large stumps, tractors cannot and ploughing in an improperly cleared field can damage a tractor. Once a field is cleared and fenced the labour problem becomes less critical although the cash expenditure remains a significant limiting factor.

34 The state is not a monolithic entity lacking in inconsistencies between its various agencies and representatives but an analysis of Botswana in these terms is beyond the scope of this paper.

35 See Hill (1986: 83–94) for a cogent critique of the uncritical acceptance of debt as a negative phenomenon.

REFERENCES

Alverson, H. 1978. *Mind in the Heart of Darkness*. New Haven: Yale University Press.

Berry, S. 1984. 'The Food Crisis and Agrarian Change in Africa'. *African Studies Review* 27 (2): 59–112.

Bohannan, P. 1959. 'The Impact of Money on an African Subsistence Economy'. *The Journal of Economic History* 19(4): 491–503.

Bonte, P. 1975. 'Cattle for God: An Attempt at a Marxist Analysis of the Religion of East African Herdsmen'. *Social Compass* 22(3–4): 381–96.

Bonte, P. 1981. 'Ecological and Economic Factors in the Determination of Pastoral Specialization,' in *Change and Development in Nomadic and Pastoral Societies*, J. Galaty and P. Saltzman, eds., pp. 33–59. Leiden: Brill.

Botswana, Republic of. 1977. *National Development Plan*. Gaborone: Government Printer.

Botswana, Republic of. 1989. *Budget Speech*. Gaborone: Government Printer.

Botswana, Republic of. 1991. *National Development Plan*. Gaborone: Government Printer.

Botswana, Republic of. 1993. *Budget Speech*. Gaborone: Government Printer.

Bush, R. 1988. 'Hunger in Sudan: The Case of Darfur'. *African Affairs* 87(346): 5–24.

Carstens, P. 1983. 'The Inheritance of Private Property Among the Nama of Southern Africa Reconsidered'. *Africa* 53(2): 58–70.

Colclough, C. and S. McCarthy. 1980. *The Political Economy of Botswana*. Oxford: Oxford University Press.

Comaroff, J. 1985. *Body of Power, Spirit of Resistance*. Chicago: University of Chicago Press.

Comaroff, J.L. 1982. 'Class and Culture in a Peasant Economy: The Transformation of Land Tenure in Barolong', in *Land Reform in the Making*, R. Werbner, ed., pp. 85–113. London: Rex Collings.

Comaroff, J.L. and J. Comaroff. 1987. 'The Madman and the Migrant: Work and Labor in the Historical Consciousness of a South African People'. *American Ethnologist* 14(2): 191–209.

Drèze, J. 1990. 'Famine Prevention in Africa: Some Experiences and Lessons', in *The Political Economy of Hunger, Vol. 2*, J. Drèze and A. Sen, eds., pp. 123–72. Oxford: Clarendon Press.

Drèze, J. and A. Sen. 1990. *The Political Economy of Hunger, Vol. 2*. Oxford: Clarendon Press.

Ferguson, J. 1990. *The Anti-Politics Machine: 'Development', Depoliticization and Bureaucratic Power in Lesotho*. Cambridge: Cambridge University Press.

Franke, R. and B. Chasin.1980. *Seeds of Famine*. Montclair, NJ: Allanheld, Osmun and Company.

Galaty, J. 1981. 'Introduction: Nomadic Pastoralists and Social Change — Processes and Perspectives', in *Change and Development in Nomadic and Pastoral Societies*. J. Galaty and P. Saltzman, eds., pp. 4–26. Leiden: Brill.

Gasper, D. 1993. 'Entitlement Analysis: Relating Concepts and Contexts'. *Development and Change* 24(4): 679–718.

Glantz, M. 1987. *Drought and Hunger in Africa*. Cambridge: Cambridge University Press.

Good, K. 1992. 'Interpreting the Exceptionality of Botswana'. *Journal of Modern African Studies* 30(1): 69–95.

Gray, R. and P. Gulliver. 1964. *The Family Estate in Africa*. London: Routledge.

Hay, R. 1988. 'Famine Incomes and Employment: Has Botswana Anything to Teach Africa?'. *World Development* 16(9): 1113–25.

Hitchcock, R. 1987. 'Socio-economic Change among the Basarwa in Botswana: An Ethno-historical Analysis'. *Ethnohistory* 34(3): 219–55.

Hill, P. 1986. *Development Economics on Trial*. Cambridge: Cambridge University Press.

Hobsbawm, E.J. and T. Ranger. 1983. *The Invention of Tradition*. Cambridge: Cambridge University Press.

Holm, J. and P. Molutsi. 1989. *Democracy in Botswana*. Gaborone: Macmillan.

Holm J. and R. Morgan. 1985. 'Coping with Drought in Botswana: an African Success'. *The Journal of Modern African Studies* 23(3): 463–82.

Izzard, W. 1979. 'Rural-Urban Migration of Women in Botswana'. Gaborone: Central Statistics Office.

Kuper, A. 1966. 'Kinship and Politics in a Kgalagari Village', unpublished PhD dissertation, Cambridge University.

Kuper, A. 1970a. *Kalahari Village Politics*. Cambridge: Cambridge University Press.

Kuper, A. 1970b. 'The Kgalagari and the Jural Consequences of Marriage'. *Man* 5(3): 466–82.

Lofchie, M. and S. Commins. 1982. 'Food Deficits and Agricultural Policies in Tropical Africa'. *Journal of Modern African Studies* 20(1): 1–25.

Molutsi, P. 1989. 'Whose Interests do Botswana's Politicians Represent?', in *Democracy in Botswana*, J. Holm and P. Molutsi, eds., pp. 120–32. Gaborone: Macmillan.

Molutsi, P. and J. Holm. 1989. 'Introduction', in *Democracy in Botswana*, J. Holm and P. Molutsi, eds., pp. 1–7. Gaborone: Macmillan.

Molutsi, P. and J. Holm. 1990. 'Developing Democracy When Civil Society is Weak: The Case of Botswana'. *African Affairs* 89(356): 323–40.

Ortner, S. 1989. *High Religion*. Princeton: Princeton University Press.

Peters, P. 1984. 'Struggles Over Water, Struggles Over Meaning: Cattle, Water and the State in Botswana'. *Africa* 54(3): 29–49.

Peters, P. 1992. 'Manoeuvres and Debates in the Interpretation of Land Rights in Botswana'. *Africa* 6(3): 413–34.

Pitt-Rivers, J. 1954. *The People of the Sierra*. London: Weidenfeld and Nicolson.

Sahlins, M. 1972. *Stone Age Economics*. Chicago: Aldine.

Scott, J. 1972. 'The Erosion of Patron-Client Bonds and Social Change in Rural Southeast Asia'. *Journal of Asian Studies* 32(1): 5–37.

Scott, J. 1985. *Weapons of the Weak*. New Haven: Yale University Press.

Sen, A. 1982. *Poverty and Famines: An Essay on Entitlement and Deprivation*. Oxford: Clarendon Press.

Shipton, P. 1990. 'African Famines and Food Security: Anthropological Perspectives'. *Annual Review of Anthropology* 19: 353–94.

Solway, J. 1979. 'People, Cattle, and Drought'. Rural Sociology Unit Report. Gaborone: Government Printer.

Solway, J. 1987. 'Commercialization and Social Differentiation in a Kalahari Village, Botswana'. Unpublished PhD dissertation, University of Toronto.

Solway, J. nd. 'Taking Stock in the Kalahari: Class, Culture, and Resistance in Southern Africa'. Unpublished manuscript.

Solway, J. and R.B. Lee. 1990. 'Foragers, Genuine and Spurious: Situating the Kalahari San in History'. *Current Anthropology* 31(2): 109–46.

Swift, J. 1993. 'Understanding and Preventing Famine and Famine Mortality'. *IDS Bulletin* 24(4): 1–16.

Valentine, T. 1993. 'Drought, Transfer Entitlements, and Income Distribution: The Botswana Experience'. *World Development* 21(1): 109–26.

Vaughan, M. 1987. *The Story of an African Famine*. Cambridge: Cambridge University Press.

de Waal, A. 1990. 'A Re-assessment of Entitlement Theory in the Light of Recent Famines in Africa'. *Development and Change* 21(3): 469–90.

Watts, M. 1987. 'Drought, Environment and Food Security: Some Reflections on Peasants, Pastoralists and Commoditization in Dryland West Africa', in *Drought and Hunger in Africa*, M. Glantz, ed., pp. 171–211. Cambridge: Cambridge University Press.

Wilmsen, E. 1989. *Land Filled with Flies: A Political Economy of the Kalahari*. Chicago: University of Chicago Press.

Wolf, E. 1966, *Peasants*. Englewood Cliffs, NJ: Prentice-Hall.

World Bank, 1992, *World Development Report 1992: Development and the Environment*. New York: Oxford University Press.

Wylie, D. 1989, 'The Changing Face of Hunger in Southern Africa', *Past and Present* 122: 159–99.

Part III
Vulnerability and Control

Culture and Control of Climate

13

Rain-Shrines of the Plateau Tonga of Northern Rhodesia

Elizabeth Colson

The Social Structure

This is a preliminary report on the social and political significance of the rain-shrines as an integrating force in Tonga society.[1] In a sense it is a misnomer to refer to them as rain-shrines, for they are also appealed to on any occasion of general community disaster, such as epidemics or cattle plagues, but to the Tonga themselves the dominant aspect of the shrines is their efficacy in ensuring the proper rainfall.

The Tonga inhabit the railway belt on the Northern Rhodesian plateau. They are affiliated linguistically and culturally with the Ila of the Kafue plain, who have been described by Smith and Dale (1920). Although the Tonga probably number over 80,000 persons, they are little known except for an occasional short note in mission journals or government reports.

On the whole the Tonga might be defined as culturally a have-not group. They have never had an organized state. They were unwarlike and had neither regimental organizations nor armies. They were and are equally lacking in an age-grade set-up, secret societies, and social stratification of all kinds. The Tonga would not even attract those fascinated by the intricate rules of lineage organization, for while they have clans and smaller matrilineal kin-groups, they have them in a characteristically unorganized fashion which leaves the investigator with a baffled, frustrated desire to rearrange their social structure into some more ordered system. It is only in the rain-rituals and their associated shrines that the Tonga show a half-hearted grouping towards the establishment of a larger community than that which existed in the village or in the ties of kinship. I think it important to study this nexus, not only because of the numerical importance of the Tonga themselves but because theirs is probably an extreme variant of a general Central Bantu culture type which seems characteristic of much of Northern Rhodesia, except where the more highly developed systems of the Congo or South Africa have impinged upon it. I should say that it is a type of society based on shifting cultivation with unlimited land and so little variation in land values in terms of the culture that no given spot possesses particular attractions.

From the earliest times of which we have any knowledge the Tonga have been settled in scattered villages throughout this countryside. A man might settle where he would, with his matrilineal kin, with his father's matrilineal

The Anthropology of Climate Change: An Historical Reader, First Edition. Edited by Michael R. Dove.

group, with his wife's relatives, with a non-related fellow clansman, with a stranger, or by himself. Hence neither clan nor kin-group could be localized. The clans have no internal structure, no leaders, and no common rituals. Their only attributes, besides the bearing of a common clan name, seem to be the prohibition of marriage within the clan and the obligation of hospitality to all fellow clansmen. The kin-group was the unit which exacted vengeance, paid the fines for its members' misdeeds, functioned in inheritance, and gave assistance towards paying bridewealth or in meeting other emergencies. This kin-group is a small unit of matrilineally related persons, plus their former slaves, plus the matrilineal descendants of their female slaves. It differs from the clan in that all its members feel themselves to be related, although in some cases the genealogical connections have been forgotten, while the clan is composed of unrelated groups. Because the members of the kin-group lived scattered wherever they pleased to settle the kin-group was not a corporate entity in daily life. Equally it tended always to lose its members and to shrink back to a numerically tiny unit of perhaps two to three generations' span above its older members. Many shrank even further, as the dispersed members and their descendants moved farther and farther apart until finally all memory of them vanished. The diminished cells of kin began a new proliferation which in turn would give rise to dispersal and the vanishing of kin solidarity with its attendant responsibilities. Very probably a kin-group rarely numbered more than a hundred adult individuals at any one time, some of whom would be the attached slaves or their descendants, who within a generation or so were indistinguishable from the lineal descendants of the group. Not all of these people would live closely enough together for them to act effectively as a unit. The descendants of the slaves shared in the common right to move where they would and to live with whom they would, so that not even the slaves owned within a kin-group formed a stable core lasting more than a generation.

The village in which a man lived thus included some members who recognized no responsibility for his actions and for whom in turn he felt no responsibility. His loyalties to the village head were tenuous ones, based perhaps on gratitude for favours received, perhaps on kinship, perhaps on the ties which bound a slave to his owner. Only over his slaves did a village leader have the coercive authority which would force them to continue to live with him. When the head of a village, or indeed any other family man who still had effective kin, died, a successor was chosen for him, usually from among his matrilineal kin. It was quite possible, however, that the successor would come from among the slaves or even from the sons. The declared principle was always, 'We choose the one who will keep the people best.' The heir did not necessarily succeed to any position of ascendancy over village or kin. His adherents might split into small groups, move off on their own, or go off to seek other kin more congenial to them.

The kin-group theoretically was held together by the recognition of its common ancestral spirits, who had power to affect them all and only them. In practice this recognition had only a limited power to stabilize their relationships. Any adult woman, and any man who had a wife capable of brewing beer, could approach the ancestors; it was not canalized through one representative of the whole group. Moreover, the power of the ancestors was effective no matter where a man might settle. At the present time the ancestors follow the labour migrants to the Union of South Africa, and the spirits of those who die there have no difficulty in returning to demand the attention of the stay-at-homes.

Since the village is composed of unrelated elements, the ancestral spirits of any one kin-group cannot affect all of its members. I have been unable to discover traces of ritual which might symbolize the village as a corporate unit.

The Rain-Shrines

Into this anarchy, some semblance of order is infused by the rain-rituals, which effectively organize small groups of villages for corporate activity, and which are able to impose sanctions on offences affecting its organization. The districts over which a particular rain-shrine or group of shrines held sway seem usually to have contained only a few square miles and four or five villages. One or two shrines, such as that of Monze, had a wide reputation, and probably drew people from a much larger area. But there was no hierarchy of shrines organizing the various separate cult districts into a country-wide system.

On the two or three days of the year when the rain-rituals were being enacted, a general district peace was imposed in the name of the shrine, which overrode the customary rights of the kin-groups to exact compensation for offences against their members' persons or property. In some districts this peace was instituted through a ritual licence at this period, and the district refused to recognize any offence save murder itself as culpable. In other districts the customary code of behaviour remained in effect through these days, but offences against it were subject to a fine paid to the shrine or to the community as a whole through its elders, instead of to the injured person or his kin. At the Monze shrine the ground is covered with hoe-blades paid by those who chose the ritual period to fight or commit adultery or take their neighbour's goods. To the north-west, adherents to shrines in the Mwanacingwala area were fined a fowl, a beast, or tobacco, which was then distributed among the elders of the area. At the present time the ritual peace has disappeared under the general peace maintained by the Administration. Breaches of the peace or civil cases arising during the ritual period now find their way into the Native Authority Courts, where they are treated on the same plane with cases arising at any other period of the year.

It has also become impossible for the elders to fine a man who fails to attend the ritual, though once the ritual was coercive upon all who lived within the cult area. Thus, in former days a man knew that at least once a year he must co-operate with his neighbours in a common ritual to ensure a benefit equally desired by all and which was for the common good. If feuds split the neighbouring communities so badly that it was impossible for them to co-operate in the ritual, the dissident sections could move into another district where they could co-operate, or the whole area might expect the visitation of drought, famine, epidemic, or other pestilence. The shrines were therefore effective in keeping the internal differences in the communities which they served within reasonable bounds. Occasionally during the rest of the year the community of the rain-shrine might be called into existence and its integrity reaffirmed. Disrespect towards the shrines even on non-ritual occasions might ensure general disaster for the community, unless the offenders were punished and a ritual cleansing performed. If it is discovered that someone has cut wood in the immediate vicinity of a shrine he is ordered to pay a black chicken. This is killed and eaten by the assembled elders of the shrine community. These fines are still exacted, although I have never heard of such a case coming into the Native Authority Courts.

In the Chona country, on the edge of the Zambezi escarpment, the highest hill in the area is regarded as a shrine. It is strictly forbidden for anyone to gather roots or wood in its vicinity, or to burn the grass upon it before the head of the rain-ritual announces that there is to be a communal hunt upon its slopes. In 1946 a foolhardy youth burned a portion of the hill by accident and then came to ask permission to burn the rest. He chose a day when the headmen of the area were assembled for another purpose. They screamed their rage at his impudence, put him in handcuffs, and threatened to make all the men in the district who were of his approximate age pay chickens for a general purifying feast. Finally, however, they agreed to fine only the offender. They maintained wholeheartedly their right to punish him. His act had endangered the whole community, and especially themselves as its leaders. If no retribution were made to the shrine, they could only expect when the rains finally came that their huts would be destroyed by lightning. During the same year others offended against this shrine. Finally the headman of the nearest village decided to move to a new site to escape the drain on his chickens and goats caused by the trespass of his people. The other people of the district were beginning to look at them askance and murmur that they were responsible for the prolonged drought which was endangering the crops. In the

Mwanacingwala area the same drought was attributed by the chief to a sacrilege which he discovered—a pile of firewood at the base of a very large and very hollow fig-tree which is associated with the shrine of Luanga. He marched into the nearest village, which happened to have immigrated into the area a year or two before, asked its inhabitants about the matter, and told them that if they offended again they must pay a beast to the district. However, on this occasion they were let off with a warning, for they apologized profusely and said they had not realized that this particular tree was sacred though they knew and carefully respected three other trees which had been pointed out to them.

This same shrine of Luanga figures in the last big feud in this country. About 1910 some men found women cutting thatching-grass close to the sacred spot. The women came from a village only about three miles from Luanga, but belonging to the cult district of Ciboya. When the Luanga men told them they were trespassing at a sacred spot the women cursed them heartily in terms which left no doubt as to their opinion of Luanga and its people. The leader of Luanga went to demand an apology from his peer at Ciboya, who promptly spat in his face. The next step was an armed raid by the Luanga people in which several men were killed and the followers of Ciboya were driven from their homes. European intervention ended the hostilities, and the Ciboya people drifted back to their former sites. It is clear from the last two cases that those who live outside the cult area owe the shrine no reverence as a holy place for all Tonga, though any irreverence will be actively resented by the adherents to the particular place. It is regarded as an insult to the whole cult community. Once people move within the cult area they are expected to observe the taboos which surround the particular shrine of the place.

Description of the Shrines

Since the largest social group ever mobilized by the Tonga is never of imposing size, we cannot expect that the shrines which symbolize the groups will be pretentious structures. And they are not. We can classify them into two general types, though both are referred to by the Tonga under the one term *malende*. One type has already been mentioned briefly. This consists in natural objects which have become sacred, though there is usually nothing to explain to the untutored eye why they have been selected out of the general landscape. Large hollow fig-trees are very apt to be sacred, and I know of one place where the sacredness has now been transferred to the hole in the ground left when such a tree was blown down. The Tonga of the area said, 'We can't lose our rain-shrine like this', and with no more ado continued to visit the hole. Such hollow trees are understandably sacred, for they are regarded as being dwelling-places of the spirits responsible for the rain. When the little huts at the man-made shrines lose their roofs, through wind or rot or the depredations of the browsing cattle, the spirits take refuge from the rain in the tree-hollows. It is more difficult to discover the motive for sanctifying other spots, such as the hill in Chona country, or another shrine in the same area, a spring which trickles out of the broken face of a rock above a watercourse which holds a small pool of water through the dryest spells of the dryest years. In Ufwenuka area also one of the rain-shrines is a spring. There the ritual is said to consist of throwing a tortoise into the spring's depths. I have not been able to discover that these particular shrines are connected specifically with any spirit, or in what the mechanism of their power consists. In the Chona country the same spirits are supplicated at both these shrines and at the artificial ones of the area, and the people claim they visit the spots at the present time because the first Chona visited them when he was alive and they are simply following the procedure he taught them to use in obtaining rain. But they do not seem to hold that the spirit of the first Chona dwells at either of these spots.

The other general type of shrine is man-made, and consists of small structures called *kaanda* (plur. *twaanda*), which literally means 'small hut'. They are not pretentious and are all built on the same general pattern—a circle of supports capped with the mushroom-like thatched roof which tops all structures in this area. In the Chona area I have seen two which have supports of upright slabs of stone. The later of these could not have been built after about 1880. The others I have seen are built with poles. Slender twigs are used in the Mwanacingwala area, where the shrines are tiny affairs, perhaps only thirty inches high when complete with thatched roof, and the shrines may vanish completely from one year to the next under the trampling of the cattle. Elsewhere substantial poles are used, which last from year to year, and some of the huts are large enough to admit an adult through the doorway, though no human being does enter once the shrine is built. Since the supports are set well apart, the contents of the shrine are always in clear view and there is little protection from the weather. However, the only permanent furniture are the pots, placed upside down near the doorway. These are the ordinary black pots of the area, indistinguishable from those used within the villages at the present time, though some of them may be of pre-European date. Two such pots should be found at each shrine, one for beer and one for the food for the communion feast which is a major portion of the ritual. Sometimes there are many more. They remain always at the site. Should they be broken, the ritual leader must seek new ones. My information indicates that he may acquire them in any way he likes, and they become ritualized only when placed within the shrine. They are found, incidentally, only at the hut-shrines. Nothing is placed at the natural ones. Another feature of most hut-shrines is the circling cluster of trees, planted when the shrine was first made. This is all. Through much of the year the shrines lie quiet and undisturbed, and their paths are overgrown.

The Role of the Spirits

The hut-shrines are connected with definite spirits who are thought to have power over the rain through their intervention with Leza, the god who controls all things. Their origin, however, is diverse. Some shrines are said to be those of former *ulanyika*, or leaders. Formerly, when a man moved into an unoccupied section of the country he was regarded as having authority over those who followed him into the area. Just what his authority amounted to it is difficult to say, but probably it amounted to little more than an expectation that he would be listened to with respect. His status was essentially that of a first among equals, rather than that of a chief in the usual sense of the word. When he died, his kin and his neighbours might decide to honour him by building a shrine at his grave, but this did not necessarily become a cult-centre, and it might soon decay and be forgotten if the village moved away. If, however, the area suffered drought or other disaster within a few years of the leader's death, a diviner or prophet might announce that the spirit of the dead man was angry because the people had forgotten him though he had looked after the community during his lifetime. The people would then attempt to rectify their mistake and would either rebuild the shrine or begin a new one for him. Thereafter they might carry out the rites at the site each year before the beginning of the rains, simply to be on the safe side, or they might return to it only in another period of emergency.

Other shrines were initiated by people whom we may call prophets, or rain-makers, though there seems to be no specific term for them in Tonga. They are subject to possession by spirits. Through the prophets, the spirits make demands on the people, lecture them for their misdeeds, and demand the institution of new rituals or the better conduct of the old. Usually when the rain-maker becomes possessed for the first time he calls the people together and tells them to build a shrine for his spirit, which may be that of some important former member of the community, or sometimes of someone completely foreign to the area. Such spirits are called *basangu* and are regarded as distinct from the *mizimu* or ancestral spirits, though there are cases where the same spirits are addressed as *basangu* at the rain-shrines and as *mizimu* at the private household rituals. This is true if the *musangu* is that of a former member of the community. Such a man is honoured as a *muzimu* by his kin-group, who approach him, as they do any other ancestor, in private rites. He is honoured as a *musangu* by the community, which includes his kin-group, but as a *musangu* he is concerned with community affairs and not with the narrow sphere of individual or private matters. While the ancestral spirits affect only their own kin, the *basangu* can possess anyone they choose to enter without regard for the proprieties of kinship regulations, and they can afflict communities with drought, cattle epidemics, disease of epidemic proportions, or any other disaster of a general nature. I have never heard of *basangu* sending sickness to only one individual, or crop failures to the fields of only one person. For such misfortunes one accuses either the ancestors or the witchcraft of one's enemies. The only exception is that the *basangu* punish with illness individuals who violate their shrines and do not make restitution.

Aside from the regulation of the rain-ritual, the *basangu* seem chiefly concerned with combating the cultural changes which European contact and other influences are producing in Tonga country. In 1946 one announced that among the sins of the people to which the drought was due was the failure to follow the customs used by the *basangu* when they were living individuals, and it cited specifically the building of kimberley brick houses instead of the old-style mud-and-pole type. The Tonga, however, are not slavishly obedient to the whims of the *basangu*. This statement was met by jeers and the declaration that the times had changed. Another *musangu* announced that the previous rain-ceremonies had been ineffective because the ritual leader had bought the beast he sacrificed instead of killing one from his own herd. This was listened to more respectfully.

When the rain-maker is first possessed by his spirit, if it is a foreign one or one which does not already have a shrine within the immediate vicinity, he calls upon the people to build a hut-shrine for it and to participate in ritual at the spot. This will be carried out annually. After a few years it becomes institutionalized if it seems to be effective in producing rain. When the rain-maker dies the shrine continues to be visited under the leadership of some member of the rain-maker's kin-group. Another shrine may be built at the grave of the rain-maker, for he too is now considered to have become a *musangu* and to have power over the area. Such a figure was the Monze whom Livingstone described as 'the chief of all the Batonga'. Occasionally the original shrine of the *basangu* and the shrine of the dead rain-maker may coalesce, and both will be appealed to at the same place and at the same time.

A few other figures have been honoured in this manner. The first Chona is said to have had powerful magic, although he was never possessed by any spirit. He seems to have been able to dominate a fairly large section of the escarpment country. His shrine is still visited, though the cult community has now shrunk to three villages.

The hut-shrines are thus dedicated to two different types of spirits: the spirits of former leaders in the area, some of whom were rain-makers and some of whom had a secular and largely personal influence, and the foreign spirits who announced themselves through the rain-makers. All are called, in their public aspect, *basangu*. In some cases rain-makers in different districts have been possessed by the same *musangu*, but this imposes no connection between the resulting shrines.

Once the original figure is dead all such cults pass into the hands of his kin-group. The shrines of the indigenous leaders are in the hands of their own kin; the shrines of the foreign spirits are in the hands of the kin of the rain-maker whom they possessed. While the shrine affects the entire local community, the kin are regarded as the proper medium through which appeals may be made to the indwelling spirit. As officials of the cult they are the living representatives of the community itself, but their power and responsibility begin and end in the ritual sphere. They choose one from their group to decide when the rituals will be held and to direct the activities of the ritual period. Such a person may himself be possessed by *basangu*, but often he is an ordinary person without supernatural assistance. If he wishes to do so he may delegate his work to some more energetic person. His role is that of director and chief participant, but he can do nothing by himself. As far as I know, he never visits the shrines by himself, nor on any save the public occasions. Even then the *basangu* are only invoked for the public good.

The Ritual

The ritual itself is simple, and may be stripped even beyond the generally recognized proprieties of the occasion. I shall describe the rites as I saw them performed in the Chona area, as they followed the general pattern described to me by informants. However, each cult area slightly varies the basic pattern of the rites. Shortly before the first planting-rains of the year began in earnest, the man in charge of the ritual announced that it was time to prepare the beer for the *luinde*, as the ceremony is called. When the beer was ready the inhabitants of each village spent the eve of the ceremony dancing and singing rain-songs. They moved from house to house, calling out pleas to the *basangu* to send the rain. This is supposed to continue until the rain comes, and on this occasion the rain came conveniently about eleven in the evening so that we could get some sleep to prepare for the activities of the next day. The following morning the people set out equipped with axes, hoes, a chicken, and some meal. The women hoed the site to clear it of grass, while the men repaired the thatched roof which had fallen into disrepair during the previous year. The chicken was killed by striking its neck against the doorway of the hut-shrine, and it was roasted while the meal was being cooked. Then all joined in a communion meal, in which each received only a mouthful. Meantime the leader addressed the spirit at the shrine: 'Send us rain and good crops and health. We have done all the things you told us to do. We are still living in the way you showed us. We have not forgotten what you told us. We have not forgotten you. Send us rain. Help us.' When the meal ended the people returned to their villages. Since in this neighbourhood there are several shrines bound together in a common cult, each village went to the shrine nearest to it to perform this part of the ceremony.

When the people returned to their villages the dancing and singing continued intermittently through the day and evening. Early the following morning, when the light was first coming, young men carried a pot of beer to each shrine and left it there. Later in the morning the people went through the village again, singing the rain-songs and led by the drums. Then in a body they went to the village of the chief officiant and circled this village in the same manner. Each village in the area had sent beer to this central point, and here they spent the morning drinking the beer and eating the groundnuts which they had levied at each hut which had failed to make beer for the occasion. About noon the procession re-formed. It now contained people from all the villages in the district, and it moved from one shrine to the next in a specified order and manner, pausing at one point to circle a barely discernible mound, at another so that all participants could rub themselves with a white clay from the river bank. At each shrine they divided the beer, and as they drank they addressed the shrine with appeals for rain in formulae similar to that given above. An offering of beer was poured over the doorway of the hut, the people saluted the spirits by clapping, and then they danced around the shrine to show the spirits that they were happy. Such dances were either solos or by twos and threes, and were pantomimes explained by the words of the accompanying songs. Some are intelligible to us as sympathetic magic, as when a woman sings and dances that she is delighted with her fine harvest. Others are less obvious. I am frankly puzzled as to why some are considered appropriate to the occasion. In one the dancer limped sadly about the shrine with a bundle on her head and sang that she had a case which she had taken to the District Officer and that she was heartily weary because all the district officials kept sending her from one to another. Other dances were obscene. After perhaps half an hour at each shrine the participants rushed down to some pool to bathe, and then returned to their villages to drink the remainder of the ceremonial beer.

This ended the ceremony. But if the rain fails to come, or does not last, then the shrines are visited again and again. At Chona the people went five times in 1946–7, although they did not always visit the same shrines. In desperation they consulted oracles. A woman was possessed by the spirit of the first Chona and ordered an innovation in the rite, the purchase of a black cloth to be placed on his shrine. On another visit a beast was killed. Finally, on the last appeal to the *basangu*, when no rain-clouds had appeared, one of the women turned just as they left the shrine of Chona I and announced: 'Look, your children will all starve and we will all die. You don't care for us. Now we are through with you.'

Participation in the rites is general. All members of the community are allowed to attend. At the Chona shrine the leaders urged all the people and not just the kin-group which controls the shrines to dance and pray for rain. They said: 'The rain falls on your fields as well as on ours. You must all dance.'

In this area a widening of participation is also obtained through a sharing of the control of the shrines with the kin-group to which the father of the person to whom the shrine is dedicated belonged. Thus for the full ceremony at all five shrines of the district six kin-groups must co-operate. This custom, however, does not appear to be typical of all Tonga country. But everywhere, whether the control of the rites is vested in one kin-group or several, the whole community should participate in carrying them out—men and women, people of all ages. Today elders complain that the rites are failing in efficacy because the missions have ordered schoolchildren and Christians not to attend.

One shrine is atypical of the area—that of Luanga in the north-west. Here only men and boys visit the shrine itself. Elsewhere there is no exclusion of women from any part in the ritual. Women as well as men have been possessed by *basangu* and become rain-makers, and some have shrines dedicated to them.

If you press the Tonga for an explanation as to why they perform the ritual thus, they say that they are carrying out the instructions of a particular *musangu*, or a particular rain-maker. They do not seek to draw any logical connection between particular acts and the end they desire to induce. It is enough for them that they are following traditional methods. If they do as their predecessors did at this spot, then they can expect the desired result. Yet strangely enough, none of the shrines is regarded as deriving its sanctity from great antiquity, and myths concerning the first establishment of rain-ceremonies in the land are conspicuously absent. This is characteristically Tonga. They are a non-historical people in every aspect of life, and never more so than when they deal with the symbols of their ephemeral communities. None of the shrines I have traced can be said to have been in existence prior to 1850, although we do know that the Monze who apparently instituted one particular rain-cult was alive in 1855. The shrines which seem more important today are all associated with the cults of men and women remembered by those now living, or who lived only a generation earlier.

Since it is improbable that the rain-ritual and the associated shrines are recent cultural innovations, we must assume that shrines as well as their originators are mortal, and that in each generation some disappear while new ones are created. One can only guess at the causes of their extinction: their adherents may have been dispersed through epidemics or through the wars which devastated this area during the nineteenth century. Or the casual shift of villages over the land as fields became exhausted may have scattered the original village members to such an extent that they no longer represented a local community, and the shrines were too distant to be visited. Since rainfall is extremely localized, those who move some distance might well have good reason to doubt the efficacy of the rituals they had formerly practised. It may be noted that only in one area do the shrines tie their adherents to their immediate vicinity. The kin-group which controls the Chona cult assured me that if the leaders moved even four miles from the shrines the spirits would punish them with sickness. Elsewhere people laughed at the idea and said they could move where they would, though they should return each year to celebrate the ceremony. Thus in general the shrines imposed their peace on those who lived within the vicinity but had no effect in building up a permanent group tied to the area.

In other cases shrines probably disappeared as they were eclipsed by rivals instituted by new men, possessed by new spirits who clamoured for a chance to control the destinies of rain, pestilence, and famine. In some places we can see this process at work today. At Chona a first rite is performed at the shrine of Nangoma, mother of the present ritual leader. Hers is the most recent shrine now in existence. If after this all goes well with the rain, well and good. The Tonga are not people to concern themselves with rites for ritual's sake. It is only when the rains do not fall in a manner to satisfy them that they are driven to perform the rites at the four other shrines associated in this particular complex. Over a series of good years it is possible that the older shrines might be so neglected that they would vanish from the complex. But so long as they are remembered, it is always likely that in times of desperation the diviners will call upon the people to go back to the sites, rebuild the huts, and perform the rituals again. Or through a series of very bad years the attitude of 'You don't care for us, and we won't care for you' may swell into a general disgust and the shrine be abandoned as useless.

Very occasionally a shrine may survive its community. When the Mwansa people moved into their present area they found the decayed remains of an old shrine. They did not know to whom it belonged, or to what spirit it was dedicated, or indeed anything concerning the community which had performed its rites there, for that had vanished completely. But the Mwansa leader for some reason chose the ancient shrine as the centre of the rain-ritual which is now performed for his area, and today it is still visited, although the Mwansa people have since acquired two new shrines through more conventional methods.

Many districts at the present time thus have a number of shrines integrated into one cult. I am not satisfied entirely with my data on this point, but I think that this proliferation of shrines is characteristic today only in the cults controlled by the families of the chiefs and that other cults centre about only one shrine, or at least only one hut-shrine. This would suggest that the growth of such complexes is due to the conditions imposed by European administration. When the Europeans entered the country they tended to recognize the rain-makers or other such leaders as the Tonga authorities and to invest them with the status of chiefs, and thenceforth made this status hereditary in the matrilineal lines of the chiefs. Thus several successive members of the same family have been recognized and supported as leaders in a particular area, whereas in pre-European days it is quite possible, as it is today, that the next rain-maker who could speak with the voice of the *basangu*, and not merely as a shrine custodian, would appear in a non-related family, of a different clan, and in a slightly different area. His appeal would be to those who lived around him, and his influence or that of his shrine after his death might well serve to detach some of his neighbours from their alliance to an older cult centre and to draw them into a new community. Since the Tonga population was a shifting one, the new shrine which would represent the particular concatenation of population at this particular time could more truly represent the community than the old one. Control of the two cults would be vested in two different kin-groups and there would be little chance of their being integrated into a common complex unless the two communities which they represented were one and the same.

Conclusion

The system described above, overlaid though it is by the creation of hereditary government chiefs, is still the fundamental element in Tonga social structure. Authority in the last analysis rests on personal qualities, but some continuity is given to the society through the recognition that a man who has led the community continues after his death his interest in its welfare, and that his kin-group is the proper channel through which to approach him. But this recognition is only a sea-anchor in the society, which slows the drift and does not stop it. It creates a small community within which the rudiments of community law can be discerned and which forces its members to remember occasionally that they belong to a wider unit than the village or kin-group. Such communities endure for only a moment in time and then re-form themselves into new units. Unsatisfactory though it seems to us, it appears to be the only guise in which the Tonga could visualize authority. Indeed, at the present time, battered as the rituals are by the attacks of the missions, most modern chiefs tend to identify themselves with the rain-shrines as a prop to their authority. Monze, whom the Government accepts as Senior Chief of the Tonga, refuses to admit that any rain-shrines except those of the Monze line exist in all the Tonga country. Other chiefs refer only to those in the hands of their kin-groups and ignore completely the other shrines of the neighbourhood. They maintain that villages in all directions and in great numbers attend their rituals, although a little casual questioning among the surrounding peoples will narrow the influence of their particular cults down to only a fraction of their claims.

Since the Government pays little attention to the rain-rituals and seems unconscious of their role in Tonga life, this attitude must reflect a deep-seated tendency among the Tonga to equate rain-rituals with political integration.

NOTE

1 This chapter was read before the Royal Anthropological Institute in January 1948.

REFERENCE

Smith, E.W., and Dale, A.M. 1920. *The Ila-Speaking Peoples of Northern Rhodesia.* London, Macmillan & Co.

14

El Niño, Early Peruvian Civilization, and Human Agency

Some Thoughts from the Lurin Valley

Richard L. Burger

The role of El Niño in the rise and fall of early Andean civilizations has attracted increasing attention over the past two decades as more has become known about the role of climate change in human history. It is no coincidence that this greater sensitivity to climate change in the archaeological past has emerged as we have become increasingly anxious about global warming and the way it could affect our future. As science has focused more intently on global climatic change, new methods and theories have been developed that allow us to reconstruct past climates and to appreciate the degree of variability that has existed in the Holocene climate, of which the El Niño phenomenon is but one small piece.

The past two decades also saw two major El Niño events. As a consequence, many archaeologists working in Peru have experienced, either firsthand or through media coverage, the devastation that a major El Niño event can produce. By contrast, in 1980, only the most senior archaeologists had personally experienced a major Niño event, and most academics had to rely on accounts of the 1925 El Niño to imagine what its effects were like. Thus, historical happenstance has placed scholars in a situation where there are now both personal experience and the academic predisposition to take El Niño seriously in the archaeological modeling of civilizational trajectories in the distant past. Such was not always the case. In the late 1960s, for example, both Edward Lanning (1967) and Luis Lumbreras (1969) found it possible to write syntheses of Andean prehistory with barely a reference to the El Niño phenomenon, and the immediately previous generation of scholars, such as Bushnell (1957) and Bennett and Bird (1960), ignored El Niño entirely. Such an approach has now been

The Anthropology of Climate Change: An Historical Reader, First Edition. Edited by Michael R. Dove.

largely supplanted, as evidenced by the work of Michael E. Moseley (1992) and James Richardson III (1994), in which El Niño figures prominently as a possible contributing factor to the emergence, expansion, reorganization, and demise of multiple Peruvian cultures, including those of Chavín, Moche, and Chimu.

Recent archaeological literature on the possible effect of El Niño on Andean prehistory has usually focused on the role of El Niño in the evolution of Andean civilization. A classic example was a 1981 article by David Wilson, in which he posited the El Niño phenomenon as a limiting factor in the development of an early maritime civilization in the Central Andes because of the unpredictable but radical reduction in maritime carrying capacity along the coast during major El Niño events. In a more recent 1999 synthesis, Wilson (1999: 352–356) updated his earlier argument and suggested that the stresses caused by El Niño could help explain how a primarily maritime-oriented people might accept agriculture as an alternative strategy, thus creating the conditions for the emergence of complex society. In the models proposed by Wilson and others, El Niño is seen as shaping a long-term evolutionary trajectory as cultures become adapted to their environmental conditions; the role of human actors and their strategies is seen as secondary to larger evolutionary processes. Not surprisingly, such models have been criticized as treating humans as fundamentally passive and as constructing social change merely as a process of reacting to natural phenomena, such as natural disasters or long-term changes in climate. Although such models have merit, it also is important to consider whether the peoples of pre-Hispanic Peru anticipated the dangers posed by El Niño events and whether they were able to develop strategies to mitigate them. In adopting this second approach, we recognize that human agency played an important role in determining cultural stability and change in the past, just as it does in the present.

In the modern world, major disasters are most successfully dealt with at the level of the nation-state or international community. For example, 90% of the $14 billion in aid to the victims of the 1994 California earthquake came from the federal government of the United States rather than from local or state sources, and in Honduras, virtually all aid to alleviate the devastation wreaked by Hurricane Mitch has come from governmental and charitable sources outside of Central America (Davis 1998). But what disaster strategies were employed prior to the emergence of such overarching social and political structures?

Possible Pre-Hispanic Responses to El Niño Events in the Central Andes

In the absence of state-based systems, one way of dealing with recurring environmental disruptions such as El Niño is for families or small social units to develop links with distant communities that are less likely to be affected by the environmental perturbation. In the case of the northern and central coast of Peru, for example, longstanding links with adjacent highland communities would have facilitated a prehistoric alternative to current disaster relief, perhaps under the rubric of fictive kinship obligations or gift exchange. At the 1982 conference on Early Monumental Architecture at Dumbarton Oaks, I suggested that in combination with the dietary needs of highlanders (e.g., for iodine and salt), the danger presented by El Niño events would have favored the establishment of ties between highland and coastal groups that could be mobilized in times of disaster. Llama caravans could have brought highland agricultural produce down to communities where an El Niño had devastated both the year's crops and maritime productivity (Burger 1985: 276).

A modern version of such a strategy was mentioned in Robert Murphy's description of the 1925 El Niño, in which food shortages on the central coast were solved by "mutton on the hoof" driven down from the high grasslands (*puna*) and by pack trains of llamas, horses, and burros from the highland valleys carrying potatoes and other foodstuffs (Murphy 1926: 46). The continued viability of agricultural systems in the northern highlands during the last two major Niño events supports the plausibility of this idea.

Moreover, recent research by Ruth Shady (1997) at Caral in the Supe Valley and by

Shelia and Thomas Pozorski (1992) at Huaynuná and Pampa de las Llamas in Casma has reinforced our appreciation of the Late Preceramic and Initial Period links between the highland societies involved in the Kotosh Religious Tradition and their contemporaries in centers on the coast. Unfortunately, the hypothesis of highland economic assistance to coastal settlement during El Niño events has yet to be tested on a microlevel by studying examples of a Late Preceramic or Initial Period center that coped with an El Niño event. It would be fascinating to know if the survival strategy of such a site was characterized by refuse that included a sharp increase in the amounts of highland meat and agricultural produce to compensate for the disruption of marine and lower-valley agricultural resources.

In Andean archaeology, one of the rare instances of a local-level analysis of a prehistoric community struggling to deal with an El Niño event is the study of two Chimu settlements in the Casma Valley by Jerry Moore (1991). Moore not only used archaeology to document the occurrence of a fourteenth century A.D. El Niño event, he also explored some of the cultural responses to it. He concluded that immediately following a powerful El Niño, the Chimu state established a complex of ridged fields in order to reclaim waterlogged soils, and an adjacent community to house agricultural workers. This subsistence system was apparently maintained for no more than a few years, while the normal farming system was restored, after which time the site was abandoned. According to Moore, by shifting to the cultivation of fields that were not irrigated, along with exploiting El Niño-resistant species of shellfish, it was possible to support the continued occupation of the Casma Valley and its administrative center at Manchan despite the devastation wreaked by a major El Niño event. This case is particularly interesting because it illustrates how one pre-Hispanic group consciously combined two of many possible strategies in order to cope with the conditions created by El Niño. In this case, the rains from El Niño may have created the opportunity for successful dry-farming in this section of the normally arid coast.

Given recent history, it should not be news to anyone that El Niños produce occasional opportunities as well as serious problems. One has only to recall the scandal that occurred during the 1982–1983 El Niño, when several high-ranking military men were accused of using army cargo planes to fly cattle down from Panama to graze on the vast pasturelands that had appeared in Peru's Sechura Desert. The appearance of more robust *lomas* vegetation, the migration of new kinds of fish, the short-term availability of new land for rainfall farming, and the sprouting of new pastures may be only small consolation when weighed against the enormous losses occasioned by an El Niño event, but such factors may have been crucial for crafting locally based survival strategies in pre-Hispanic times. Additional studies along the lines of the Casma research should yield greater insight into the significance of these alternatives. It should be noted that the strategy posited by Moore for Casma involved the intervention of state institutions, which were absent in much earlier prehistoric times.

Pre-Hispanic Human Agency, El Niño, and the Manchay Culture

Thus far I have explored some of the possible short-term responses to the effects of El Niño events in the pre-Hispanic Central Andes. However, as geographer Kenneth Hewitt has observed, "Most natural disasters are characteristic rather than accidental features of the places and societies where they occur" (cited in Davis 1998: 52). In the longer-term perspective, humans can be considered agents that, with the accumulated knowledge of their landscape, either learn to anticipate potential disasters and avoid them or choose to ignore the dangers and place themselves in harm's way. Mike Davis's book, *The Ecology of Fear* (1998), provides an excellent illustration of this perspective. He shows how flawed human decisions acted to turn tectonic and climatic forces into major dangers in the course of human settlement in southern California.

Following in this line of thought, I want to consider here whether the coastal societies of Peru during the second millennium B.C. (known as the Initial Period) perceived the possible threat posed by the El Niño phenomenon, and if they did, what actions they may have taken to protect themselves. In the Central Andes, this period of time is of particular interest to those interested in the relationship between El Niño and the appearance of complex societies because it was the time of the emergence of the region's earliest civilizations. Among the accomplishments of the coastal cultures of the Initial Period were the creation of abundant monumental architecture, the production of sophisticated public art, breakthroughs in metallurgical techniques, and the building of extensive irrigation systems. This Initial Period culture, characterized by massive civic-ceremonial centers with a U-shaped layout, extended along the central coast from Chancay Valley on the north to Lurin Valley on the south. Investigations of these U-shaped centers of Manchay culture by myself and Lucy Salazar (Burger 1992: 60–75; cf. Silva and García 1997) have focused on the southernmost of these valleys, which is located immediately south of contemporary Lima, and the following commentary is based on our ongoing research.

During the second millennium B.C., the population in the lower Lurin Valley gradually increased, as reflected in the founding of civic-ceremonial centers that served as the focus of small-scale social units. Only one such center is known to have existed in 1800 B.C., but by 1000 B.C., at least six such centers appear to have been functioning in the lower valley, and several more in the middle valley (Figure 14.1). These centers appear to have been autonomous and were not organized by an overarching state apparatus; the latter appeared only much later in the prehistory of the central coast. The population supporting these centers survived on a mixed subsistence system based on irrigation farming of crops such as squash, peanuts, beans, red pepper, guava, pacae, and lucuma, as well as yet unidentified tubers (sweet potato?), manioc, and a small amount of maize. These domesticated crops were supplemented by the collection of wild plants, the acquisition of fish and mollusks from the Pacific shore, and the hunting of deer, camelids, vizcachas, and birds from nearby *lomas* and riverine environments (Benfer and Meadors 2002; Burger and Salazar-Burger 1991; Umlauf 2002).

The largest of the valley's centers, Mina Perdida, was occupied for about a thousand years (calibrated C14 years) without evidence of hiatus or abandonment (Burger and Salazar-Burger 1998, 2002). If one accepts the proposition that the pattern of El Niño events was established by 5,800 years ago (Rollins et al. 1986), the Manchay culture in Lurin presents an example of cultural continuity and resilience for at least ten centuries in the face of the major El Niño events that must have occurred during this time. Even if one accepts that El Niño had a longer recurrence interval until 3200–2800 cal B.P. (Sandweiss et al. 2001), the centers of the Manchay culture would still have experienced numerous major El Niño events, a fact confirmed by the research I shall describe below.

Considering the dangers posed by El Niño events, the choice of location for Lurin's U-shaped centers was not auspicious; in fact, to use Mike Davis's phrase, it could be said that the groups living in the lower Lurin Valley chose to place their centers in harm's way. The public complexes were generally built at the mouth of deeply cut ravines (*quebradas*). These *quebradas* were normally bone dry, but they sometimes carry water during El Niños. Such locations may have been chosen because they provided expanses of relatively level land adjacent to the valuable irrigated bottomlands of the valley. Moreover, the nearby rocky ravines and barren valley sides offered ample building materials, including stone blocks and lenses of clay suitable for mortar and adobes. Unfortunately for the inhabitants of these centers, the loose rock, rubble, and earth in these *quebradas*, which make such good building materials under normal circumstances, are incorporated into landslides and debris flows when heavy rainfalls occur in the lower Lurin Valley during major El Niño events.

Manchay Bajo and its monumental wall

Judging from the results of fieldwork in 1998 and 1999 by the Yale University Lurin Valley Archaeological Project at Manchay Bajo, the Initial Period occupants of Lurin not only were aware of the danger posed by an El Niño event,

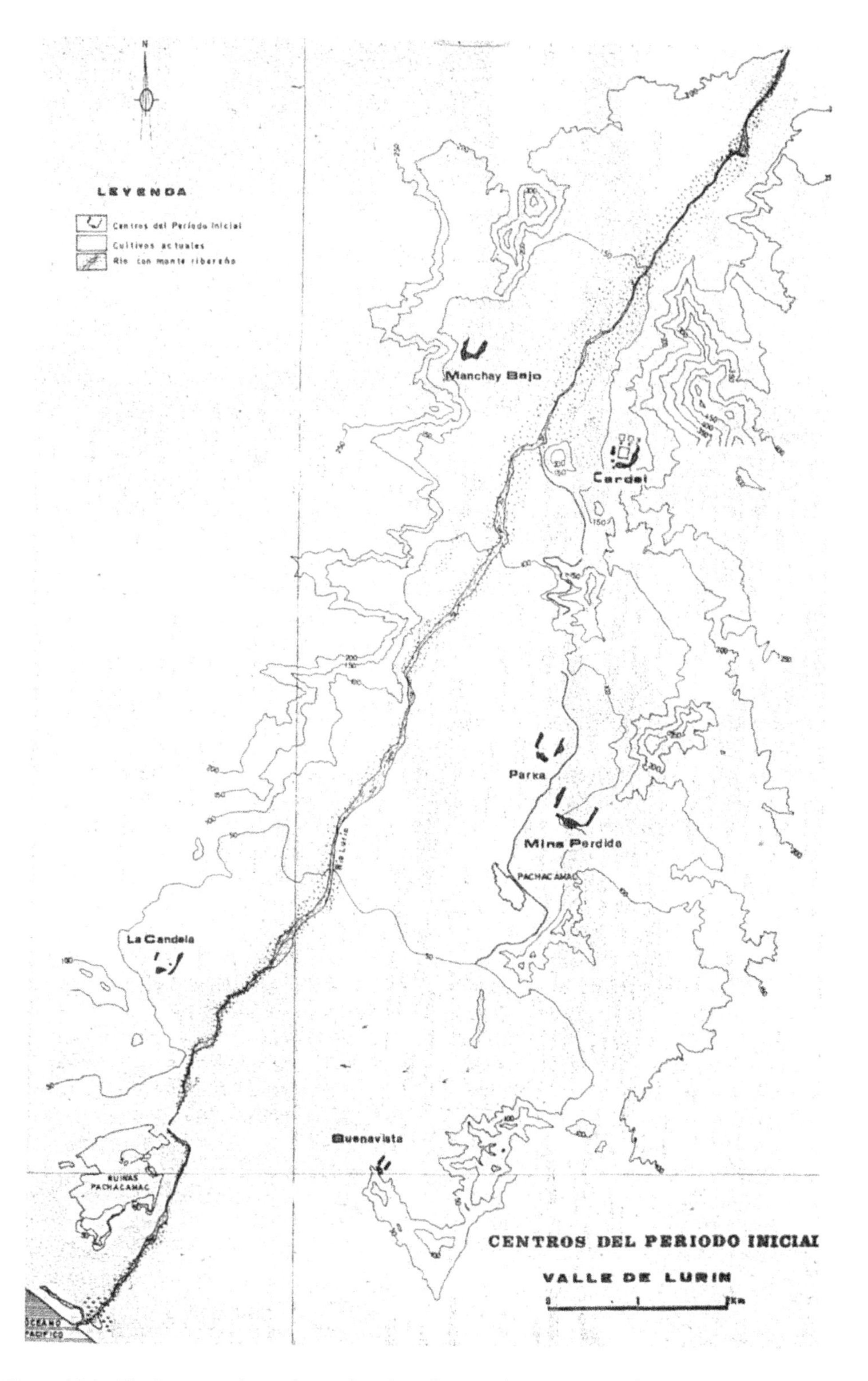

Figure 14.1 *The location of Initial Period U-shaped pyramid complexes in the lower Lurin Valley, Peru. (Map drawn by Bernadino Ojeda.)*

they consciously worked to protect themselves against potential disasters. Manchay Bajo (PV48–147) is located in the lower valley, 12 km inland from the Pacific, at 140 m above sea level (masl). In contrast to the U-shaped complexes of Mina Perdida and Cardal, Manchay Bajo is on the northern bank of the Lurin River, only 800 m from the current course of the river.

Manchay Bajo is, in most respects, a typical U-shaped complex. The archaeological complex is dominated by a terraced, flat-topped central pyramid and the site is oriented to the northeast. The pyramid, found at the apex of the U, measures 100 × 75 m at its base and rises 13 m above the current level of the valley floor. Two elongated lateral mounds, one of which is attached to the main pyramid, flank a central plaza that is 3 hectares in area (Figure 14.2). Although lower than the main mound, the lateral mounds are of considerable size, with heights of 11 m and 8 m, respectively. The area of the site, roughly 20 hectares, and the scale of the monumental architecture are slightly larger than at Cardal, which is located 1.7 km away on the other side of the river. The excavations at Manchay Bajo revealed a long history of construction at the center that included a minimum of nine superimposed central stairways, three superimposed atria, each with multiple renovations, and at least nine major building episodes. Thus, the large public constructions seen today were the result of repeated constructions that spanned at least six centuries.

Most of the ceramics associated with the monumental constructions date to the late Initial Period (approximately 1200–800 cal B.C.). This is consistent with an AMS measurement of 3010 ± 60 (AA 3442), which, when calibrated, has a 2σ range of 1404–1052 B.C.; the specimen tested comes from a fiber bag (*shicra*) used to hold stone in the fill covering the site's middle atrium. Since this measurement dates the closing of this structure, and since an even older atrium exists below this one, Manchay Bajo must have been founded significantly before this time. Although most of the construction episodes at Manchay Bajo date to the late Initial Period, the uppermost levels yielded a distinctive ceramic assemblage that dates to the Early Horizon; no hiatus is indicated. The preliminary stylistic interpretation of the pottery is consistent with two AMS C14 measurements of 2560 ± 50 B.P. (Beta-122683) and 2600 ± 50 (AA 34441) from the final period of construction which, when calibrated, produced a 2σ range between 815–525 B.C. and 894–539 B.C., respectively. Thus, the available evidence indicates that whereas Manchay Bajo was contemporary with Cardal and Mina Perdida during the final centuries of the Initial Period, it continued to function as a civic-ceremonial center for a century or more after the others were abandoned (Figure 14.3). Manchay Bajo is situated at the mouth of two *quebradas* that were incised into the Andean spur separating Lurin from the Rimac drainage. The larger of these, known today as the Quebrada Manchay, is a dry tributary valley located to the north of Manchay Bajo. It is separated from the valley through which the Lurin River flows by a massive rocky spur (246 masl). The smaller of the two *quebradas* is located to the northwest of the site and is unnamed; it extends only for about a kilometer. The Quebrada Manchay was used as a natural corridor between the Lurin and Rimac in prehistoric times, and this route continues to be used today despite the poor state of the unpaved road. Given the nature of the topography and Manchay Bajo's location, a major El Niño could have triggered landslides through the large Quebrada Manchay, which would have buried the site's central plaza in stone rubble. A debris flow from the shorter, unnamed lateral *quebrada* would have had a strong effect on the western lateral platform of Manchay Bajo. Undeterred debris flows from either or both *quebradas* would have had the greatest impact on the residential zone covering the flatland to the north and northwest of the public architecture.

The potential danger posed for prehistoric occupations and public spaces by landslides and debris flows from the small lateral *quebrada* was highlighted by the excavations at the site of Pampa Chica by archaeologists from the Pontificia Universidad Católica (PUC), Lima, under the direction of Jalh Dulanto as part of the Proyecto Arqueológico Tablada de Lurin, directed by Krzysztof Makowski. Pampa Chica, a small site located up the smaller of the two

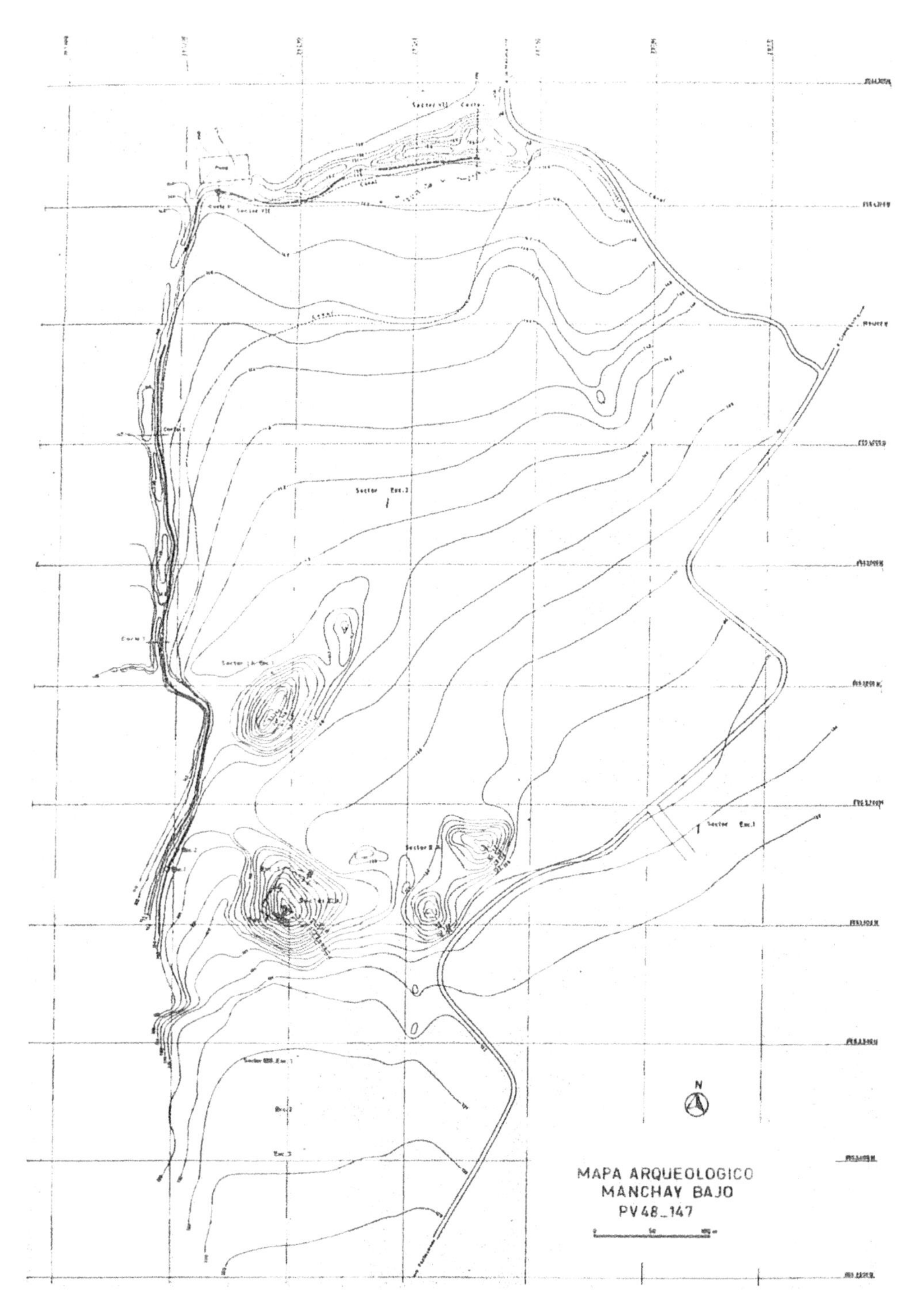

Figure 14.2 *Topographic map of the Manchay Bajo complex indicating the location of excavation units and the monumental wall extending along the western and northern extremes of the site. (Map drawn by Bernadino Ojeda.)*

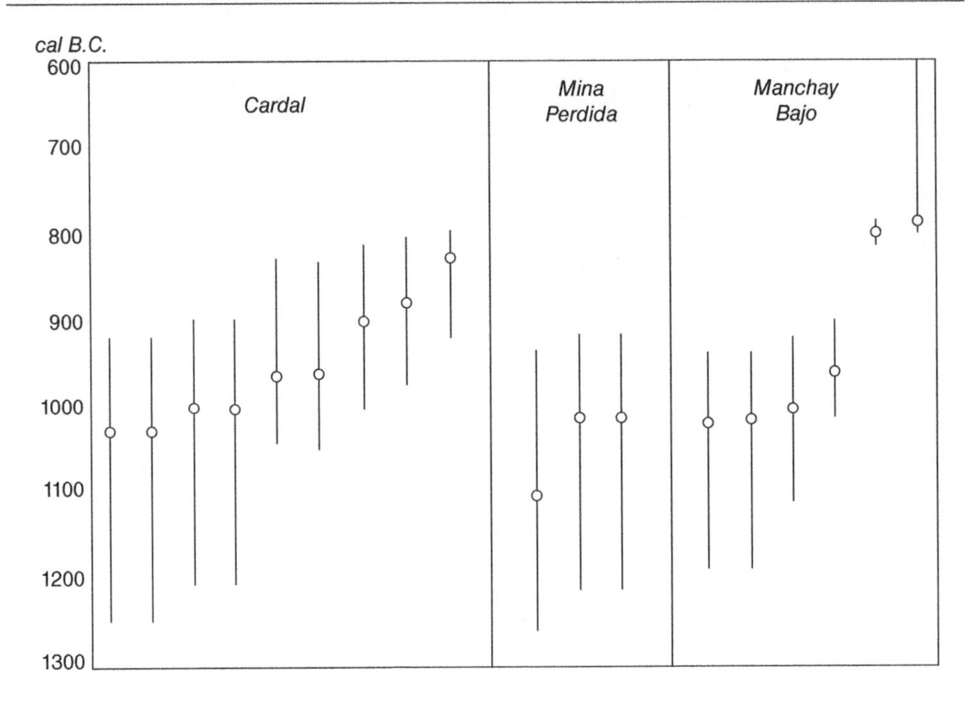

Figure 14.3 *Radiocarbon measurements from the three U-shaped complexes excavated in the Lurin Valley demonstrate both the general contemporaneity of the centers and the continued occupation of Manchay Bajo after the abandonment of the other two sites. (Chart by George Lau.)*

quebradas at 180 masl, had been covered by landslides. Occupied between the Early Horizon and the Middle Horizon, investigations at the site found evidence of repeated debris flows in prehistoric times (Dulanto et al. 2002).

In addition to the danger posed to Manchay Bajo by flash floods and debris flows coming out of the Quebrada Manchay and the smaller lateral *quebrada*, a threat also was posed by a 138-m-high rocky outcrop (278 masl) immediately to the west of the main mound (Figure 14.4). It is covered with large stone boulders and loose stone rubble, and this unconsolidated material would have become unstable during an El Niño event or an earthquake.

Prior to 1998, no topographic map of the entire archaeological site of Manchay Bajo existed. However, Harry Scheele (1970: 179–190) had carried out test excavations at the site in 1966 and produced a sketch map of the central portion of the site. In subsequent years, many visitors, including Alberto Bueno Mendoza, members of the PUC Pampa Chicha team, and me, have examined Mancho Bajo and been intrigued by a large wall that rings its western and northern perimeter. Our investigations included a detailed mapping of the entire site, and it was determined that the massive wall begins at a rocky outcrop near the southwest corner of the northwest lateral platform and runs in a northerly direction for some 460 m. The wall then turns eastward and runs for another 240 m (see Figure 14.2). Unfortunately, its final section was destroyed by the construction of a modern road, but it appears that the eastern end terminated 45 m away, at the large rocky outcrop that defines the eastern edge of the Quebrada Manchay. Both of the two extremes of the monumental wall appear to have been engaged with natural topographic features anchoring this remarkable cultural feature to the landscape. The total length of the original wall is estimated at 745 m.

During the mapping and surface survey in 1998, masonry retaining walls could be seen at various points along both the north–south and

Figure 14.4 *Rocky outcrop immediately to the west of the Manchay Banjo complex. Traces of the monumental wall at the foot of the outcrop were visible prior to excavation and can be seen near the stadia rod in this 1998 photograph. (Photograph by Richard L. Burger.)*

east–west segments of the perimetric wall. In those sections where it has been exposed, the original wall can be seen to be double-faced, with a core of unconsolidated stone, gravel, and earth. In both segments there is evidence of at least one and in some cases two renovations of the wall; this was done by adding new walls separated from the previous walls by a layer of fill. These additions would have widened the wall substantially while reducing strain on the walls incorporated into the core. This same pattern of growth was determined for the terrace walls on the central pyramid of the Manchay Bajo complex. Like the Manchay Bajo's platform constructions, construction of the original monumental wall and its subsequent renovation were carried out with medium-sized blocks of roughly dressed stone from the nearby slopes. Clay mortar was used at the junctures, but during the surface reconnaissance, no surviving evidence of surface plastering was encountered. The width and height of the wall varied, and in many places its limits are completely hidden under collapsed or accumulated material. Nevertheless, the topographic map suggested that the north–south segment had an average width of about 12.5 m and an average height of at least 5 m, and portions of the east–west segment were still more massive. Surface ceramics were encountered at various points on the summit of the wall; they were particularly common near its eastern end because of disturbance from the modern construction of a small chapel on top of the wall. Based on vessel forms and decoration, all of the pottery can be dated to the late Initial Period. It included numerous shallow open bowls and neckless ollas. Some of the former had the vessel interior decorated with broad incisions. The surface ceramics found on the wall in 1998 were indistinguishable from those found in the excavations of the main mound at Manchay Bajo that same year. No later ceramics were on or near the wall. Based on the masonry style of the surface architecture and the ceramic evidence, a preliminary conclusion was reached that the monumental wall had been constructed during the Initial Period and was roughly contemporary with the adjacent U-shaped public architecture.

Following the mapping and study of the monumental wall just described, we encountered another masonry feature at the foot of the steep rocky outcrop to the west of the main pyramid. Cut by a modern canal, ancient floors and fills were exposed, and these were reminiscent of features such as circular courts such as those found at Cardal. However, clearing and excavation in this area revealed that the remains actually corresponded to another massive wall running for at least 105 m, with a width of 5 m and a height of 5 m. The stonework and construction were similar to those already described, and there was evidence of two episodes of renovation in which additional walls were added. In the 1998 excavations of a small section of this wall, late Initial Period pottery was recovered from intact floor surfaces along the wall's eastern face. This evidence, combined with the construction style and technique, lends support to the conclusion that this and the other monumental walls at Manchay Bajo are coeval with the platform complex and were built by the same population. It is hypothesized that this wall was built to protect the area of the central mound from debris slides coming from the steep slopes of the rocky outcrop above it. Thus, the total extent of the monumental perimetric wall at Manchay Bajo must include this feature as well, bringing the total length of the wall constructions to some 850 m. If a rough calculation is made of the volume of earth and stone moved to construct these walls, it produces a figure in excess of 30,000 m^3.

In 1999, during the second field season, archaeological excavations were carried out in a small section of the monumental wall (Sector VIIA, Excavation 3) in order to clarify its date, construction history, and the building techniques utilized. The work in this sector was supervised by Marcelo Saco (PUC), and technical assistance in the interpretation of the stratigraphy was provided by the Polish sedimentologist, Krzyzstof Mastalerz. The excavated units were located along the wall's north–south section, which crosses the mouth of the small lateral *quebrada* to the west of the site. Initially, 7 m of the eastern face of the wall was cleared. This revealed that the southern half of this section was well-preserved, while the northern half had collapsed after the site's abandonment. Subsequent excavations in the area focused on the intact portion of the wall; the zone investigated had an area of 49 m^2. This included a 1-m-wide trench perpendicular to the wall face. At the conclusion of this excavation, a 17-m east–west transect of the monumental wall complemented the horizontal exposure of the wall's eastern face.

Judging from the excavations, the original monumental wall in this area was trapezoidal in cross-section. The hearting of the wall consists of loose soil, gravel, and stones. The wall was built on a sloping surface created by ephemeral sheet flows that predated the occupation of the site. Both faces of the wall consisted of roughly quarried medium-size stones (e.g., 40 × 38 cm) set in mud mortar. Both sides of the wall cant inward for greater stability, and as a result, the upper section of the wall is approximately 2 m in width and nearly 3 m wide at its base. The upper section of the original wall was missing. It was feasible to reach the wall base on the western face, and it can be demonstrated that the original wall was over 2 m in height.

Later in the history of Manchay Bajo, the wall was widened by stone retaining walls built parallel to the faces of the original wall. Along the eastern face the new retaining wall was terraced. The lower terrace was 1 m in height, and 1.2 m remains of the upper terrace wall. Along the western face, sterile fills of gravel and stone were added, completely burying the original wall. The massive layers piled against the wall's original western face were studied in terms of their sorting and position, to determine whether they were man-made construction fills or the result of slumps or debris flows from the lateral *quebrada*. These layers include loose, fragmented material ranging from angular boulders to muddy, coarse-grained sand. Sedimentologist K. Mastalerz (1999) concluded that they were man-made deposits piled against the western side of the original wall. These fills added at least 1 m in height and 4 m in width to the monumental wall, bringing the total scale of the wall in this section to over 9 m in width and over 3 m in height.

Interestingly, the floor articulating with the western face of the original wall showed evidence of caliche-like cementation due to the precipitation of soluble compounds from groundwater. Mastalerz (1999) believes that such a layer was probably the result of the pooling of water from El Niño rains against the monumental wall. Significantly, this cementation was not encountered along the eastern face of the wall. Little evidence survived of the new western face of the expanded monumental wall due to the narrowness of our trench (1 m); only a limited portion of what remained could be exposed. However, the base of the wall (Muro 6) and its associated floor was identified. Surprisingly, the wall was made of stone-filled *shicra* bags covered with mud mortar. This technique of wall construction was rare at the U-shaped complexes in the Lurin Valley, but it had been identified previously at Mina Perdida (Burger and Salazar-Burger 2002). It was possible to date the fiber used in the *shicra* in order to get an idea of the age of the monumental wall's renovation. The AMS measurement on this sample produced a date of 3020±40 B.P. (calibrated 2σ range of 1389–1129 B.C.). This result confirms the overall contemporaneity of the monumental wall with the U-shaped civic-ceremonial complex and the associated residential constructions at Manchay Bajo. The date suggests that the original monumental wall was built early in the site's history and renovated at least once during the late Initial Period. Judging from the section excavated, that renovation may have involved as much labor as the original construction itself. Finally, it would appear from the caliche layer that a minimum of one major El Niño event occurred after the wall was constructed and while the site was still occupied. It is reasonable to hypothesize that this El Niño event may have stimulated the enlargement of the original wall, since the addition covers the cementation.

There are two other massive layers of gravel and stone that post-date Muro 6. According to Mastalerz, these, like the strata they cover, also are man-made deposits still in their original position. A possible explanation of these strata is that they represent a subsequent second phase of enlargement after the collapse or dismantlement of the *shicra* wall (Muro 6). This enlargement to the west could have involved a retaining wall whose traces have disappeared completely, or, alternatively, as Mastalerz (1999) suggests, the final outer western surface of the monumental wall could have been left as an unfaced embankment. At the end of this hypothetical third construction phase, the monumental wall would have reached 12 m in width and increased in height by at least another 50 cm to 3.5 m. We have no way of directly dating this third episode of wall construction; we suspect that it could date to the final Early Horizon occupation of the public center.

The location of the walls, their massive width, and their substantial height all suggest that they were built as a dam to protect the civic-ceremonial complex from land and rock slides coming off the rocky outcrops and out of the dry *quebradas*. It is significant that walls do not exist to the east or south of the Manchay Bajo complex, where there is no danger of such disasters. Moreover, there is evidence that the walls served their intended purpose with some success. In all four of the transects that we documented in 1998, the surface level outside the wall (i.e., the exterior facing the potential source of debris) was significantly higher than inside the wall (i.e., the interior facing the plaza or platform mounds). It appears that in some areas, 1–2 m of material had accumulated against the wall, presumably from one or more debris flows provoked by El Niños. In one deep cut to the north of the wall made by modern builders, this pattern of debris flow evidently recurred on several occasions both before and after the wall's construction. Judging from our excavations within the wall's perimeter, Manchay Bajo's monumental wall or dam stopped the entry of stone rubble from debris flows, as was intended. In no area inside the wall did we encounter deposits of boulders or large stones carried by landslides or other disasters. The dam also appears to have protected the civic-ceremonial center from floods during the Initial Period and Early Horizon occupation of the site.

Nevertheless, the problem posed by large quantities of flood water blocked by the monumental wall appears to have presented a serious problem. Our investigations revealed that deep layers of water-borne deposits cover most of the site, with the exception of the elevated public architecture. For example, an excavation in the Manchay Bajo's open plaza area (Sector IV, Excavation 1) revealed that the central section of this space featured a low, stone-filled platform at least 1 m in height. This Initial Period construction was buried by over 2 m in flood deposits, which, according to Mastalerz (1999), were the product of six El Niño episodes whose character varied in size and duration. Some layers of sediments were the result of flash floods, while others were produced by powerful floods followed by stagnant water conditions. In one period the rains were sufficient to stimulate in-channel fluvial processes and the resulting deposition of sand and gravel bars at Manchay Bajo. The repeated floods documented by these deposits came primarily from the Quebrada Manchay, and it would appear that the northern section of the dam was breached on numerous occasions during the last 2000 years following the center's abandonment. Considerable numbers of Initial Period artifacts are mixed in with some of these flood deposits, and it is clear that these floods destroyed some of the upper layers of the site's Formative settlement. While there is compelling evidence of destructive floods following Manchay Bajo's abandonment, at the present time there is no evidence that floods disrupted the Initial Period or Early Horizon occupation of the civic-ceremonial center of Manchay Bajo.

Agents and Environment

The evidence summarized here suggests that (1) the people of Manchay Bajo perceived a threat to their center and adjacent agricultural lands from El Niño-related landslides; (2) they were able to generate a solution to the problem using available technology and materials; (3) they were able to mobilize enough labor to complete a dam large enough to protect them from El Niño debris slides; and (4) during some six centuries of occupation, they were able to bring together enough manpower to renovate the dam on at least two occasions by encasing the original wall within new fills and retaining walls. The monumental walls succeeded as bulwarks against the feared landslides, and they are still capable of doing so. These findings highlight the importance of human agency in shaping a culture's destiny; clearly, the actions discussed here were pre-emptive, anticipating potential threats from unpredictable future El Niño events. The population employed a knowledge of environmental risks to formulate a strategy, and they were able to implement this strategy even though it involved thousands of person-days of labor without immediate short-term benefit.

The case of Manchay Bajo provides a good opportunity to reconsider some of our preconceptions about the ability of different kinds of societies to cope with environmental variability. It has often been assumed that states are uniquely well suited to deal with disasters because of their coercive capacity, managerial apparatus, and ability to marshal resources from a wide area. Nevertheless, the continuity and duration of the Manchay culture for a millennium are a clear demonstration of the resilience and flexibility of its social forms in the face of mega-El Niños and other disasters that must have occurred. In this respect, the Manchay culture's lack of centralization and hierarchy may have been an asset rather than an obstacle. The mobilization of labor for efforts like the monumental wall should not surprise us, since even greater projects were undertaken during the second millennium to obtain water through gravity canals. In fact, the creation of new canals was intimately linked to the establishment of the agricultural lands needed to support newly established social units and their public centers. Other corporate labor efforts were undertaken to obtain supernatural favor through temple construction.

The ability of the Manchay culture's subsistence economy to withstand short-term climatic disruptions is comprehensible, since its continued dependence on a range of maritime resources, hunting, and wild plants would have served the people well during an El Niño event. Moreover, the social institutions underlying the impressive public constructions of the Manchay culture would have been an asset in times of crisis. In times of emergency, the annual mobilization of public labor usually used to refurbish the U-shaped pyramid complexes could have been turned to repairing the relatively short canals that irrigated their fields and that would have been damaged by major El Niño events, or to renovate the monumental dam protecting the site.

As already noted, the construction technique, the masonry style, and the pattern of episodic renovations of the dam differ little from that used in the temple. In many respects, the challenge of building a long linear feature like the Manchay Bajo dam is analogous to the construction of a gravity canal. Contemporary communities construct and maintain canals without state intervention by dividing the required labor between the family units or communities that benefit from the irrigation water, with participation in the cooperative labor effort a prerequisite for continued community membership (i.e., access to land and water). Such cooperative labor practices have been documented for pre-Hispanic times, and may have been in place by the Initial Period (Burger 1992; Moseley 1992). Considering these factors, it is worth considering whether the pre-state societies of the Initial Period may have been as well as or, perhaps, even better equipped to deal with mega-El Niños than the more fragile complex societies of later times.

REFERENCES

Benfer, R., and S. Meadors. 2002. Adaptaciones de la dieta humana a nuevos problemas y oportunidades en la costa central del Perú (1800–800 A.C.). *In* Arqueología del Valle de Lurín, R. L. Burger and K. Makowski, eds. Lima, Peru: Imprenta de la Pontificia Universidad Católica.

Bennett, W., and J. Bird. 1960. Andean Culture History. New York: American Museum of Natural History.

Burger, R. 1985. Concluding remarks: Early Peruvian civilization and its relation to the Chavin Horizon. *In* Early Ceremonial Architecture in the Andes, C. Donnan, ed., pp. 269–289. Washington, D.C.: Dumbarton Oaks Research Library and Collection.

——. 1992. Chavin and the Origins of Andean Civilization. London: Thames and Hudson.

Burger, R., and L. Salazar-Burger. 1991. The second season of investigations at the Initial Period center of Cardal, Peru. Journal of Field Archaeology **18**(3): 275–296.

——. 1998. A sacred effigy from Mina Perdida and the unseen ceremonies of the Peruvian Formative. RES: Anthropology and Aesthetics 33: 28–53.

——. 2002. Investigaciones arqueológicas en Mina Perdida. *In* Arqueología del Valle de Lurín, R. L. Burger and K. Makowski, eds. Lima, Peru: Imprenta de la Pontificia Universidad Católica.

Bushnell, G. 1957. Peru. New York: Praeger.

Davis, M. 1998. The Ecology of Fear. New York: Metropolitan Books.

Dulanto, J., L. Caceres, M. I. Velarde, and L. F. Villacorta. 2002. Investigaciones en Pampa Chica. *In* Arqueología del Valle de Lurín, R. L. Burger and K. Makowski, eds. Lima, Peru: Imprenta de la Pontificia Universidad de Católica.

Lanning, E. 1967. Peru Before the Incas. Englewood Cliffs, N.J.: Prentice-Hall.

Lumbreras, L. G. 1969. De los pueblos, las culturas y las artes del antiguo Perú. Lima, Peru: Francisco Mancloa Editores.

Mastalerz, K. 1999. Sedimentary Processes and Modifications of Unconsolidated Deposits in the Manchay Bajo Excavation Area. Unpublished report prepared for the Yale University Lurin Valley Archaeological Project.

Moore, J. D. 1991. Cultural responses to environmental catastrophes: Post-El Niño subsistence on the prehistoric north coast of Peru. Latin American Antiquity **2**(1): 27–47.

Moseley, M. E. 1992. The Incas and Their Ancestors. London: Thames and Hudson.

Murphy, R. 1926. Oceanic and climatic phenomena along the west coast of South America during 1925. Geographical Review **16**: 26–54.

Pozorski, S., and T. Pozorski. 1992. Early civilization in the Casma Valley, Peru. Antiquity **66**: 845–870.

Richardson, J. B. III. 1994. People of the Andes. Washington, D.C.: Smithsonian Press.

Rollins, H., J. Richardson III, and D. Sandweiss. 1986. The birth of El Niño: Geoarchaeological evidence and implications. Geoarchaeology **1**(1): 3–15.

Sandweiss, D., K. A. Maasch, J. B. Richardson III, and H. B. Rollins. 2001. Variation in Holocene El Niño frequencies: Climate records and cultural consequences in ancient Peru. Geology **29**(7): 603–606.

Shady, R. 1997. La ciudad sagrada de Caral: Supe en los albores de la civilización en el Perú. Lima, Peru: UNMSM.

Silva, J., and R. García. 1997. Huachipa-Jicamarca: Cronología y desarrollo sociopolítico en el Rimac. Boletín del Instituto Frances de Estudios Andinos **26**(2): 195–228.

Umlauf, M. 2002. Restos botánicos de Cardal durante el Período Inicial, Costa Central del Perú. *In* Arqueología del Valle de Lurín, R. L. Burger and K. Makowski, eds.. Lima, Peru: Imprenta de la Pontificia Universidad Católica.

Wilson, D. 1981. Of maize and men: A critique of the maritime hypothesis of state origins on the coast of Peru. American Anthropologist **83**: 93–120.

——. 1999. South Americans of the Past and Present. Boulder, CO: Westview Press.

Climatic Disasters and Social Marginalization

15

Katrina

The Disaster and its Doubles[1]

Nancy Scheper-Hughes

One hundred years later, the Negro lives on a lonely island of poverty in the midst of a vast ocean of material prosperity. One hundred years later, the Negro is still languishing in the corners of American society and finds himself an exile in his own land. [...] I have a dream that one day even the state of Mississippi, a state sweltering with the heat of injustice, sweltering with the heat of oppression, will be transformed into an oasis of freedom and justice.

Martin Luther King Jr, 28 August 1963

The non-black population was just as devastated, but apparently they were able to get out, while the black population wasn't able to get out. So maybe maybe New Orleans has a half-decent mass transit people and some of these people don't need cars.

Radio talk show host Rush Limbaugh, 1 September 2005

If people know year after year that a natural disaster occurs in a particular place and if people continue to build there and want to live there, should they bear the responsibility of buying insurance or should everyone else bear that responsibility?

Republican Senator Jon Kyl, 4 September 2005

New Orleans is dead, man. It's dead. But [it] was already gone long before the storm hit.

Cyril Neville, youngest of the Neville Brothers, at a benefit performance for victims of Hurricane Katrina, Madison Square Garden, 20 September 2005

At times of crisis and catastrophe, people seek an explanation for what happened. Even a bad explanation can seem better than none at all. As Geertz pointed out many years ago, the one thing many humans seem unable to live with is the idea that the world may be deficient in meaning and that human existence might be absurd.

The Bush administration's spin doctors, especially Karl Rove,[2] rushed to attribute the swathe of deaths and destruction on the Gulf Coast – some 1200 lives lost – to an act of nature, to God, to inept local Democratic officials who failed to act despite multiple pleas for help, and, finally, to the stubbornness of those (mostly Black and poor) New Orleanians who were too slow and too late getting themselves and their families out of harm's way.[3] Ultimately, then, the stragglers had only themselves to blame for being turned into a population of

The Anthropology of Climate Change: An Historical Reader, First Edition. Edited by Michael R. Dove.

pitiful 'refugees', a term briefly used by news media and by some public officials before it was quickly picked up and criticized for its unconscious racism, its failure to recognize the dispossessed fleeing in rubber dinghies and rickety rowing boats as bona fide citizens of the USA. Politically correct TV commentators intervened to scold and to instruct viewers that the people fleeing were *not* to be called refugees: 'These are Americans! Not Bosnians, not Kosovars, not Bangladeshis!' I'll return to this collective slip of the tongue at the end of this editorial.

Of course, individuals' exit plans were largely determined by race and class. The poor, heavily concentrated in low-lying districts, were more exposed to high water and had few opportunities to escape. Many did not own roadworthy cars, or any cars at all. Lacking personal computers, they were dependent on TV reports (until the electricity failed) and on radio (until the batteries ran out). Both media were slower than the internet and email in sounding the alarm. Consequently, many poor residents were stranded in their one-storey homes and on their roofs waiting to be rescued.[4] New Orleans newscaster Dave Cohen captured the poor people's dilemma: 'We got amazing phone calls: a woman in her house with a two-year-old on one shoulder, a five-year-old at her side, no formula, no food. "What do I do?" What can I tell her? I'm just a guy on the radio!'

The wealthy residents of New Orleans live in sturdy two-storey homes in higher-lying districts. A front-page story in the *Wall Street Journal*[5] the day after Katrina hit captured the difference immediately: 'Ashton O'Dwyer stepped out of his home on this city's grandest street and made a bee-line for his neighbor's pool. Wearing nothing but a pair of blue swim trunks and carrying two milk jugs, he drew enough pool water to flush the toilet in his home.' The affluent had access to early warnings via fax and internet. They could jump into their tank-like 4-wheel-drives and well-stocked recreation vehicles. They had access to fast cash with their high-end credit and debit cards, and they could mobilize extensive and well-equipped personal and public support systems. Finally, the wealthy residents of New Orleans hold insurance policies that will allow them to return and to rebuild if they so wish.

No Exit

Once the more 'beloved communities'[6] were safely evacuated, leaving behind the riff-raff thousands who took shelter in the Superdome, the rumours of mass death – the mayor of New Orleans predicted 10,000 deaths; the Federal Emergency Management Agency (FEMA) had ordered 25,000 body bags – of riots, rape and anarchy circulated wildly in the media. The National Guard was deployed to control what was left of New Orleans by military means and to protect private property. Abandoned people who tried to organize among themselves to obtain and distribute water, food, medications and shelter were dispersed at gunpoint by the Guard, who were under orders not to distribute their own water to the desperate. Four days after the hurricane hit, and with basic government aid still delayed, President Bush advised the stranded to seek help from private charities such as the Salvation Army.

Two San Francisco paramedics, Larry Bradshaw and Lorrie Slonsky, who were trapped in New Orleans with the abandoned poor of the city, wrote a chilling report published in the leftist press,[7] describing police and the National Guard blocking desperate evacuees as they tried to cross the Greater New Orleans Bridge to safety. It evoked a scene reminiscent of Alabama police attacking the Selma to Montgomery Freedom Marchers as they tried to cross the Edmund Pettus Bridge in 1965. Here is what the Katrina evacuees, including some with babies in pushchairs, injured people on crutches, elderly people clasping walkers, others in wheelchairs, met as they approached the bridge they had been told was a route to safety:

> Armed sheriffs formed a line across the foot of the bridge. Before we were close enough to speak, they began firing their weapons over our heads. This sent the crowd fleeing in various directions. As the crowd scattered and dissipated, a few of us inched forward and managed to engage some of the sheriffs in conversation.

> We told them of our conversation with the police commander and the commander's assurances. The sheriffs informed us that there were no buses waiting. The commander had lied to us to get us to move. We questioned why we couldn't cross the bridge anyway, especially as there was little traffic on the six-lane highway. They responded that the West Bank was not going to become New Orleans, and there would be no Superdomes in their city.

The police and National Guard made sure that hundreds of abandoned New Orleanians were prevented from fleeing the city on foot. Contrast this violent scene with the evacuation of thousands of ordinary people from Lower Manhattan in the wake of 9/11, when Mayor Giuliani dispatched his top assistant, Rosie, clad in a fluorescent jacket and holding a megaphone, to lead panicked people across a bridge and into the safety of Queens – a beautiful and historic moment.

Self-Blame and National Shame

In the wake of a disaster people tend to ask the question: Why me? Why us, oh God, of all people? Victims collude with those who are all too willing to blame them for their misfortune. Making sense of suffering is a dicey game, a two-edged sword. In his essay on Holocaust survivors, 'Useless suffering', Immanuel Levinas goes so far as to see the search for meaning in catastrophic human suffering as a potent source of evil in the world. Conversely, those who escape a catastrophe experienced by others (especially their own loved ones) tend to ask the opposite question: Why was I spared? Why did I live? – an equally devastating experience of self-blame.

My particular perspective on the Katrina disaster derives from the 18 months (1967–1968) that I spent living and working in Selma, Alabama and its rural environs as a civil rights worker investigating hunger and malnutrition among Black sharecroppers. The reports I wrote for the Southern Rural Research Project (SRRP) – 'Black farm families: Hunger and malnutrition in rural Alabama' and 'The extinction of Black farm families' (both of them scathing attacks on the perverse relations between Black farm families and local agents of the US Department of Agriculture) – were based on a survey of 243 households in several Blackbelt counties of Southwest Alabama. The reports were used in a class action suit: 'Peoples v. the US Department of Agriculture' (US District Court, 23 March 1967). We brought three busloads of undernourished adults and children – 130 Black Alabamans ranging from 7 weeks to 75 years old – into that Washington, DC court room, along with a team of doctors (including Robert Coles and Charles Wheeler) to verify the shocking nutritional disorders, ranging from paediatric marasmus and kwashiorkor to the pellagra suffered by many of the adults.

SRRP lost its case against the US Department of Agriculture in the courts but won in the media as newscasters from ABC, CBS and NBC, and reporters from the *Washington Post* to the *New York Times* expressed alarm at the possibility of widespread hunger among the rural Black poor of the American South. Dr Wheeler continued to work with us in documenting the effects of chronic malnutrition on Alabama's sharecroppers. A CBS team came to Selma, Alabama in 1968 to film a segment of the 90-minute documentary 'Hunger in America'. I accompanied the team to the home of a large family of sharecroppers where Dr Wheeler interviewed a 14-year-old boy named Charles. Of all the images of hunger in America, this one tore at the collective conscience of the American public.

Wheeler asked the 14-year-old sitting across him on a bed covered with a tattered bundle of rags, the only seat in the shack:

> 'Do you eat breakfast before school?'
> 'Sometimes, sir. Sometimes I have peas.'
> 'And when you get to school, do you eat?'
> 'No, sir.'
> 'Isn't there any cafeteria food there?'
> 'Yas, sir.'
> 'Why don't you have it?'
> 'I don't have the 25 cents.'
> 'What do you do while the other children eat lunch?'
> 'I just sits there on the side.'
> [Here Charles turns his face away from the cameras]

But Dr Wheeler continues:

> '*How do you feel* when you see the other children eating?' 'I feels ashamed'. [Charles' voice breaks]

Raymond Wheeler asks incredulously:

> 'You feel *ashamed*?' 'Yas, sir.'

After the CBS documentary was aired hundreds of letters bearing small cheques arrived at our 'Freedom House' in Selma, Alabama. I answered them all. No American child, these concerned citizens argued, should feel ashamed because they had nothing to eat. And no child should sit by empty-handed while his or her schoolmates ate lunch. This one CBS documentary had enormous impact, leading, ultimately, to Congressional action. Consequently, Charles got his school lunch free, as did thousands of other rural Southern kids like him.

Many years later, when Governor Cuomo made his memorable nominating keynote speech at the 1988 Democratic Convention, he recalled that scene from 'Hunger in America'. Though he did not recall his name, over 20 years later Cuomo evoked the burning sense of misplaced shame in that one hungry American child. For shame, America! was Cuomo's message. Since that time the US has agreed to put an end to welfare (as we knew it), thereby putting an end to childhood (as we knew it). A raw deal replaced the New Deal, contributing to the dangerous material decline of poor urban (mostly African-American) communities, including the quality of transport, public housing and public schools in New Orleans and its environs, anticipating the shameful scenes of public neglect of victims and survivors of Hurricane Katrina. As Illinois senator Barack Obama put it: 'The people of New Orleans weren't just abandoned during the hurricane, they were abandoned long ago.'

Vulnerability of the Poor

As Eric Kleinenberg demonstrated in his masterful study of the Chicago heatwave of 1995,[8] the poor are vulnerable to 'natural' disasters and other catastrophes not because of geography and climate changes (although these set the stage) but because of political lassitude, racism and entrenched poverty, all of these exacerbated by the dismantling of social welfare by both Democratic and Republican administrations that have left them stranded.

In marked contrast to the public response to Katrina, the response to the 9/11 World Trade Center attack was immediate: private companies and public agencies swooped down and into action. Necessary supplies and equipment were put into place with or without contracts. A sense of solidarity united bureaucracies, NGOs and political units. True, there was that long pause, the endless 90 seconds or so that it took George W. Bush to get what had just happened, his deer-in-the-headlights paralysis that was captured so painfully in Michael Moore's film. But this time the presidential paralysis was days and weeks long. No one in the government seemed ready to push a panic button, despite advance warning from the National Weather Service, which declared Katrina a major hurricane likely to make the targeted area 'uninhabitable for weeks, perhaps longer' (quoted in the *New York Times*, 2005).

The amazement with which people around the world greeted the stark images of dead bodies in the lethal sewage of post-Katrina New Orleans contrasted sharply with the 'What do you expect from sub-citizens who refuse to follow orders, who are looting and shooting and raping and killing?' attitude of Fox TV and its associates. Could it be that while white bodies count, black bodies are merely counted? What explains the absurd miscalculations of 20,000, then 40,000 presumed deaths in the wake of the killer hurricane? The body counts, like the exaggerated reports of mayhem, circulated like an urban legend, based on what? A subconscious wish that it be so, a genocidal fantasy?

Today, the attention of the country and the press is focused on reconstruction and on the 'golden opportunity' afforded to developers by the destruction of New Orleans. Congressman Richard Baker of Baton Rouge greeted the devastation with evident glee: 'We finally cleaned up public housing in New Orleans … We couldn't do it, but God did for us.' Today the media are preoccupied with debates about architectural preservation vs economic development.[9] There is talk of allowing certain low-lying sections of Black New Orleans to be 'let go' permanently. In one of his columns conservative pundit David Brooks opined that 'people who lack middle-class skills' should not be allowed to resettle in the city. 'If we put up new buildings and allow the same people to move back into their old neighborhoods, then urban New Orleans will become just as run down as before.'

Will New Orleans be rebuilt with higher levees and fewer African Americans? Will the French Quarter be transformed into a permanent watery theme park for college students on holiday? Will African Americans, Creoles, Cajuns and other Louisiana cultural minorities ever again account for two-thirds of New Orleans' population and for nearly 100% of the city's distinctive culture and social history?

***

Katrina may have tapped into the collective unconscious, pointing to something that Americans need to confront about themselves and their nation. The 'refugee' Freudian slip might be seen as a feeble step toward acknowledging what Michael Harrington recognized decades ago in his book, *The other America* – that is, the reality of two Americas, one bona fide, the other a step-child nation, the un-American America, refugee America, apartheid America. The term 'refugees' implies that there are American-born Americans without a symbolic passport, without a president, without protection, who live and die outside the political circle of trust and care. Perhaps this is why anthropologist Susanna Hoffman[10] suggested that humanitarian efforts for the victims of Katrina might be understood as 'aid', a term most often associated with people living in other countries (as in USAID).

Perhaps the designation 'refugees' is an unformed way of suggesting that 'normative' America (Amerika?) owes something to the displaced victims of American apartheid, something akin to Jacques Derrida's call for a cosmo-politics based on open cities of refuge and a politics of hospitality based on *human* rights, since the Black and poor population's *civil* rights seem to have so utterly failed them.

NOTES

1 The 'double' refers to the social and political responses to the catastrophe that amplify its disastrous effects to the extent that it is difficult to say which is worse – the killer hurricane or the national response to it.

2 On Friday 5 August the *New York Times* reported that Karl Rove and White House communications director Dan Bartlett had 'rolled out a plan … to contain the political damage from the administration's response to Hurricane Katrina.' The core of the strategy, the *Times* report stated, was 'to shift the blame away from the White House and toward officials of New Orleans and Louisiana'.

3 It was not the rains, torrential as they were, that caused the death and destruction. New Orleanians pride themselves on 'toughing out' major storms. It was the breach of the levees and the government's breach of promise to the city of New Orleans that caused the catastrophe.

4 See, for example, D. Gaines, 'The one that's left behind', *Pacific News Service* 8 November 2005, http://www.alternet.org/story/25114/.

5 C. Cooper, 'Old-line families escape worst of flood and plot the future: Mr. O'Dwyer, at his mansion, enjoys highball with ice; Meeting with the mayor', *Wall Street Journal*, 8 September 2005.

6 This is an ironic reference to Martin Luther King's search for the 'beloved community' as the Kingdom of God on earth, his vision of a non-racialized egalitarian society. (See Smith, K.L. and Zepp, I.G., Jr., *Search for the beloved community: The thinking of Dr. Martin Luther King, Jr.* Valley Forge: Judson Press 1974). The Katrina catastrophe showed that only part of the afflicted New Orleans community was 'beloved' and appropriately mourned by the US commander-in-chief in the first days of the tragedy.

7 'Trapped in New Orleans by the flood – and martial law', *Socialist Worker*, 9 September 2005: 4–5.

8 See Kleinenberg 2002.

9 See, for example, J. Shafer, 'Don't refloat: The case against rebuilding the sunken city of New Orleans'. Posted by Slate, 7 September 2005. http://www.slate.com/?id=2125810&nav=tap1.

10 S.M. Hoffman, 'Katrina and Rita', *Anthropology News*, November 2005: 19.

REFERENCES

Bode, B. 1990. *No bells to toll: Destruction and creation in the Andes*. New York: Scribner.

Colson, E. 1971. *The social consequences of resettlement: The impact of the Kariba resettlement upon the Gwembe Tonga*. Manchester: Manchester University Press.

Davis, M. 1999. *Ecology of fear: Los Angeles and the imagination of disaster*. New York: Vintage Books.

Erikson, K. T. 1976. *Everything in its path: Destruction of community in the Buffalo Creek flood*. New York: Simon and Schuster.

Kleinenberg, E. 2002. *Heat wave: A social autopsy of disaster in Chicago*. Chicago: University of Chicago Press.

Linenthal, E. 2001. *The unfinished bombing: Oklahoma City in American memory*. New York: Oxford University Press.

Petryna, A. 2002. *Life exposed: Biological citizens after Chernobyl*. Princeton: Princeton University Press.

SSRC (Social Science Research Council) Katrina site: http://understandingkatrina.ssrc.org/.

16

"Nature", "Culture" and Disasters

Floods and Gender in Bangladesh

Rosalind Shaw

In September 1988, three-quarters of Bangladesh was submerged in the most devastating flood in living memory. This and other recent disasters have generated worldwide concern about the human impact on the environment which, in relation to Bangladesh, has been focused overwhelmingly upon the theory that recent damaging floods are the direct outcome of deforestation in the Himalayas. While deforestation is clearly of crucial importance in relation to such other issues as global warming, its applicability as a unilinear explanation of particular floods is highly questionable (e.g. Currey 1984; Park 1981). More seriously, the most popular solutions generated by this explanation (popular with national governments and international aid organizations), namely large-scale afforestation projects upstream and large-scale construction of embankments downstream in Bangladesh, rely upon one-dimensional conceptions of the interaction between people and the environment. In the case of Bangladesh, such narrowly defined understandings obscure the relationships between floods and those who live with them. It is with these relationships, in particular how they vary for men, women and others with differential access to material and symbolic resources and consequently different 'environments', that this chapter is concerned.

A fundamental aspect of the people–environment relationship is emphasized by geographers researching natural hazards. Thus 'Floods', wrote Gilbert White, 'would not be a hazard were not man tempted to occupy the floodplain' (1974: 3). Such writings commonly begin in this way. Part of a reaction against environmental determinism, geographers were careful to incorporate human agency into the definition of 'hazards', raising a quite different issue from the adverse environmental impact of, e.g., pollution or deforestation. The question here is not what *causes* events but their *conceptualization*.

Yet such research has all too often failed to live up to the promise of its humanized definitions of hazards, since in actual usage in the same writings the term often becomes interchangeable with geophysical events themselves – floods, droughts or earthquakes (see Torry's criticism of this semantic slippage (Torry 1979a: 370)).

The Anthropology of Climate Change: An Historical Reader, First Edition. Edited by Michael R. Dove.

Indeed, this primacy given to an apparently autonomous physical environment is embedded within the term 'natural hazards' itself, in which such hazards are attributed to nature. Although the delineation of 'natural' and 'man-made' events is highly problematic (see Turton 1979), a view polarizing 'nature' and 'culture' is characteristic of thought patterns concerning hazards in many western institutional contexts – international-aid agencies, the environmental movement, natural hazards research and even anthropology.

This polarization has apparently marginalized research on hazards within anthropology since they are ascribed to nature. Anthropologists recognize that the 'nature' versus 'culture' dichotomy is socially constructed and of restricted cultural provenance, so it is ironic that the definition and practice of 'cultural anthropology' reifies it. Our conceptions of disciplinary boundaries appear to be protected from those insights about the 'nature–culture' opposition which our own discipline has developed. Even cultural ecology and ecological anthropology, explicitly concerned with the environment, have themselves been pervaded by this polarity in the form of biological and cybernetic models of culture's 'interaction' with or 'adaptation' to a distinctly conceived 'natural environment' (e.g. Bateson 1972; Rappaport 1979). A recent exception, however, is Ingold's critique from within ecological anthropology of the latter's tenet of cultural 'adaptation' to the environment (Ingold 1986). Prior to this, the major alternative to such neo-functionalist models within anthropology has been the work of the French Marxist school of economic anthropology, in which drought and famine in the Sahel are situated both ecologically and historically, in terms of the history of colonialism, cash-cropping and the creation of dependency as well as the climate (e.g. Meillassoux 1974; Copans 1975; 1983).

Apart from this, the nature–culture distinction has been used to define disciplinary boundaries, confining hazards to the 'proper' domain of geography. Hazards research has usually, until recently, been approached within geography in such a way as to eclipse the social and cultural nature of hazards by the 'natural' nature of hazards. In such research, investigations of human involvement have usually been limited to such questions as how accurately people appraise risks, how they cope with and adapt to 'the hazard' (the cyclone or the earthquake, in the definition contours of which people are now excluded), and/or how personality factors influence both of these. Causal priority is given to nature. People merely respond and adjust on the basis of 'subjective' knowledge. Attention has been given to people not as social actors but as bearers of idiosyncratic qualities such as 'personality differences'. Khan (1974), for example, relates responses to cyclones in Bangladesh to such constructs as 'optimism and pessimism' and to 'superstitious' beliefs.

In their 'adaptationist' assumptions, these approaches run parallel to ecological anthropology (Watts 1983: 235). The dominant perspective which they represent, however, has been challenged by radical critiques within geography which both parallel and draw from the analyses of the French neo-Marxist anthropologists (see esp. Hewitt 1983). Watts (1983) argues that the escalation of disasters over the past 50 years, and their increasingly 'Third World' locations, can be much more clearly attributed to domination and dependency than to climatic change. Similarly, Susman *et al.* (1983) formulate a theory linking disasters to the process of underdevelopment, in which marginalized people are forced into 'marginal', often hazard-prone areas. Slum-dwellers of Guatemala City, well aware of this, referred to the 1976 earthquake as a 'classquake' (ibid.: 277). Since 'natural hazards', then, are as much products of human agency and power as of geophysical events, 'we would be right to replace the term *natural* with the more appropriate term *social* or *political* disaster' (Richards 1975).

Such approaches entail the dissolution of two polarities implicit in most natural hazards research. The first is that of the natural and human worlds: instead of discrete entities in interaction, they are a people/environment mutuality in an '*inner*action' which is internally differentiated according to social relations and inequalities (Watts 1983: 234, following Sayer 1980). The second polarity is that of 'normal' versus 'abnormal' events, which misrepresents the precariousness and instability of everyday life for those whose marginality makes them vulnerable to hazards. As Hewitt (1983) has observed, the 'forces of nature' which furnish the basis of explanation in such studies of natural hazards are typically represented as discrete, sharp discontinuities, distanced from everyday life by their 'un'-ness as *un*predictable, *un*precedented, *un*certain, *un*managed situations. In our mental maps, a 'disaster archipelago' is drawn:

> In the technocratic style of work there is a structure of assumptions, and a use of science and management that always situates natural calamity beyond an assumed order of definite knowledge, and of reasonable expectation. More importantly it places disaster outside the realm of everyday responsibility both of society and individual. More important still, it makes assumptions about everyday life – about its being 'normal', 'stable', 'predictable' – that are in turn debatable. (Hewitt 1983: 16)

Although hazards are thus viewed as largely removed from 'ordinary' human action, they are represented as potentially amenable (to some degree) to prediction and control by specialists: planners, managers, scientists and engineers. We thus have a hierarchy of agency, underpinned by a conception of hazards as 'nature' which favours the authority of (usually western expatriate) experts and institutions and their scientific and technical knowledge, and which tends to discount the 'culture' of those who live with hazards.

Floods and Bangladesh

To those in Bangladesh the 1988 flood was clearly a hazard. But far from being distanced from everyday life, many of its features (understood as a people/environment 'inneraction') were constituted by social processes and structures. In Bangladesh, highly variable floods are a regular feature, and experiencing some as 'hazards' does not depend merely upon the presence of people who live in the floodplain, but upon how those who do so *utilize* floods and, most importantly, who has access to such use. Thus, the 'hazardous' nature of the 1988 flood was differently constituted for men and women, for urban and rural dwellers and for poor and wealthy.

Being on the delta of three major rivers, Bangladesh is flooded annually, no more than a third of the country being submerged at any one time. Because of its fertile alluvial deposits, floods are resources, enabling three harvests per year. The cropping pattern spreads the risk of flood or drought by the cultivation of different rice varieties on different elevations of land at different times of year (Ralph 1975: 46–50). The rice variety *aus*, which cannot survive flooding, is planted in the dry season in February/March and harvested in June, while the flood-tolerant *aman* variety is also planted during February/March, grows in the monsoon floods from June to late September and is harvested in October. Sometimes both are sown together in the same field: *aus* will survive if the water level is low, *aman* if flooding is greater. Since the *aman* harvest is the main one, it might be said that an adjustment has been turned to an advantage so that farmers have become dependent on floods.

Other crucial resources provided by rivers and floods (the few available to the poor landless) include fish, shellfish, snails and turtles, the major (and often only) source of animal protein available to poor families (Bangladesh Agricultural Research Council 1989: 4–5). While not everyone in rural areas has access to land, the surface floodwater is a common property resource to which the poor have equal access, the reduction of which would itself constitute a hazard for the landless.

The construction of villages and houses are part of everyday 'flood-practices'. Villages are built on mounds (*bhiti*) whose heights indicate the usual levels of flooding, and houses are further raised on mud plinths, which are renewed after every flood, if possible, although this is a very labour-intensive task. Some houses also have a 'false roof' (*kar*), where goods can be stored and people can inhabit if necessary. If water enters the house, the bed (*choki*) becomes the living area: households live, cook and store belongings on it, raising it on bricks. Food, clothes and even chickens are hung from the roof in jute nets (*shika*). Cattle and goats are placed on bamboo platforms (*machan*) or taken to a road upon an embankment.

These are just a few examples of the varied (and differential) ways in which people live with and use floods. This is not to suggest a 'traditional' perfect environmental homeostasis, a long coexistence in a benign equilibrium. Damaging floods have characterized this delta for centuries. An account of a flood in 1787–88 could describe the 1988 flood: the rivers rose to such a height that they flooded the country 'to an extent never remembered by the oldest inhabitant', boats sailed through the flooded streets of Dhaka, and those in rural areas took refuge on rafts and bamboo platforms (Taylor 1840: 301, cited in Ralph 1975: 10). Sixty thousand people were estimated to have died, either in the flood itself or in the resulting famine (ibid.).

Most floods are not so extreme but are still very variable in timing, duration and magnitude. They also effect changes in parts of the river system from year to year: different areas are flooded, and rivers change their course. Flooded land is resurrected in the form of highly-fertile new islands (*char*), and form the basis of fresh settlement. So farmers are continually adjusting cropping strategies as floods vary, perhaps losing their land through erosion, migrating to new areas and exploiting new islands (immediately colonized and controlled by the most wealthy landowners). Farmers also negotiate ritually with powers behind the river, praying in the mosque and sometimes throwing offerings into the water.

There are two related points here: first, the duality of floods as resources and hazards – even the 1988 flood resulted in excellent harvests in some areas, although taking the country as a whole, a third of the *aman* harvest was estimated to have been lost. Second, given this duality of floods (and countless other contingencies), 'everyday life' for most Bangladeshi farmers is unstable, unrepetitive, but characterized by a series of calculated risks and strategies.

The duality of floods finds expression in Bengali terminology. Whereas the English term 'flood' connotes an abnormal phenomenon, in Bengali there are different terms, which vary from region to region, with different connotations. Around Dhaka, *borsha* and *bonna* are used. *Borsha* is used generally for the monsoon season, and specifically for rain and the annual inundation of the river, thus associating flooding with a certain degree of regularity and seasonality. *Bonna*, like 'flood', implies extremity, but refers to a different range of contexts from the English term. Instead of a sharp dichotomy between a predictable 'normality' and an unpredictable 'hazard', the distinction between the Bengali terms suggests the crossing of a line on a continuous scale beyond which the damage of the 'flood-as-hazard' outweighs the benefits of the 'flood-as-resource'. The line will not be the same throughout society: the asymmetries of power and interests between men and women, landlords and the landless, rural and urban dwellers will mean differentiated environmental 'inner-actions' and very different experiences. When these asymmetries change through time, the distinction between *bonna* and *borsha* changes with it. Although it is questionable whether extreme floods have increased in the recent past, it is certain that people have become more flood-vulnerable, largely due to processes set in motion during the colonial era.

Prior to British colonialism, Bengal was a prosperous province of the Mughal Empire, famous for its cotton industry and agriculture. In the mid-eighteenth century, the British East India Company established control and acquired through coercion and violence large quantities of cotton cloth at minimal cost from Bengali weavers, which they sold at considerable profit. Subsequently, the cotton industry was eliminated to create a monopoly for imported cotton from Manchester. Towards the end of the eighteenth century, under British colonial rule, agriculture suffered due to changes in land tenure. Formerly, the Mughal rulers had collected land tax from Bengali farmers via an elite known as *zamindars*. Under British rule, the *zamindars* were made the owners of the land from which they collected taxes, as part of the common British strategy of creating a loyal elite. The *zamindars* (mostly Hindu) then moved from predominantly Muslim East Bengal to Calcutta in West Bengal, becoming absentee landlords who used the rents from their tenant farmers (inflated via intermediaries who subleased the land) to invest in Calcutta. Thus agriculture stagnated and the tenant farmers, forced to become indebted to moneylenders in order to pay their rents, grew increasingly poor. Jute cultivation was encouraged as a cash crop, but its processing took place outside East Bengal. There was a second era of colonialism in 1947 when East Bengal became East Pakistan. While Bengalis formed the majority in Pakistan, they were greatly under-represented in the government, the civil service and the military, and their jute exports were controlled by and invested in West Pakistan. When Bangladesh became independent in 1971, it had a colonial legacy of over two hundred years in which its surplus had been channelled out to finance development elsewhere. Today, a third of rural households are landless, rising to 48 per cent if those households owning less than half an acre are included, the latter considered functionally landless (Chen 1986: 60).

Landless households do not, then, have the same access to the flood-as-resource as landowners. For poor households (unable to maintain the *bhiti* of their house, unable to afford a bed, without reserves of rice and/or having been forced to move into more marginal and flood-prone areas), the flood becomes disastrous considerably earlier than for resource-rich households, whose flood practices have not been undermined.

Gender, danger and floods

For men with land in rural areas, cultivation focuses not only one's relationship with the physical environment, but with the spiritual world. 'A good Muslim is a good farmer' is a common Bengali expression. The origins of agriculture, Islam and flooded rivers in the region are linked historically. The shift of the Ganges delta from West to East Bengal was accompanied by the extension of Mughal rule and economic exploitation of the area, leading both to intensive rice cultivation, and to large-scale conversion to Islam (Eaton 1985).

Bonna, accordingly, is perceived by farmers in terms of disruption to cultivation. According to Ralph's study of floods in Char Bhabanathpur village, southeast of Dhaka:

> The villagers describe abnormal flood to be those times when the water rises one and a half to two times the normal height on the fields (fifteen–twenty feet on *aman*, as opposed to the normal eight–nine feet). A few call abnormal those floods when water comes on the *bhiti* and into the homestead area. (Ralph 1975: 71)

Since the focus of Ralph's study was primarily agriculture, the respondents to the questionnaire were male (ibid.: 61–5). Whereas only a few men in Ralph's sample defined *bonna* in terms of disruption to the homestead (*bari*), this disruption is much more central to women's perceptions: a flood is *bonna* when it enters the home.

For most Bangladeshi women, the experience of the environment is centred upon purdah, as well as concepts of pollution which have become incorporated into Bengali Islam. Purdah ranges from women's seclusion in the home (except when wearing the *burqa*, a long black garment completely covering head and body) to simply placing the end of the sari over the hair and avoiding eye contact with male affines and strangers. These observances are stricter among Muslims than among Christians, among young brides than older women, and may vary for the same person in different circumstances (Ahmed and Shamsun Naher 1987: 53–60).

Many scholars have also observed the status associations of purdah practices, particularly their more complete observation among the wealthy. Deserted or poor women in landless families are forced to face the shame (*lojja*) of working outside the home to survive. Elite women, however, are often able, by for example having had a western education or a professional occupation, either to abandon purdah practices or to redefine them as the 'inner' observance of modesty (ibid.: 55, 57). For women in between, work is usually confined within the homestead.[1] However, such women are responsible for storing and germinating the seed rice and threshing, winnowing, parboiling and husking the paddy after the harvest. They also grow vegetables in small plots near the home, while men tend larger plots further away in lowland fields (ibid.: 67–8). However, such work is not considered to be 'work' by men, and is undervalued and discounted (see Chen 1986; Hartmann and Boyce 1983).

Purdah is underscored not only by status considerations, but also by concepts of pollution. Purity and pollution have, as Blanchet (1984: 30–50) has argued, become transformed in the Muslim Bengali context into auspiciousness and misfortune. Women are defined as more subject to pollution than men, particularly with regard to childbirth and menstruation, and this pollution is construed as potentially dangerous for crops, cows, the river and the granary – all of which as Blanchet (1984: 48) observes, form the basis of the rural economy. At the same time 'the perfect wife' (defined in terms of thrift, modesty, cleanliness and a disinclination to quarrel) is linked with fertility and prosperity. Thus, auspiciousness is maintained and pollution guarded against by purdah observances restricting women's contact with the environment.

The latter relationship, however, is more complex than such restrictions might suggest. According to Blanchet, 'Women have more to do than men with … spirits of the land, for women are associated with fertility' (1984: 14). Moreover, Blanchet argues that by performing rituals involving local spirits who may either disturb or protect them when menstruating, pregnant or giving birth, and as mothers, 'women have maintained separate pre-Islamic and pre-Brahmanistic traditions which give meaning and value to their specific roles and functions in society' (Blanchet 1984: 16).

This argument appears, however, to overstate the positive aspects of women's relationships with land and river spirits. Purdah restriction is central to such relationships. In Blanchet's research area in the northeast of Bangladesh, considerable ambivalence is apparent in the relationship between women and Kwaz, an Islamicized guardian spirit of the river (ibid.: 44–8). Kwaz is honoured at marriages, during pregnancy and protects children from attack by ghosts (*bhut*) during their first 5 or 6 years. Sometimes Kwaz is attracted to women and their sexuality, sometimes pulling a woman bathing naked into the depths of the river; conversely menstruating women are forbidden to bathe in the river because he is angered by female pollution which brings calamity. Kwaz's anger manifests itself by river erosion, in which he 'eats' the river bank, aided by invisible manual workers.

The relationship between women and another group of beings is, however, unambiguous. River banks, bamboo groves and road junctions are inhabited by *bhut*, ghosts of those who have not died peaceful deaths: suicides, murder victims, stillborn children and women who died in childbirth. Blanchet has also collected traditions which claim that *bhut* were once masters of the land prior to the 'great religions' (1984: 54). Everyone is vulnerable to possession or illness from *bhut*, especially women and children. *Bhut* are especially attracted to women when they are at their most polluting, and to foetuses and young children. When *bhut* are most active (usually dawn, sunset, midday and midnight), and during menstruation, pregnancy, and up to seven days after giving birth, women should stay indoors. The best protection is afforded by purdah practices; non-observance endangers a woman and makes her children more vulnerable. There is considerable overlap between *bhut* and the specifically Islamic *jinn* and *pori* ('fairies'), as well as a more impersonal and indistinct category known as 'bad wind' (*batash*), protection from all of which is secured by remaining inside the home and by wearing Islamic amulets (*tabiz*). Since, then, *bhut* attack both sexes, but are particularly feared and avoided by women, can women really be said to be their custodians?

Blanchet argues that the ambivalence of women's sexuality in relationships with these 'spirits of the land' derives from the different chronological layers of religious ideas, in which female sexuality is valued positively in relation to the original, local spirits and negatively in relation to Brahmanized deities and to Islam (1984: 47–8; 61–3). However, the static spatial image of layers may obscure the extent to which originally distinct sets of ideas merge and mutate in 'inneraction' with each other. Thus 'whether women are interacting with high spirits who are repulsed by their pollution, or low *bhut* who are attracted to it, the end result is the same; a woman is always safer "inside"' (Blanchet 1984: 48). There is a contrast between harmful autochthonous spirits of the 'outside' which endanger women, and beneficent Brahmanized spirits of cultivation which are endangered by women. But in both cases, purdah practices are protective. There is overlap in the case of the river, whose power either enables or destroys cultivation, but here, too, the association of women with fertility and prosperity in relation to the river depends upon its insulation from direct contact with female sexuality. The interfusion of earlier traditions with ideas from Brahmanical religion and Islam has given rise to contradictions in women's relationships with the spirits, sometimes being custodians, sometimes victims. (See Blanchet 1984: 69–123 for an account of birth rituals.)

Women's experience of floods, then, is shaped by the necessity to maintain purdah and pollution-removal practices. Floods may cause women to temporarily abandon purdah and pollution-removal practices, as Ahmed and Shamsun Naher relate:

> I have been to the fields twice, once when flash floods suddenly hit upon us, the crops would have been ready for harvesting in two or three days, but with the flood waters rising there was no way we could wait. Kamlas [labourers] could not be found and we couldn't even wait for them to be hired, so then I accompanied my husband to the field and did the harvesting along with him. No, no one spoke bad of me because it was such an emergency situation and then, every one else was in the same boat. But villagers refer to it as the bad times (*durdin*), when *even* women had to go to the fields. (Ahmed and Shamsun Naher 1987: 68)

When not forced out by such contingencies, women tend to be confined more than usual by floods. During *borsha*, the mobility of women may be curtailed, since their visits to each other are impeded by mud or water (Sushila Zeitlyn, pers. com.). During a severe flood (*bonna*), female mobility is severely restricted, since women's work is confined to the dry space of the bed. Flooding can also reduce restrictions on women since boat transport is facilitated: this is the traditional occasion for the *naior*, a wife's visit to her natal home, where she usually has much more freedom of movement and is fed better food than she would eat in her husband's household. Thus, for women and men there is a duality to floods. A particularly damaging flood, however, has more serious consequences for women than for men. In addition to women's work being confined and undervalued by men, wives in poor homes are perceived as a burden and finally deserted by their husbands during the severe impoverishment, which is the long-term consequence of flood disasters for the very poor.[2] The downward spiral of impoverishment, then flood-vulnerability, increasingly damaging floods and further impoverishment, has thus been steeper for poor women.

The Flood of 1988

In the summer of 1988, the river Buriganga suddenly flooded alarmingly so that two-thirds of Dhaka (later three-quarters) were underwater. Relief camps for flood victims were set up by the government, wealthy businessmen and various organizations. I regularly visited a camp in Azimpur, in the southeast of the city and also visited an island on the outskirts of Dhaka (Kamranghir Char), from which most victims in the camp had come. Kamranghir Char is a settlement of landless migrants from rural areas who came to Dhaka to find work (mostly as rickshaw pullers), and who live in the Char because rents are cheap owing to the low-lying area's vulnerability to flooding.

The government-run camp was directed by a city Commissioner, and the daily organization was managed on a volunteer basis by young men (mostly relatives of the Commissioner) from two organizations whose membership overlapped: the junior branch of the ruling Jatiya party, and a cultural group which performed Bengali plays. There were over 1,200 people in the camp, divided into 260 families, using six bathrooms into which the water flowed for about half an hour in the morning and an hour in the evening. Each adult received a half kilogram of uncooked rice (almost) every day. Food was cooked by individual households on their own stoves (*chula*) (see Shaw 1989).

Most people in the camp had already passed through two major stages: living on a raised bed in a flooded house followed by living on the roof. Once outside on the roof, people extended their usual flood strategies: bamboo platforms were built, and some people lived on top of their bed raised by bricks on the submerged roof itself. They carried as many possessions as possible on to the roof with them, and built shelters out of anything they could find. In one case two households had a jute sack stretched over a couple of square yards of roof space, which provided some privacy for women. Rickshaw pullers, all men, could still work as rickshaws and boats were the only viable forms of transport in much of the city. They worked in Dhaka, bought food and went back at night to their families on the roof.

For most families, considerable pressure was required before they left their homes for the relief camp. The risk of having their possessions stolen or squatters in their houses was thought to be worse than staying on the roof. Staying put meant not having a secure food supply or clean water, becoming ill without access to medicines, being bitten by the snakes which competed with people for shelter, and the attentions of 'muscle men' (pirates, *mastans*). Moreover, the reduced privacy, and thus purdah for women, would be completely lacking in the camp, compared with the shelters, a situation both wives and husbands were reluctant to have. Since Kamranghir Char is a community of migrant workers with kin and affines in different parts of the country, only very few women were able to make the *naior* to their natal home (which probably would also have been flooded). For many families it was the illness or near-accident of a child (usually falling in the water) which precipitated the move to the camp, as Muhammad, a rickshaw puller, recounts:

> As we were living in an open space my children became very sick. Then I asked other people about a better shelter. They told me about this camp. We stayed on the roof for two days. Living on the roof is very dangerous. I couldn't go to work, leaving my children on the roof. I was worried about their health, problems with washing; and of course there's always a chance the children may fall in the water.

In the camp at Azimpur, households occupied a single mat each in the rooms and corridors of the school, their stove in front of them and their salvaged possessions (cooking pots, knife, a change of clothes and a couple of quilts) demarcating them from neighbouring households. Most men worked pulling rickshaws during the day, often while weak and ill. The women mostly stayed in the camp, queuing for rice and medicines, cooking, caring for their children and visiting each other, the worst aspect being the difficulty in maintaining purdah. To be seen by strangers while washing, sleeping and especially eating (since a wife is defined as a provider, not a consumer, of food) caused them great shame. The latter was perceived by all as the cost of physical survival, but with far-reaching consequences extending beyond shame.

For instance, living in the camp (or on the roof of a submerged house) entailed exposure to the 'invisible hazards' of *bhut* and *batash*. All of the households in the camp had had at least one sick child, whose illness was usually attributed both to the camp's dirty conditions and to attack by a *bhut* or *batash*. Also, the camp was situated next to a graveyard, a favourite habitat of these entities.

The predicament of women who gave birth during the flood, and were unable to perform rituals to remove pollution and thereby protect themselves and the babies from *bhut* and *batash*, was perceived as particularly serious. A young mother in the camp, Halina, had given birth in her flooded house, and remained there with her husband for 3 days while her mother-in-law searched for a camp. Her mother-in-law, Momataz, was a midwife (*dai*), and described her problems in not being able to carry out birth rituals after the birth of Halina's baby:

> It was in the morning when my grandchild was born. I was standing in knee-deep water in the room during his birth. As soon as it was over I came to Dhaka to look for shelter. After three days, when I went back to get them,

it was impossible for them to get out of the house. The roof had to be taken off. As I knew many people in this neighbourhood, they gave me shelter in this camp....

There are so many people here in this camp, she can't wash herself, or eat, or take rest. A new born child is not to be taken outside the house for the first seven days, but we had to bring the child out. Then the child is vulnerable to a *batash*. If a child is possessed by a *bhut* or *batash* he will stop taking any food and will be very irritated all the time.... Usually, we bury the placenta and the umbilical cord in the earth so that no *jinn, bhut* or *batash* can put their eye on it. When my daughter-in-law had the child I tied his placenta to a rock so that it wouldn't come up to the surface. After seven days the child's hair is shaved off. This hair too, we bury it. I have kept the child's hair. When we go home I will bury it in the ground. If this is not done, the child will get a very severe cold which will last a long time. If the mother doesn't follow the customs (*achan*) properly and does things which she is not supposed to do, this harms the child. The child becomes sick, and it may even cause the child's death. A new baby and the mother are not allowed to go out of the house for the first seven days, except for going to a doctor or a *fakir*. Because at this time they are unclean (*napak*): this will harm cattle, crops and other things.

Thus the flood was far more dangerous for Halina and her baby than for other women and children. Purdah and pollution beliefs transform the whole character of floods as 'hazards' for women, and this varies in degree not only between different women (poor or wealthy), but for the same woman at different times. This gendered nature of hazards also intersects with rural and urban contrasts.

For both men and women in the urban environment of Kamranghir Char, the duality of floods as both resource and hazard has disappeared: the men are not farmers who depend upon certain levels of flooding, while the women, usually far from their natal homes, cannot visit during a flood. A further contrast lies in men's perceptions of *bonna*, which rickshaw pullers in Dhaka do not, of course, define in terms of disruption to rice cultivation but, like women, in terms of water entering the house. Additionally, since the river, crops and cattle are no longer central to the economy of the residents of Kamranghir Char, women's 'ideal' behaviour and ritual observance no longer enhances fertility. It is interesting, however, that this relationship between women, cattle and crops was drawn upon by Momataz in explaining the negative consequences of the flood's disruption of women's birth rituals. But even this underscores the fact that for women, the negative consequences of uncontrolled contact with spirits of the land are a feature of urban as well as rural life, while the positive aspects of their relationships with certain of these spirits are not.

Conclusions

The deforestation theory of disastrous floods presumes a balanced 'natural' environment which does not include human culture and which is upset by (in this case local) human agency. Both of the favoured solutions – large-scale afforestation and embankment-building – imply that the forces of nature thus unleashed can only be returned to their presumed former equilibrium by applying specialized ecological and technical knowledge by western-trained experts. Such one-dimensional explanatory schemes, single-tracked upon a straight path, are based upon a perception of 'nature' and 'culture' as dichotomized entities which have 'an impact' on each other in mechanical Newtonian cause-and-effect sequences without being *instantiated* in each other. Thus the social and cultural nature (and consequently the heterogeneous, differentiated nature) of 'natural hazards' is disregarded. This chapter shows that floods not only have varying consequences for rich and poor, men and women, and rural and urban dwellers, but that their very nature as hazards is *constituted* by these and other forms of human social difference.

By overlooking this heterogeneity, large-scale afforestation projects and high-tech embankment projects fail to take account of their own differential consequences, in which they may themselves present hazards for the most vulnerable people. Certain afforestation projects, for example, have benefited industry rather than the rural poor (see Shiva *et al.*

1982) and, by failing to alleviate poverty, may in fact be contributing to further deforestation. As Currey observes, it is because the poor urgently need resources that the tree resource is being reduced, yet '[t]he same fences which keep the goats and cattle from reafforestation projects may also be the boundary lines for poor farmers who formerly grazed their goats on the scrubland' (Currey 1984: 10). Similarly, the construction of embankments can reduce the soil moisture and fertility (Hossain *et al.* 1987:36; Rasid and Paul 1987:164), leading to deteriorating rice harvests for those farmers who cannot afford irrigation. Embankments can also reduce surface water and the latter's resources, upon which, as already seen, the poor particularly rely. Also, because husbands desert their wives and children during extreme hardship, any further impoverishment generated by such projects has more serious consequences for women and children than for men.

Setting aside such considerations, in a river system as vast, complex and shifting as that of Bangladesh, the practical problems of flood control by large-scale building of embankments are immense, and perhaps insuperable. In the end the vast sums of money may be 'poured into the water', a high-tech equivalent of the ritual offerings traditionally thrown into rivers. For Bangladeshi farmers, rickshaw pullers and most women the powers behind the rivers are considerably easier to negotiate with than the powers behind the World Bank, which are unamenable to local knowledge and established strategies. For many western aid organizations, however, disasters such as the 1988 flood are important resources ('We'll get three more jobs out of this flood', crowed an official from one such organization in Dhaka), as large-scale projects ('offerings') maximize the flow of money ('fertility'). This might be described as a new duality of floods, in which the agency of those in charge of mega-projects is augmented while the agency of the majority of Bangladeshis in their relationship to floods is impaired.

The perceived remedies are, to say the least, uncertain. Many geographers in Bangladesh are advocating alternatives to embankment-construction, for example, the development of local warning schemes, zoning and the support of indigenous strategies towards floods at village level (e.g. Islam 1986; Rasid and Paul 1987). It is even more important to recognize that since floods are constituted as environmental disasters by poverty and, additionally for women, by ideas of female pollution, they can be radically ameliorated by attacking poverty and empowering women.[3]

NOTES

1 This is less true of Christian women. Ahmed and Shamsun Naher observed Christian women – especially older ones – working in the fields beside their husbands (1987: 61–2).
2 The Association of Development Agencies in Bangladesh's Flood Disaster Report (1984) states: '[o]ne striking phenomenon is that no less than eleven organisations have reported increased instances of abandonment of wives and children by menfolk' (1984: 2).
3 Female empowerment is likely to be, if anything, undermined by patronizing attempts to 're-educate' women about purdah and pollution. Development programmes which increase poor women's access to economic and social resources, on the other hand, can sometimes transform women's attitudes and relationships along with their economic security and opportunities (Chen 1986).

REFERENCES

Ahmed, R. and M. Shamsun Naher. 1987. *Brides and the Demand System in Bangladesh*. Dhaka: Centre for Social Studies, Dhaka University.
Association of Development Agencies in Bangladesh (ADAB). 1984. ADAB's Flood Disaster Report. *ADAB News* XI (6): 2–3, 6.
Bangladesh Agricultural Research Council. 1989. Report on Floodplain Agriculture. Dhaka.
Bateson, G. 1972. *Steps to an Ecology of Mind*. New York: Ballantine.
Blanchet, T. 1984. *Meanings and Rituals of Birth in Rural Bangladesh*. Dhaka: University Press Limited.
Chen, M. A. 1986. *A Quiet Revolution. Women in Transition in Rural Bangladesh*. Dhaka: BRAC Prokashana; Cambridge, MA: Schenkman Publishing Co.
Copans, J. ed. 1975. *Sécheresses et famines du Sahel*. Paris: Maspero.

—— 1983. The Sahelian drought: social sciences and the political economy of underdevelopment. In *Interpretations of Calamity*, K. Hewitt, ed. London: Allen & Unwin.

Currey, B. 1984. Fragile mountain or fragile theory? *ADAB News* XI (6): 7–13.

Eaton, R. 1985. Approaches to the study of conversion to Islam in India. In *Approaches to Islam in Religious Studies*, R. C. Martin, ed. Phoenix: University of Arizona Press.

Hartmann, B. and Boyce, J. 1983. *A Quiet Violence. View from a Bangladesh Village*. London: Zed Press.

Hewitt, K. ed. 1983. *Interpretations of Calamity*. London: Allen & Unwin.

Hossain, M. A. T. M., A. T. M. A. Islam, and S. K. Saha. 1987. *Floods in Bangladesh. Recurrent Disasters and People's Survival*. Dhaka: Universities Research Centre.

Ingold, T. 1986. *The Appropriation of Nature: Essays on Human Ecology and Social Relations*. Manchester: Manchester University Press.

Islam, M. A. 1986. Alternative adjustments to natural hazards; implications for Bangladesh. Presidential Address, 11th Annual Bangladesh Science Conference, Rajshahi University, March 2–6, Section V: Geology and Geography.

Khan, A. A. 1974. Perception of cyclone hazard and community response in the Chittagong coastal area. *The Oriental Geographer* XVIII: 1–25.

Meillassoux, C. 1974. Development or exploitation: is the Sahel famine good business? *Rev Afr Polit Econ* 1.

Park, C. C. 1981. Man, river systems and environmental impacts. *Progress in Physical Geography* 5: 1–31.

Ralph, K. A. 1975. Perception and adjustment to flood in the Meghna flood plain. MA thesis, Department of Geography, University of Hawaii.

Rappaport, R. 1979. *Ecology, Meaning and Religion*. Richmond: North Atlantic Books.

Rasid, H. and B. K. Paul. 1987. Flood problems in Bangladesh: is there an indigenous solution? *Environmental Management* 11: 155–73.

Richards, P. ed. 1975. *African Environment: Problems and Perspectives*. London: International African Institute.

Sayer, A. 1980. *Epistemology and Regional Science*. Falmer, Sussex: School of Social Science, University of Sussex.

Shaw, R. 1989. Living with floods in Bangladesh. *Anthropology Today* 5 (1): 11–13.

Shiva, V., H. C. Sharatchandra, and J. Bandyopadhyay. 1982. Social forestry: no solutions within the market. *The Ecologist* 12 (4): 158–63.

Susman, P., P. O'Keefe, and B. Wisner. 1983. Global disasters: a radical interpretation. In *Interpretations of Calamity*, K. Hewitt, ed. London: Allen & Unwin.

Torry, W. I. 1979a. Anthropological studies in hazardous environments: past trends and new horizons. *Current Anthropology* 20: 517–31.

—— 1979b. Hazards, hazes and holes: a critique of *The Environment as Hazard* and general reflections on disaster research. *Canadian Geographer* XXIII: 368–83.

Turton, D. 1979. Comment on W.I. Torry, Anthropological studies in hazardous environments: past trends and new horizons. *Current Anthropology* 20: 532–3.

Watts, M. 1983. On the poverty of theory: natural hazards research in context. In *Interpretations of Calamity*, K. Hewitt, ed. London: Allen & Unwin.

White, G. F. ed. 1974. *Natural Hazards: Local, National, Global*. Oxford: Oxford University Press.

Part IV

Knowledge and its Circulation

Emic Views of Climatic Perturbation/ Disaster

17

Typhoons on Yap

David M. Schneider

Introduction

Typhoons are common events in the Western Caroline Islands. They are well known and a chronic threat. How do a people faced with such a recurrent threat handle the situation? How do they prepare for typhoons? What do they expect from them? What do they do until the typhoon comes? What do they do during one and afterwards?

Four typhoons struck Yap on November 2nd and 10th, December 23rd of 1947, and on January 13th of 1948. These were not the first typhoons to strike Yap, nor are they likely to be the last. They are, however, the ones I was able to observe myself. The material for this paper is based on field work on Yap which took place before, during, and after these four typhoons.

There were no deaths from any of these four typhoons, and only a very few minor injuries. Property damage, in houses and canoes demolished, was considerable, while damage to food resources was significant but not devastating. This pattern of low casualty rate, for either impact or post impact period, is characteristic of high islands in the Western Carolines which are not overpopulated with respect to food resources. Where population density is such as to press food resources, or where the island is a low atoll (perhaps five feet above sea level) the risk of devastation is much greater. The low islands can be swept by deep water so that only a very few escape, while even shallow inundation by the sea ruins vegetable foods. A further source of risk for both high and low atolls is to canoes at sea. Yap has close ties with the islands to its east and southwest and these ties depend on long overseas voyages. Typhoons and severe storms account for the loss of many sailors.

Despite the relatively low casualty rate, typhoons are classed by the Yap people themselves with disasters or catastrophes. Typhoons, epidemics, tidal waves, earthquakes and a phenomenon with which we are unfamiliar, thunder falling to the ground (which is said to be fatal for anyone in the area) are all grouped together.

Typhoons have a characteristic pattern. First, warning is always there but it is difficult to predict just how bad the typhoon will be. The sky darkens, the sea becomes angry and rises, wind velocity increases. How high the wind velocity will get, or how high the water will rise is not, however, clearly predictable. There is always hope that any given moment is the worst and that the next period will see the storm abate. The clear warning with the slow buildup time permits certain kinds of anticipatory action. There are often two to three hours at least within which something can be done, and often more time than that. Even if the

The Anthropology of Climate Change: An Historical Reader, First Edition. Edited by Michael R. Dove.

typhoon comes at night there are these warnings and, although it may be more difficult to do as much at night, it is still possible to act.

Second, the impact is not instantaneous but stretches over a period of time which may vary from one to three or four hours.

Third, the dangers are fairly clear-cut and predictable. These dangers are from flying material (coconuts, fronds, branches, etc.), collapsing houses, and rising water. Since the first danger comes from rising water and can be anticipated, canoes are the first objects of attention. They are of course quite valuable and some pains are taken to protect them. They will be taken into sheltered water or drawn far inland and protected. After this, refuge is taken inside a house. This is good protection against flying objects and falling trees. The danger from collapsing houses is actually fairly small, because of the type of construction. Yap houses consist of stout uprights set deep in the house foundation with horizontals set across them, the whole of which supports the bamboo framework of the roof. All connections are made by tying with coconut fibre cord; nails and pegs are not used. Coconut fibre cord expands when damp, and even when dry has a certain amount of "give." The house is thus far from a rigid brittle structure. I was in a number of different houses which collapsed and in each case the collapse was slow, steady, gradual. At no time was it necessary to dive for the nearest exit; instead the house slowly settled as one cord after another first stretched, creaked and groaned, and finally broke. Since the pattern of tying the horizontals to uprights involves a good deal of cord, there is enough time for all occupants to get outside. Finally, most of the houses have neither doors nor walls, so that there is no real difficulty in discovering means of egress.

If I have given the impression that typhoons on Yap are neither very dangerous nor very destructive by our standards, then I have given precisely the impression intended. Yap is not a small, low island where a rise in water level can sweep the entire population into the sea. It is a high island with good protection. It is true that waves, sometimes called tidal waves, can follow a typhoon and batter their way far inland, cutting pieces out of the island itself, drowning and devastating what they touch. But Yap is big enough so that only a small portion of the island is affected.

On the other hand, typhoons on Yap are terrifying, and were so both to me and to the Yaps. They do destroy houses, and houses are hard to build and take much time and labor. Further, the destruction of a house on Yap, as elsewhere, means more than the mere destruction of a piece of material; it is a home and the locus of a family, and its destruction entails the drastic revision of not only habit patterns, but emotionally significant social relationships. Americans have a tendency to measure disasters in terms of figures: casualty rates and property damage costs. Perhaps these are good indices of the extent to which social relationships are upset. I only invite attention to the fact that the measure of the disaster of a typhoon on Yap is not in its casualty rate, which is low or non-existent, but rather in terms of the disruption of social and emotional relationships which it entails.

Social History of Typhoons on Yap

A few words might be in order now as to the social history of a typhoon from the time the warning comes—an ambiguous warning it always is, for it is difficult to tell how bad it will get.

When the warnings are fairly clear and the probability is that it is not just a bad storm, the village chief will usually instruct the magician or magicians available to go to the sacred places of the village to try to contact the supernaturals involved and to ask that the typhoon be halted. Such action is a public responsibility of the village chief and the village magicians. At the same time, chiefs and magicians at the district level, having more prestige and more powerful magic, will be concentrating their efforts to the same end.

At this time each individual takes what precautions he can on his own behalf. If he is very close to the beach he goes inland, for he knows the water will rise and may become dangerous. If he has any especially valued possessions, he will take these with him if he can. He will get help and bring his canoe as far inland as he can and secure it. If he is inland, he will get into his house and stay there. If, for any reason, he must be outdoors, there are certain magical manipulations available to protect himself against flying coconuts and branches. If he stays indoors, he will chew his betel and wait out the typhoon. If his house shows serious signs of giving under the stress, he will strengthen the houseposts with magic and communicate with the ancestral spirits which live under the floor, asking their aid in seeing that the house weathers the storm. He knows that he must not abandon the house. Until the house actually collapses—or is within a few feet of total collapse—he must not leave it, for if he leaves the building it will surely go down. By leaving the building he abandons his ancestral spirits, which live there too, and if he abandons them, they will abandon the house and abandon him as well. At the same time, he knows that his house and roof will protect him against wind-propelled objects.

The individual besides doing what is sensible and reasonable, like going inland and remaining indoors, making such magic as he can and asking the assistance of his ancestral spirits, also "gets religion." He puts himself on his good behavior with respect to all rules which have any supernatural sanction behind them. It is ordinarily wrong for young men and old men to share food or drink, yet in everyday life I found old men willing to share my food and drink. Since Yaps "drink a cigarette," where we "smoke a cigarette," cigarette smoking ordinarily falls within the prohibition, but on ordinary days old men would explain that it was not a breach of regulations to take a cigarette out of my pack since I was not a Yap. But during typhoons, while I found old men as willing as ever to smoke my cigarettes, I had to give them a whole unopened pack and a whole unused pack of matches. They would not share a light or a cigarette at that time.

After the typhoon, too, there will be visits by the magicians to the sacred place and prayer to prevent further typhoons. Each village will also be represented at the district sacred place, and one particular sacred place, which may be especially powerful, will have representatives from the whole island to ask that its magician implore the spirits to cease and desist from further events of this sort.

During the typhoons the gusts of wind, the shaking of a house, or any sudden activity will elicit the characteristic Yap "startle" reaction. This is very much the same sort of reaction observed in combat casualties during the war. The sudden gasp or exclamation with a kind of pulling together or closing into one's shell which lasts just an instant or two but leaves the person in a state of apprehension. The significant difference between combat casualties and Yaps, however, is that the startle reaction is both normal and universal for the Yaps. Any sudden action at any time—during the normal daily routine—will elicit this startle, from adults and children alike. It is perhaps most closely akin to our stereotyped "Oops" on losing balance or slipping or falling. Yaps slip or fall without exclamations, but any sudden event will make them cry out "Ehh."

After the typhoon is over there may be some slight attempt to pick up the debris around the house, to visit close kinsmen in other houses and see how they fared, or just simply to get reorganized—cooking food, getting "kinks" out of one's legs, and so on. If the house has gone down, there will be a few simple salvage and shelter problems. Valuables inside the collapsed house will be dug out, assembled and put in some safe place. But Yap people don't break out into highly purposeful, coordinated efforts at reconstruction at high rates of activity. There is a lot of sitting around and talking, of moving in apparently random ways, doing this, doing that, never quite finishing one job. However, such inaction cannot be interpreted as either daze or apathy, since it is the normal manner of living and housekeeping on Yap. Meals are often cooked in the morning, but often not. Food is eaten when one is hungry, not at regular meal times. Only rarely, if ever, is there any highly organized pitching in at the first possible moment to get a job done.

Similarly, although the last typhoon occurred in January, when I left at the end of June only a few houses had been rebuilt, most were still much as they were the day after they went down. But here again, this is less a sign of apathy or daze than it is of the normal Yap conception of the world and of life as lacking any time-urgency. There are only a few things on Yap which demand imperative action at a particular time. Things get done eventually and no one is ever in much of a hurry.

There is, however, one very notable reaction to the typhoons, and that is the highly vocal, clearly reiterated affirmation that there is no more food left and that starvation is their fate. During the typhoon people kept saying to me "Alas, alas! Now there will be no food! No coconut, no taro, no yams, no bananas, nothing to eat, nothing to drink." They said this to me and to each other during the typhoon and after, and they said this repeatedly to the commander of the civil administration in Yaptown. They said it so convincingly to him that he radioed Guam, and relief supplies in the form of tinned beef and gravy and rice were sent immediately and distributed to the disaster-stricken Yaps.

There are two important points here. The first is that there were really two possible foci of concern. One of these was the food resources and the other the housing situation. More than 60% of the houses were totally destroyed by the end of the typhoons, while 100% of those left standing suffered more or less serious damage—particularly to the roofing, which meant that even the standing houses were more like sieves than houses. Canoes, on the other hand, in very short supply before the typhoons, had suffered very little damage. Of the two possible bases for concern it was food, not housing, which was selected.

The second point is that the literal content of the concern over food supply was demonstrably unrealistic and grossly out of proportion to the real extent of damage and deprivation. I was very puzzled when people told me on all sides, "Alas, alas, no food," and I spent a good deal of time after each typhoon, and particularly the last one, surveying the extent of damage and destruction both to housing and food resources. I am thus in a position to say that for all practical purposes, there was *some* destruction of food resources, but there was no danger that anyone would go hungry. The really serious damage was to supplemental foods, not staples. Bananas were badly hit, but bananas play only a small part in the Yap diet. Coconuts were hit, but the shortage would not be felt for three to six months after the last typhoon. Even then, the shortage would not be too serious. Fish were disturbed, but hardly seriously, and the supply of taro was practically unaffected. The relief supplies of rice and canned beef were thus a delightful luxury which the Yaps enjoyed while they lasted, but this relief was quite unnecessary. The commanding officer who requested them took the Yaps at their word, and his belief in their sincerity was not due to misunderstanding. They really did mean it when they said that they would starve, that there was no food left, and that they had been destroyed. As sincerely as they meant it, however, it was nonetheless an unrealistic conviction on their part.

Food plays an extraordinary part in Yap culture. It is involved in and somehow pervades almost every form of activity—ceremonial and mundane. Neither a lovers' tryst nor a funeral is complete without some exchange of food; a husband's role is defined in terms of what food he should contribute to the family as is the wife's; there are special sacred places which have supernatural control over the fertility of food plants and women, and other ones for assuring the abundance of fish. Activities which we separate from eating are considered analogous to eating by the Yaps, as the previously mentioned "drinking" of the cigarette. And of course there is a good deal of food anxiety in daily life, concern with the possible failure of food supply. Lovesickness is seen as a form of stomach trouble, for it makes for upset stomach and pains in the stomach, and so on.

It is, therefore, reasonable to expect that in a situation of stress food will be pushed to the forefront; it is involved with almost all activities. Yet this hardly explains anything; it is only a statement of a kind of statistical order—food crops up all the time, it will probably crop up here too. What is the context, what role does it play?

"Meaning" of Typhoon on Yap

In order to understand the unrealistic concern over the food supply and in order to answer the questions posed at the outset, how do a people face a chronic disaster situation, how do they "train" for it, how do they prepare for it, we must turn to another order of data. That is, simply, what does a typhoon "mean" on Yap?

Americans view typhoons, tornadoes, hurricanes as "natural" phenomena—events which have natural causes. Yaps class typhoons, epidemics, earthquakes, tidal waves, the falling of thunder to the ground, and sorcery as events which are caused by people who know how to manipulate supernatural forces. There is no need to go into a detailed discussion of Yap theology here. Suffice it to say that a magician who knows how can bring about one of these catastrophes. But a magician who knows how can never work on his own initiative except when he gets out of hand and proceeds with some piece of personal vengeance. In such a case, however, he will not bring a catastrophe, but rather will focus trouble on his particular enemy. The catastrophes—sorcery included here—are brought on by a magician when he has been ordered to do so by some chief. A chief, usually only a district chief, is only supposed to order a catastrophe when in his judgment people have deserved such serious punishment for failing to heed his advice and counsel. A typhoon is, therefore, a punishment meted out to the people by some chief whom they have neglected to heed.

This is a difficult problem. A good chief never orders—he has no authority. But he is expected to give wise counsel, and his people are supposed to recognize the wisdom of his counsel and follow it. Yet a bad chief is one who does not heed the desires or wants of his people and who works against his people, trying to make them do what he wants instead of leading them along the ways they want to go. And a bad chief can be killed by his people or, nowadays, since killing a man brings trouble from foreign authorities, a bad chief may be deposed. When a chief and his people part company, each is convinced that the other is wrong and each is convinced that right lies on his side. A chief has, by common consent, the right to punish his people for failing to "follow" him; yet his people, by common consent, have the right to take drastic measures if their chief fails to represent them and lead them as they want to be led.

The typhoon is thus not an open and shut case of clear-cut wrong and just punishment. It is a half-right measure of force and power which can overweigh the question of ethical right. The people may be right, but they would rather not have a typhoon or epidemic than stand too firmly on those rights.

Such force at the command of a chief, if it were unlimited, would give him considerable power indeed. Particularly if he could really bring on typhoons and epidemics at will. An important balance in the system, which makes the typhoon or epidemic a measure of last resort, is that no magician can ordinarily call forth more than two disasters in his lifetime. It is only the rarest and most unusual magician who can bring on three in his lifetime, and he never knows until he tries it. But no magician in the history of Yap—I was told—had ever brought four disasters and lived. Usually if a magician brings on two disasters and then tries a third, he himself will die immediately thereafter.

Hence, any given district or alliance chief is confined by the fact that his magician can only invoke two disasters. Chiefs and magicians look forward to long lives and they are not particularly keen on bringing those lives to abrupt endings. A chief will therefore keep the disasters in reserve—he will not loose such a bolt unless it is vitally important to him, for he has not many to waste.

It is a fact that the magician who was said to have brought on the four typhoons of 1947–48 was found dead the morning after the fourth typhoon of apparently natural but undeterminable causes. There was no autopsy, but there were no visible signs of foul play.

A typhoon on Yap is, therefore, supernaturally determined, but its presence is the result of some human's action on these supernaturals.

Supernaturals being what they are, if one man can cause a typhoon, another man, approaching the supernaturals from another perspective and with his own power, can try to stop the typhoon, or minimize it, or at least protect himself. Thus the question of planning and preparation follows directly from the presumed causes. We feel that we cannot control the weather, but only what the weather does to us. We therefore build tornado cellars but make no effort to halt or deflect a tornado. On Yap it is otherwise. Since the typhoon is sent by and controlled by the actions of people on supernaturals, they get right to the heart of the matter by sending deputations to sacred places to work on the supernaturals sending the typhoons.

They stay with their houses so that their ancestral ghosts will protect them and let everyone else fend for himself. A bit of magic may protect any given individual from being struck by flying materials even though other individuals have, by magic, brought the typhoon.

Correspondingly, there is a minimum—though a perceptible and significant minimum—of rational preparation. People do get away from the beaches and they do get under roofs. But they rely on magic to steady the house, when they have known for about forty years or more that ropes thrown across the top of the house and tied firmly to the ground will help considerably in keeping the house from blowing up and then down. Germans, Japanese and Americans have all used this method for keeping roofs on, and the Yaps have helped them rig these cables and have understood what they were for. Yet no Yap house to my knowledge ever had a rope thrown across its roof—or better, two to four ropes, criss-crossed, to help hold the roof down.

I should turn now to Yap behavior during the typhoon itself. In the first typhoon, which occurred about a month after I arrived and when my command of the language was almost nil, I was able to see people but not understand much of what was said. This typhoon struck at night and was over by morning. The other three typhoons occurred during the day, or partly during the day, partly during the night. In each there was that strange combination of tension without concrete or specific manifestations. In the first typhoon, I observed a young man and woman trying to get as far into a dark corner as possible and disappear beneath a collection of mats, ponchos and other coverings and, although they were quiet, it was evident that they were "necking." Yet there was never hilarity, jokes, singing or any other kind of jollity. Mothers tended to stay with their babies and just sit, doing little things but looking drawn, tense, uncomfortable. Nobody liked the typhoons, yet there seemed to be nothing to fear. They were silent more than they talked, and when they talked they talked quietly. The Yaps described the feeling as one of confinement, boredom, strained inactivity. And my own feelings were similar; cold, discomfort, confinement, an unspecified anxiety and tenseness, yet nothing clearly to be afraid of. Yet after the typhoon there was no release, no bursting forth from the confinement as there is at other times. I have remained through long days of rain and storm on Yap and at the end, when the sun comes out and the thick heat and moisture can almost be cut with a knife, people would break out of the house and *go*. Dogs would bark. Children would shout and play. But after the typhoon, they just got up and left or aimlessly puttered, or dug in the ruins in a dispirited, listless way.

It is this context, I think, which illuminates the anxiety centered on food. I am not Yap, but there is something about the awesome activity of the elements which is fundamentally disturbing to me. It gives one pause. One feels very much cut down to size. For the Yaps, however, these feelings are much more clearly focused. Where I see the typhoon as a natural phenomenon and am reminded forcefully of the overwhelming size and strength of nature, the Yaps see the typhoon as the work of people who can control those forces. It is therefore in social relationships, backed by the conscious and unconscious interpretations which people invest in them, that the meaning and the threat of the typhoon lies. The relations between individuals, the relations between the people and their chief, the relations between the old and the young, the relations between parents and children are disturbed and this disturbance is symbolized by the very awesome experience of the typhoon. Where the *exchange* of food is the symbol of good relations, it is the absence of food which symbolizes disturbed or broken relations. And when someone has sent a typhoon, relations are not too good. Therefore, the typhoon is seen as expressing a situation of broken relations, symbolized as "no food." It is at this level that the lament "Ah piri e gafago, dari e thamunamun" ("Ah, alas, no food") is meant literally. It is not at the level of concrete food at all, for they will voice this lament with food in their hands.

Thus, a typhoon on Yap is the end product of a series of purposeful actions by people through their manipulation of the supernatural. The typhoon is caused as much by the disturbance of social relations between the chief and the people as it is by the manipulations of the magician. The exchange of food is the symbol of good relations; the typhoon means that relations have deteriorated, and where social relations deteriorate, food is absent. Hence a typhoon means "no food" in the sense that the omnipresent symbol of good social relations is absent when relations are bad.

The impact of the typhoon thus has much wider and more important meaning than the mere velocity of the wind and its physical destruction. The impact of the typhoon is felt where it hurts most, as disturbed social relations, and it is this impact which is voiced by the lament over food. The impact of the wind and water represents significantly lesser issues. Indeed, in surveying the damage of the typhoon as measured in facts and figures of houses destroyed and casualties, one is measuring only the secondary, less important damage.

Conclusion

I opened with the question of what happens in a situation of chronic threat. I think that it is plain by now that the chronic threat of a typhoon on Yap is a situation in which the concrete, real situation has a much deeper, wider set of meanings built into it than the mere physical phenomenon itself. The typhoon is put in a causal context and becomes symbolic. It is invested with meaning and becomes the focus for a host of other, ancillary anxieties or concerns. Because of these meanings which are invested in the typhoon, the typhoon itself is only one element in the impact and the response to the physical impact is only part of the response. A significant portion—indeed, on Yap the most significant portion—of the response is not to the physical impact at all, but rather to the impact of the meanings with which it is invested.

There are certain other points which may be worth noting in comparing the Yap typhoon with, say an American tornado. One of these is the difference in the degree to which Yap and American cultures depend on material things and the variable effect of physical destruction on each. On Yap, houses that go down stay down for quite a while. Paths are only gradually cleared of debris. In America it is really difficult to say which has the greater impact, the tornado or the rescue and clean-up operations. The fury of a tornado is admittedly packed into a shorter time period than the fury of the fire trucks, ambulances, bulldozers and cranes which roar onto the scene in the wake of a tornado. I say that the tornado's fury takes less time, but there is every indication that the speed with which the rescue and clean-up operations are conducted is growing greater with each disaster. We are learning how to pack the greatest clean-up impact in the shortest possible time. Here is perhaps one clue to the very different meanings of the disaster in America and in Yap. The rescue and clean-up operations on Yap are conducted in the sacred places and in the political arena, for these areas represent the importantly dislocated relations. In the U.S., it is the order and efficiency of material apparatus which are seen as the important consequences of disaster.

A second point which suggests itself is the question of the disaster syndrome, particularly with respect to the dazed and stunned reaction which has been so prominently described in the literature on American disasters. It might be possible to misinterpret the Yap puttering and apparently aimless reorganization of their physical community as daze and apathy. I do not think that this is so, however. In a sense the Yap is simply normally apathetic with respect to getting things done in a hurry—they just never do. My own feeling is that Yaps show neither daze nor apathy after a typhoon and that their response is nearly rational, given their premises as to the causes and primary areas of value involved in the typhoon. Thus, if typhoons are controlled by supernaturals, it is reasonable to attempt to influence these supernaturals, and if the primary area of value hit by a typhoon is the area of intergenerational relationships symbolized by political relations, then their pre- and post-typhoon behavior aimed at restoring these relations is reasonable.

The absence of apathy and daze may partly be accounted for in terms of the fact that the disaster and its meaning are more clearly structured in certain respects for the Yaps, and hence can be met in more closely rational ways, given their view of the nature of the situation. Partly too, there is less, in their view, of a highly complex nature that needs doing. Perhaps this point can be put more clearly by saying that the extraordinarily elaborate functional differentiation of roles in American culture requires a very high degree of coordination among individuals, so that any dislocation of that scheme of coordination leaves a series of individuals who simply cannot act because their acts are neither properly cued nor fitted into the reciprocal actions of others. If the definition of the role of a casualty is that he be taken by some special vehicle to proper medical quarters, then there is only a very little that a man with a broken arm can do for another man with a broken arm beyond waiting for help and an ambulance. Similarly, it is a doctor's job to fix broken people, and practicing medicine without a license or without good training in first aid is wrong. Hence people stand around and stare not only because they don't know what to do, but also because they have been trained to "not know" and to leave these things to the expert. Perhaps equally important is the impotence that may be felt by the person who sees the need for action and is nevertheless trained not to act. He must defend himself against this, and he does so by "apathy," and he looks as if he were in a "daze." On Yap there is none of this elaborate complexity of functional differentiation. What needs doing, people do. The fact that in their view hardly anything really needs doing immediately helps to prevent overwhelming necessity for action conflicting with incapacity. Finally, of course, the absence of physical shock and physical casualties makes their task much easier.

Communication, too, is very different on Yap and in the U.S. On Yap, communication is primarily by word of mouth from one person directly to another. Howling above a high wind is a perfectly adequate mode of communication on Yap. But howling through a telephone when the wires are down just doesn't work in America. Communication, and the technical aids to communication in America are geared very closely to the elaborate functional differentiation of roles. Every man is a fireman and policeman on Yap and the medical specialists take care of only severe and refractory cases that require special supernatural assistance, so that virtually every man is his own doctor. The communication network is so much simpler on Yap that it is proportionately harder to disrupt.

Finally, let me return to the point with which I opened: what are the implications of a *chronic threat?* I would make this suggestion. Where there is a chronic threat, that threat will take on meaning far above and beyond its own real and inherent nature. The event which is threatened will have meaning in terms of causes, and all human beings are vain enough to see these causes in themselves. The unique catastrophe is very different: it is unstructured and by the time it is structured it is finished and does not occur again. But the chronic threat, the catastrophe that is long-awaited, takes on distinct meanings and provides a focus for long-standing anxieties, guilts, fears, and hostilities. One important part of the structuring of meaning is the extent to which it becomes common for the whole population. The unique catastrophe is responded to in terms of the socially structured motives of individuals. The chronic threat takes on common meanings for a wide population.

If we look forward to and plan for the thermonuclear disaster and if we look forward to it and plan for it over a sufficient time, we must take into account in our plans more than the bare physical effects of that disaster, for by that time people will not only respond to the thermonuclear detonation itself, but they will also be responding to all the things they have projected into it and built into it in terms of their primary concerns and their primary anxieties. Indeed, their response to the physical detonation and its physical effects will be largely shaped by the meanings they impute to those physical phenomena, and not simply to the physical phenomena themselves.

I cannot say at this moment what primary concerns and anxieties Americans will manifest and thereby determine their response to a thermonuclear blast. Nor can I say just what meaning they have already given such a potential event. But I would suggest that, if the Yap data provide a valid base for generalization, it would be reasonable to expect these anxieties and concerns to follow closely the meaning they impute to such a disaster, and that the response to the disaster will be composed of the response to those meanings as well as to the actual, physical explosion; just as the disturbance of social relations is as much a part of a Yap typhoon as the wind, the rain and the water.

18

The Politics of Place

Inhabiting and Defending Glacier Hazard Zones in Peru's Cordillera Blanca

Mark Carey

Glacier retreat during recent decades has threatened human populations worldwide. In the Peruvian Andes, the consequences of melting glaciers have been particularly dramatic and deadly. Glacier retreat in Peru's Cordillera Blanca has caused two dozen glacier avalanches and glacial lake outburst floods since the 1930s (Ames Marquez and Francou 1995; Zapata Luyo 2002). After a 1941 outburst flood killed 5,000 residents of Huaraz, the Peruvian government began monitoring glaciers and draining glacial lakes to prevent additional glacier disasters. Despite its efforts, in 1970 the Cordillera Blanca produced one of the world's most deadly glacier disasters when an earthquake triggered a massive glacier avalanche that buried Yungay and killed thousands. In response, the state expanded disaster mitigation programs to include hazard zoning. Experts agreed that hazard zoning would keep people out of potential avalanche and flood paths, thereby reducing their vulnerability to glacier disasters. But in every town where the authorities tried to implement hazard zoning, residents resisted government initiatives. Even though the region had just experienced a series of cataclysmic glacier disasters, people refused to move to safe areas; instead, they recolonized the areas that had been destroyed. Analysis of hazard zoning in the case of Yungay, the area most devastated by the 1970 glacier avalanche, helps to explain how hazard zoning was derailed.

Hazard zones and hazard zoning held widely divergent meanings for scientists, government officials, and local residents. To scientists, hazard zones represented paths that avalanches and outburst floods could follow. Hazard zoning was seen as the most prudent way to avoid future glacier disasters, and because the 1970 earthquake and avalanche had destroyed most of the region it did not entail moving structures or communities; rather, it required shifting reconstruction to new areas. To the avalanche survivors, however, the hazard zones were historically produced spaces with cultural, economic, social, and political meanings. Relocating to a safe place meant major compromises and significant risks

The Anthropology of Climate Change: An Historical Reader, First Edition. Edited by Michael R. Dove.

to their livelihoods, connection with ancestors, material well-being, social status, and political power. Consequently, their decisions about whether to remain or to relocate involved the ranking of risks. For those who rejected hazard zoning, there were cultural, social, economic, and political risks associated with leaving that outweighed the risk of unknown and unpredictable glacier disasters. In other words, while scientists focused on a single risk, residents contended with a host of them. By analyzing the rationality of Yungay residents' risk perception—the historical forces informing that perception, the multiple meanings they assigned to the hazard zone, and their reasons for resisting relocation—it is possible to understand why experts, policy makers, and local residents clashed over the 1970s disaster mitigation policies.

Natural disasters occur not only because environmental processes inflict damage but also because vulnerable people, property, and infrastructure get in the way. Scholars acknowledge that some people choose to live in vulnerable locations because they are wealthy enough to afford insurance or powerful enough to be guaranteed government relief (Davis 1998; Steinberg 2000). More often, and especially for the developing world, researchers suggest that people inhabit vulnerable hazard zones because power imbalances, government neglect, poverty, racism, economic development, or social injustice force marginalized populations into areas prone to flooding, hurricanes, fire, drought, and other natural disasters. To protect people, many scholars call on governments and experts (scientists, engineers, planners, etc.) to develop policies that reduce human vulnerability to natural hazards (Maskrey 1993; Alexander 2000; Wisner et al. 2004).

While marginalized populations do suffer disproportionately from natural disasters, scholarship that deflects the *reasons* for people's vulnerability to forces beyond their control can yield two problematic interpretations. First, by overlooking people's own role—even the role of marginalized or developing world populations—in influencing their vulnerability to natural disasters, researchers can deny the historical agency of these groups. Second, by calling on governments and experts to implement policies that reduce people's vulnerability to natural disasters, scholars often assume that vulnerable populations will embrace these plans. The case of Yungay in the 1970s challenges these two views.

On the one hand, when Yungay residents rejected hazard zoning, they played a vital role in determining their vulnerability to future glacier disasters. They acted in multifaceted and complex ways that stemmed from historical, social, cultural, economic, political, and ideological factors (Johnston and Klandermans 1995; Rubin 2004; Chuang 2005). On the other hand, their views of hazard zones as space and hazard zoning as policy clashed markedly with the perspectives of scientists, planners, engineers, and national government officials. These local reactions to state plans demonstrate the intimate relationship between science and power (Arnold 1993; Scott 1998; Prakash 1999; Mitchell 2002). In short, this Yungay case study helps to clarify not only the local consequences of global warming and glacier retreat but also the conflicts that can arise when governments attempt to implement scientific policy, mitigate natural disasters, and adapt to climate change.

Cordillera Blanca Glacier Monitoring and Disaster Mitigation

The Cordillera Blanca in Peru's Ancash Department contains approximately 600 glaciers that cover slightly less than 600 km^2 (Georges 2004). A half-million people inhabit the valleys and upland slopes surrounding this range. These residents consist of Quechua-speaking indigenous people, mestizos (of mixed Spanish-indigenous descent), and whites (of Spanish descent). The majority of them live in the Santa River valley, known as the Callejón de Huaylas, along the western base of the cordillera (Figure 18.1). More than 70% of Cordillera Blanca glacier meltwater drains into the Santa River, which therefore supports extensive agriculture and livestock as well as Peru's important Cañón del Pato hydroelectric station.

Peruvians living near the Cordillera Blanca have lived with repeated glacier disasters since the 1930s (Carey 2005*a*). As a result of rising global temperatures since the end of the Little Ice Age (~1350–1850), glacier melting in the

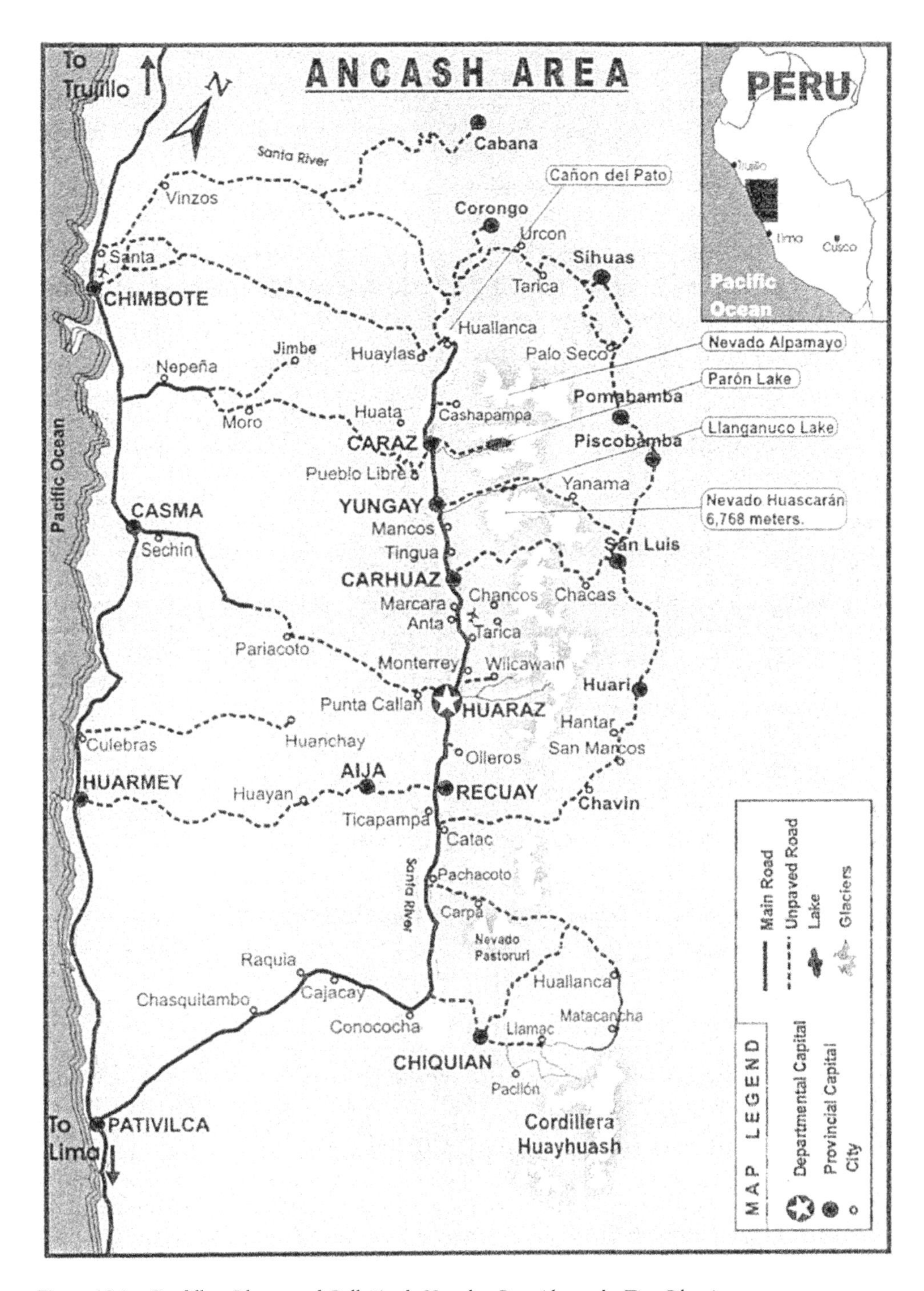

Figure 18.1 *Cordillera Blanca and Callejón de Huaylas, Peru (drawn by Tito Olaza).*

cordillera has caused two types of glacier disasters (Portocarrero 1995; Ames 1998; Kaser and Osmaston 2002). First, glacial lake outburst floods have occurred when lakes grew in the space left by retreating ice. The number of glacial lakes has increased significantly, from 223 in 1953 to 374 in 1997 (Fernández Concha and Hoempler 1953; ElectroPerú 1997). Dammed behind unstable moraines that sometimes ruptured, these glacial lakes produced 13 major outburst floods between 1932 and 1959 (Zapata Luyo 2002; Ames Marquez 2003). Three floods were particularly catastrophic: the 1941 Huaraz (5,000 deaths), the 1945 Chavín (500 deaths), and the 1950 Los Cedros (200 deaths).

A second type of glacier disaster has occurred when glacial ice thinned, fractured, and generated landslides. In 1962, for example, Glacier 511 on Mt. Huascarán caused an avalanche that destroyed the town of Ranrahirca, killing an estimated 4,000 inhabitants. The most deadly glacier disaster occurred on May 31, 1970, when an earthquake registering 7.7 on the Richter scale dislodged Glacier 511 again and the ensuing avalanche buried Yungay (Ericksen, Plafker, and Concha 1970). Most sources cite the avalanche death toll as 15,000 or 18,000 (e.g., Oficina Nacional de Información 1970; Oliver-Smith 1986; Ames Marquez and Francou 1995). Analysis of the most recent census data prior to 1970, however, indicates that in 1961 only 15,068 lived in the entire district of Yungay (República del Perú 1968). Given that many survived the earthquake and avalanche and even accounting for population growth between 1961 and 1970, the estimate of 15,000 deaths is perhaps inflated. Even so, the Yungay avalanche remains one of the world's most deadly glacier disasters.

To prevent glacial lake outburst floods and glacier avalanches, the Peruvian government has monitored glaciers and drained glacial lakes since the 1941 Huaraz flood. In 1951, after the Los Cedros outburst flood destroyed the nearly completed 50-megawatt Cañón del Pato hydroelectric station, the Peruvian government created the Comisión de Control de Las Lagunas de la Cordillera Blanca (CCLCB). The CCLCB conducted hundreds of studies and developed one of the world's first systematic glacial lake classification indexes to identify and rank glacial lake hazards. This classification system helped categorize glacial lakes so that unstable lakes could be drained and dammed before they inundated the Cañón del Pato hydroelectric station or the Callejón de Huaylas communities. The CCLCB partially drained and dammed 19 glacial lakes between 1952 and 1971, when the Peruvian Corporación Peruana del Santa (CPS), which ran Cañón del Pato, absorbed it into its own Division of Glaciology and Lakes Security (Gálvez Paredes 1970; Carey 2005*b*). Spurred partly by the 1962 Ranrahirca glacier avalanche, which pointed to ice—and not just glacial lake—hazards, and partly by its economic interest in determining the amount of water stored in glaciers, the CPS created the division in 1966 with the Peruvian glaciologist and Huaraz resident Benjamín Morales as its head (Morales 1969). By 1970, the commission and the company had carried out large-scale glacial lake control projects and extensive glacier monitoring. With the exception of declaring the 1941 Huaraz flood path off-limits to rebuilding, however, the glacier-disaster prevention agenda had not focused on reducing people's vulnerability to floods or avalanches. The magnitude of the 1970 Yungay avalanche expanded the government's disaster mitigation agenda to include hazard zoning.

Disaster Mitigation from Above

Responses to Peru's 1970 earthquake and avalanche took place in two important contexts. First, the scale of the earthquake and Yungay avalanche stimulated increased state funding for glacier-disaster prevention projects. During the first half of the 1970s, government investments in glacier monitoring, lake control, and hazard zoning surpassed those of all previous eras. Second, Peru's political situation in 1970 was unique. In 1968, General Juan Velasco had seized Peru's presidency, and his "Revolutionary Government" had begun implementing far-reaching social, economic, political, and even cultural changes, including accelerated economic growth, redistribution of income and wealth, integration of the indigenous population into mainstream society, reduction of foreign dependence, and, essentially, the fabrication of a "new society" (Jaquette and Lowenthal 1987; Contreras and Cueto 2000). While the execution of these sweeping changes generally fell short of the goal, Velasco's reforms significantly altered the traditional social landscape.

In many ways, Velasco saw the 1970 earthquake and avalanche as an opportunity to rebuild the region according to his ideals. The near-complete obliteration of the Callejón de Huaylas offered a clean slate on which to create a "new and improved" society. He was determined to create a society less vulnerable to glacier avalanches and outburst floods, and hazard zoning—along with glacier monitoring and lake control—would reduce these long-term risks. To distribute relief aid and reconstruct the region, he established the Comisión de Reconstrucción y Rehabilitación de la Zona Afectada (CRYRZA). As a centralized, bureaucratic agency run from Lima, CRYRZA did not always operate efficiently or in cooperation with the Yungay survivors (Oficina Nacional de Informatión 1970; Oliver-Smith 1986; Bode 1990). Most scholars have criticized Velasco and CRYRZA for imposing top-down reconstruction agendas and attempting social engineering. But to (mis)classify hazard zoning as just another aspect of Velasco's social experiment neglects the contribution of hazard zoning to long-term disaster mitigation—an approach that natural-disaster scholars advocate today (Maskrey 1993; Sarewitz, Pielke, and Mojdeh 2003; Wisner et al. 2004).

In June 1970, experts identified three principal problems that disaster prevention was subsequently to address: (1) unpredictable glaciers that required continuous monitoring, (2) unstable glacial lakes caused both by continued glacier retreat and by earthquake damage to previously constructed dams and drainage canals, and (3) human habitation of hazard zones, the riparian zones through which future avalanches and outburst floods could pass (Lliboutry 1970; Lliboutry et al. 1977).

First, given the unpredictability of glacier avalanches, the experts called for comprehensive glacier monitoring and research. The CPS (and its successor, ElectroPerú) maintained a distinct office for glacier research and hired dozens of new scientists and engineers after 1970. This research required access to remote Cordillera Blanca canyons in which researchers sometimes spent weeks at isolated camps above 4,000 m. A typical glacial lake was at least 20 km from and 1,500 m above the nearest road, and access to it involved climbing steep slopes and navigating boulder fields, dense brush, and marshy valleys, with mules and indigenous porters carrying the equipment. Glacier monitoring often benefited from the guidance of indigenous residents who were familiar with the high-elevation terrain from pasturing cattle and sheep or gathering plants and firewood. These locals also furnished information about recent glacier history and glacial lake formation. Glacier hazard monitoring took scientists to new heights and into unprecedented danger. In August 1973, for example, scientists spent a night at 6,400 m during an expedition to analyze a glacier crevasse on Mt. Huascarán Norte (Morales 1972; Zamora Cobos 1973).

Second, the scientists recommended effective lake control, especially at Lake Llanganuco above Yungay. Located in the province of Yungay, the upper and lower parts of Lake Llanganuco are two of the region's largest lakes. On June 5, 1970, scientists flying over the Cordillera Blanca noted that the upper lake outlet was blocked by a rock and ice dam formed by an earthquake-induced landslide that had covered part of the valley floor (*El Comercio* 1970*a*; Lliboutry 1970). By June 20, 30 men were digging frantically through long days and into the nights to open a drainage canal through the new dam (Ortega 1970). By July 1970 they had succeeded in reopening the drainage canal, thereby lowering the water level and averting catastrophe in the valley below. Through subsequent years, President Velasco allocated funds for several glacial lake control projects, including the drilling of four drainage tunnels and the digging of five drainage canals to remove water from unstable glacial lakes. Because access to these lakes was difficult, he also funded construction of 50 km of roads to glacial lakes (Morales 1970). From 1970 to 1978 the Division of Glaciology and Lakes Security drained, dammed, and secured a dozen glacial lakes (División de Glaciología y

Seguridad de Lagunas 1972; ElectroPerú 1975, 1984; Lliboutry et al. 1977).

Finally, the experts advocated the relocation of towns that were vulnerable to avalanches and outburst floods. Because glacier disasters such as the 1962 and 1970 avalanches from Glacier 511 could not be predicted or avoided, scientists proposed hazard zoning and the relocation of several towns. As Morales observed, "The catastrophes produced by avalanches from hanging glaciers such as Huascarán will be periodic events, which is to say that they will occur again in the future. This type of phenomenon has no solution; there are no measures to avoid them or control them. Consequently, populations located below glaciers with these characteristics must be relocated" (Morales 1970, 71). Experts identified many hazard zones below Cordillera Blanca glaciers and glacial lakes, including areas of Huaraz, Carhuaz, and Yungay (Lliboutry et al. 1970; Oberti 1975). Because the earthquake and avalanche had destroyed most of these towns, relocation from hazard zones did *not* involve relocation of intact communities. Rather, it generally involved the rebuilding of previously demolished structures in different, safer places. It was proposed that Yungay be moved to a site called Tingua, 15 km away, which was believed to be protected from glacier disasters (*El Comercio* 1970*e*; CRYRZA 1970). By the time authorities announced relocation plans in November 1970, however, Yungay survivors had already begun reconstruction adjacent to the avalanche path in an area they named Yungay Norte. Scientists determined that the settlement would soon spread into the Yungay hazard zone, and therefore they made efforts to move it to Tingua.

Local Resistance from Below

Despite decades of glacier disasters, the people of Yungay and other Callejón de Huaylas residents were eager to resettle the hazard zone. The Yungay elite—property owners, businesspeople, politicians, entrepreneurs, and professionals—led this resistance to relocation and gained the support of other locals. Resistance to hazard zoning should not be taken to indicate that Yungay residents were unafraid or ignorant of glacier-related hazards. Rumors circulated widely in Yungay about the instability of the Huascarán glaciers. Some believed that the Cordillera Blanca was made up of water volcanoes that could erupt at any point and send deadly floods or landslides into populated valleys below (Oliver-Smith 1986; Walter 2003). Others fled in fear of future disasters and the horrifying postdisaster reality; of the approximately 27,000 earthquake and avalanche survivors from Yungay Province, an estimated 4,000–5,000 moved to Lima immediately (Angeles 1970*a*; *El Comercio* 1970*d*). At the same time, fatalism sometimes led to inaction among residents who believed that natural disasters were beyond their control. Instead of recognizing their vulnerable location beneath unstable glaciers, they implicated others for causing natural disasters, among them God, who took revenge on sinners and controlled nature, or the French, who had detonated underwater atomic bombs in May 1970 that many believed had triggered the earthquake (Ramírez Gamarra 1971; Miano Pique 1972).

Still others decided to rebuild their community in its previous location. Trauma, grief, and a sense of place helped motivate this choice (Oliver-Smith 1982). The avalanche, after all, had buried thousands of people, and their survivors wanted to remain close to them. One of them said, "This is where we want to be. We are accustomed to dying, to losing family.... We want to be here, to die where we were born" (quoted in Bode 1990, 201). Beyond this, despite fear of disaster and hatred of the peak for causing catastrophe, residents felt attached to Mt. Huascarán and Lake Llanganuco (Flores Vásquez 1972). As one observer explained, "Most struggle to live and die on these prodigal lands because, although Huascarán and the lakes threaten us with their fractured bases, they have spirits that attract and captivate everyone who lives here and sees them" (Zavaleta Figueroa 1970, 16).

The Yungay hazard zone was also vital for economic productivity. Relocation to Tingua, residents argued, would undermine their capacity to maintain commercial relations within the province and exploit the rural labor force. As an urban area that possessed both the geographical location and the gravitational pull to attract people, trade, labor, and resources from surrounding areas, Yungay had become a "central place" within the Callejón de Huaylas (Oliver-Smith 1977*b*). Before 1970 it was the economic center of the province and attracted markets, transportation, labor, food products, agriculture, and natural resources. Yungay leaders feared that moving the town to Tingua would diminish its importance and cripple its economy.

The hazard zone held great potential for tourism as well. Yungay was the access point for the ascent of Mt. Huascarán (6,768 m a. s. l.), Peru's highest peak, and for recreation at Lake Llanganuco, a stunning turquoise glacial lake. Prior to 1970 Yungay Province had been a principal tourist destination in Peru. After the 1970 disaster, Yungay survivors believed that tourists would come not only for traditional recreation at Huascarán and Llanganuco but also to see the avalanche site (Angeles 1970*c*; *El Comercio* 1970*b*). Tourism was so important that the Yungay authorities reopened the road to Lake Llanganuco as quickly as they built schools and hospitals (Ángeles Asín 2002). The safe zone at Tingua simply did not possess the proximity to the Cordillera Blanca that made Yungay "a worldwide tourist attraction" (Angeles 1970*b*).

Beyond its economic potential, the hazard zone also helped determine social status. Yungay's town boundaries had historically divided rural and urban, upland and valley, and indigenous and mestizo populations. Throughout Peruvian history, the country's dominant social classes, including the Yungay elite, had defined the indigenous sphere as rural, highland, and close to nature (Orlove 1993, 1998). With the conflation of race, class, and geographical location in Peru, habitation of Yungay's urban area signified superior social status over rural inhabitants labeled as indigenous (Stein 1974; Walton 1974). Although historical migration, miscegenation, and shifting identities had blurred their boundaries, these categories remained prominent in Peruvians' minds from the colonial era through the twentieth century (de la Cadena 2000). The Yungay elite maintained their superior status over the rural population (67% of the population in 1967) in part because they inhabited an urban area. Further, elite anxiety about their privileged social standing was acute after the disaster because, while rural survivors descended to Yungay for aid, many urban residents fled to Lima (Walton 1974). In the first year after the disaster, the Yungay population rose from several hundred survivors to 2,000 because of an influx of rural residents (Oliver-Smith 1982). Maintenance of political positions, professional jobs, property, and control of the urban space hinged partly on clear markers to distinguish rural from urban, lower from upper class (Oliver-Smith 1977*a*; Ángeles Asín 2002), and the 1970 avalanche had destroyed the identifiable urban–rural division that was one such marker. According to urban survivors, whereas Yungay possessed remnants of these markers and historical memories of the division, Tingua had no recognizable boundaries to demarcate social standing. They therefore rejected relocation because improved safety from avalanches and outburst floods threatened to reduce their social status and control (Cabel 1973).

Rural Yungay districts did not necessarily oppose relocation to Tingua, and Yungay leaders used political incentives to gain allies in their struggle. For example, in the district of Yanama, they persuaded Mayor Isidro Obregón to reject relocation by promising to allocate funds to the building of a long-awaited road across the Cordillera Blanca (Ángeles Asín 2002). Further, to persuade the new indigenous immigrants in Yungay Norte to resist relocation, they made sure that those in the relief camp received sufficient food and shelter (Oliver-Smith 1982). Meeting their immediate needs linked residents to Yungay Norte and demonstrated the leaders' effectiveness, thereby generating political support for them and creating a popular base to resist hazard zoning.

Hazard zones became sites not only for regional power struggles but also for national politics. Even though it was scientists who had identified the hazard zones, urban survivors believed that relocation signified their subordination and loss of power to Lima. Several factors help explain why Yungay survivors saw hazard zones as political battlegrounds. First, it was CRYRZA planners and engineers, rather than glaciologists and hydrologists, who attempted to implement the Tingua relocation plan. The scientists generally worked near the glaciers, where they did not appear in survivors' daily lives, while CRYRZA officials set up their offices in survivor camps and interacted with local residents daily.

Second, the CRYRZA representatives sought to distribute aid equally among all survivors, but for the urban elite accustomed to advantages over the indigenous population equality was an insidious proposition. A Yungay woman captured the rift when she grumbled about CRYRZA's aid distribution: "The people of the heights, the Indians, never had anything, so why should they get help? On the other hand, we, the real Yungainos, have lost everything, so we should get more" (quoted in Oliver-Smith 1977*a*: 8). By helping rural and indigenous people in the Callejón de Huaylas, CRYRZA planned to eliminate elite social status and political control—an agenda the Yungay elite emphatically opposed (Walton 1974).

Third, President Velasco was threatening Yungay landowners with a vast agrarian reform program. He and his successor eventually redistributed land to 375,000 families (25% of all farm families) throughout Peru (Klarén 2000). As elsewhere, agrarian reform in the Callejón de Huaylas turned landowners against Velasco because land expropriation diminished their regional power and economic base (Barker 1980; Kay 1982). In Yungay, where the elite already felt threatened by CRYRZA's support for the downward-migrating indigenous population, the risk of government land expropriation posed another threat. Hazard zoning and government relocation plans resembled agrarian reform in that the state determined who would inhabit (or leave vacant) a specific plot of land. Yungay survivors thus interpreted the scientists' well-intentioned efforts to relocate them to safe areas as a national-government affront (Cabel 1973; Walton 1974).

Finally, Yungay residents were alienated by CRYRZA's failure to connect relocation plans with local needs. Discussions about building Tingua did not involve local people, and lack of faith in Velasco, combined with historical mistrust of the state, led many of them to doubt that the government would support the move. In November 1970, residents demanded that government planners "give more attention to future considerations, such as population, production, design of the city, and communication and transportation" (*El Comercio* 1970*b*). In subsequent weeks, the local authorities from Yungay insisted that CRYRZA present concrete plans for the development of Tingua (*El Comercio* 1970*c*; *La Prensa* 1970). People complained because official plans ignored them.

To the Yungay elite, hazard zones were literal and figurative spaces that signified much more than potential avalanches or floods. The Yungay hazard zone and the proposed relocation site at Tingua had complex meanings tied to cultural beliefs, livelihoods, material well-being, social status, identity, and power structures. The glacier avalanche had thus been much more than a natural disaster; it had also triggered cultural, economic, social, and political catastrophes. To many, and especially to the Yungay elite, recovery from these multiple disasters meant rebuilding their lives and their societies in the hazard zone. The risks of further losses of social status, economic security, political power, and cultural beliefs were far more pressing and important than the risk of a glacier avalanche or an outburst flood. The occupation of hazard zones and resistance to relocation thus reflected Yungay residents' ranking of risks.

Conclusion

Peruvian responses to the 1970 Yungay avalanche illustrate the complexities of contending with glacier retreat and mitigating natural disasters. Challenges to hazard zoning below the Cordillera Blanca glaciers emerged because distinct groups attributed different meanings to the hazard zone. Scientists saw hazard zoning as a way to reduce the risk of glacier disasters. The government recognized hazard zoning as a political opportunity to remake Peruvian society according to President Velasco's ideals. And many Yungay residents believed that hazard zoning was less important than resolution of cultural, social, economic, and political disruptions. While all three groups recognized the hazard, their perceptions of risk involved a host of interconnected historical, environmental, and human factors.

Clearly, people can influence not only their vulnerability to natural disasters but also the implementation of policy. No major Cordillera Blanca avalanche has occurred since 1970, so Yungay residents may have made the right choice. However, if global warming persists, glaciers will likely continue to retreat, thin, fracture, and possibly produce more avalanches and outburst floods. If, in contrast, cooling occurs, a new set of problems may emerge as glacier tongues advance into existing glacial lakes and threaten to displace lake water. Ironically, local people's having chosen to resettle vulnerable zones, in part to defy government authority, has made them more dependent than ever on the government's use of science and technology to protect them from glacier disasters.

REFERENCES

Alexander, D. 2000. *Confronting catastrophe: New perspectives on natural disasters.* New York: Oxford University Press.

Ames, A. 1998. A documentation of glacier tongue variations and lake development in the Cordillera Blanca, Peru. *Zeitschrift für Gletscherkunde und Glazialgeologie* 34: 1–36.

Ames Marquez, A. 2003. Chronology of ice avalanches and floods occurring in the Cordilleras Blanca and Huayhuash since the beginning of the 18th century. MS, Huaraz.

Ames Marquez, A., and B. Francou. 1995. Cordillera Blanca glaciares en la historia. *Bulletin de l'Institut Français d'Études Andines* 24: 37–64.

Ángeles Asín, L. 2002. Mientras haya un yungaíno con vida, Yungay no desaparecerá. In *Vida, muerte y resurrección: Testimonios sobre el sismoalud 1970*, ed. R. Pajuelo Prieto, 187–92. Yungay: Ediciones Elinca.

Angeles, P. M. 1970*a*. Acuerdo no. 9: Reubicación y recomendaciones sobre las ciudades y centros poblados en Yungay, 15 de diciembre. In *Lo mejor de nuestra juventud al servicio de Yungay y de los yungainos por una vida mejor a través del Centro Unión Yungay*, ed. Centro Unión Yungay y la Junta Directiva. Lima: Centro Unión Yungay.

——. 1970*b*. Carta al decano del colegio de arquitectos del Perú, Lima, 7 de setiembre. In *Lo mejor de nuestra juventud al servicio de Yungay y de los yungainos por una vida mejor a través del Centro Unión Yungay*, ed. Centro Unión Yungay y la Junta Directiva. Lima: Centro Unión Yungay.

——. 1970*c*. Carta al sr. Director del diario El Comercio, Lima, 13 de noviembre. In *Lo mejor de nuestra juventud al servicio de Yungay y de los yungainos por una vida mejor a través del Centro Unión Yungay*, ed. Centro Unión Yungay y la Junta Directiva. Lima: Centro Unión Yungay.

Arnold, D. 1993. *Colonizing the body: State medicine and epidemic disease in nineteenth-century India.* Berkeley: University of California Press.

Barker, M. L. 1980. National parks, conservation, and agrarian reform in Peru. *Geographical Review* 70: 1–18.

Bode, B. 1990. *No bells to toll: Destruction and creation in the Andes.* New York: Paragon House.

Cabel, Jesús. 1973. *Literatura del sismo: Reportaje a Ancash.* Lima: Juan Mejia Baca.

Carey, M. 2005*a*. Living and dying with glaciers: People's historical vulnerability to avalanches and outburst floods in Peru. *Global and Planetary Change* 47: 122–34.

——. 2005*b*. People and glaciers in the Peruvian Andes: A history of climate change and natural disasters, 1941–1980. Ph.D. diss., University of California, Davis.

Chuang, Y.-C. 2005. Place, identity, and social movements: *Shequ* and neighborhood organizing in Taipei City. *Positions* 13: 379–410.

Contreras, C., and M. Cueto. 2000. *Historia del Perú contemporáneo.* Lima: Instituto de Estudios Peruanos.

CRYRZA (Comisión de Reconstrucción y Rehabilitación de la Zona Afectada). 1970. La reubicación de las ciudades del Callejón de Huaylas. *Revista Peruana de Andinismo y Glaciología* 19(9): 28–29.

Davis, M. 1998. *Ecology of fear: Los Angeles and the imagination of disaster.* New York: Vintage Books.

de la Cadena, M. 2000. *Indigenous mestizos: The politics of race and culture in Cuzco, Peru, 1919–1991.* Durham: Duke University Press.

División de Glaciología y Seguridad de Lagunas, Corporación Peruana del Santa. 1972. *Estudios glaciológicos, bienio 1971–1972.* Biblioteca, Unidad de Glaciología y Recursos Hídricos, Huaraz, Doc #. I-GLACIO-015.

El Comercio. 1970*a*. Hacen estudios para que laguna de Llanganuco no siga siendo una amenaza. *El Comercio*, June 5.

——. 1970*b*. Piden que el gobierno profundice estudio sobre reubicación de la ciudad de Yungay. *El Comercio*, November 15.

——. 1970*c*. Sugieron profundizar los estudios de reubicación de pueblos en prov. Yungay. *El Comercio*, December 21.

——. 1970*d*. Yungaínos acuerdan formar colonias hogares en Lima. *El Comercio*, October 8.

——. 1970*e*. Yungay, Carhuaz, Mancos y Ranrahirca tendrán otra ubicación, señala CRYRZA. *El Comercio*, November 12.

ElectroPerú. 1975. *Memoría bienal del programa de glaciología y seguridad de lagunas*. Huaraz, febrero. Biblioteca, Unidad de Glaciología y Recursos Hídricos, Huaraz, Doc#.I-MEMORIAS-008.

——. 1984. *Información básica de la labor realizada por la unidad de glaciología y seguridad de lagunas entre los años 1973 y 1984*. Huaraz, agosto. Biblioteca, Unidad de Glaciología y Recursos Hídricos, Huaraz, Doc #. I-MEM-002.

——. 1997. *Mapa indice de lagunas de la cordillera blanca*. Huaraz, octubre. Biblioteca de ElectroPerú, Lima, Doc #. Caja 060902, No. H-10.

Ericksen, G. E., G. Plafker, and J. F. Concha. 1970. Preliminary report on the geological events associated with the May 31, 1970, Peru earthquake. *United States Geological Survey Circular* 639: 1–25.

Fernández, C. J., and A. Hoempler. 1953. *Indice de lagunas y glaciares de la Cordillera Blanca*. Comisión de Control de Las Lagunas de la Cordillera Blanca, Ministerio de Fomento, Lima, mayo. Biblioteca, Unidad de Glaciología y Recursos Hídricos, Huaraz, Doc #. I-INVEN-011.

Flores Vásquez, A. 1972. Discurso del Prof. Alejandro Flores Vásquez, pronunciado al conmemorarse el primer aniversario de la catástrofe. *Forjando Ancash (Organo del Club Ancash)* 21: 30–31.

Gálvez Paredes, H. 1970. *Breve información sobre lagunas de la Cordillera Blanca*. Huaraz, abril. Biblioteca de ElectroPerú, Lima, Doc #. 70I 27.725.

Georges, C. 2004. 20th-century glacier fluctuations in the tropical Cordillera Blanca, Peru. *Arctic, Antarctic, and Alpine Research* 36: 100–107.

Jaquette, J. S., and A. F. Lowenthal. 1987. The Peruvian experiment in retrospect. *World Politics* 39: 280–96.

Johnston, H., and B. Klandermans, eds. 1995. *Social movements and culture*. Minneapolis: University of Minnesota Press.

Kaser, G., and H. Osmaston. 2002. *Tropical glaciers*. New York: Cambridge University Press.

Kay, C. 1982. Achievements and contradictions of the Peruvian agrarian reform. *Journal of Development Studies* 18: 141–70.

Klarén, P. F. 2000. *Peru: Society and nationhood in the Andes*. New York: Oxford University Press.

La Prensa. 1970. Solicitan estudios adecuados para ubicar pueblos de Yungay. *La Prensa*, December 29.

Lliboutry, L. 1970. Informe preliminar sobre los fenómenos glaciológicos que acompañaron el terremoto y sobre los peligros presentes. *Revista Peruana de Andinismo y Glaciología* 19(9): 20–26.

Lliboutry, L., V. Mencl, E. Schneider, and M. Vallon. 1970. *Evaluación de los riesgos telúricos en el Callejón de Huaylas, con vista a la reubicación de poblaciones y obras públicas*. Paris: UNESCO.

Lliboutry, L., A. B. Morales, A. Pautre, and B. Schneider. 1977. Glaciological problems set by the control of dangerous lakes in Cordillera Blanca, Peru. 1. Historical failures of morainic dams, their causes and prevention. *Journal of Glaciology* 18: 239–54.

Maskrey, A., ed. 1993. *Los desastres no son naturales*. Bogotá: La Red de Estudios Sociales en Prevención de Desastres en América Latina.

Miano Pique, C. 1972. *¡¡Basta!! La bomba atómica francesa, la contaminación atmosférica y los terremotos*. Lima: Tangrat.

Mitchell, T. 2002. *Rule of experts: Egypt, technopolitics, and modernity*. Berkeley: University of California Press.

Morales, A. B. 1969. Las lagunas y glaciares de la Cordillera Blanca y su control. *Boletín del Instituto Nacional de Glaciología* (Peru) 1: 14–17.

——. 1970. El día más largo en el hemisferio sur. *Revista Peruana de Andinismo y Glaciología* 19(9): 63–71.

——. 1972. Comentarios sobre el memorandum del Dr. Leonidas Castro B. en el caso Huascarán. Lima, 20 diciembre. Biblioteca, Unidad de Glaciología y Recursos Hídricos, Huaraz, Doc #. I-GLACIO-005.

Oberti, I.L. 1975. Estudio glaciológico del cono aluviónico de Huaraz. Biblioteca, Unidad de Glaciología y Recursos Hídricos, Huaraz, Doc #. I-GLACIO-014.

Oficina Nacional de Información. 1970. *¡Cataclismo en el Perú!* Lima.

Oliver-Smith, A. 1977*a*. Disaster rehabilitation and social change in Yungay, Peru. *Human Organization* 36: 5–13.

——. 1977*b*. Traditional agriculture, central places, and postdisaster urban relocation in Peru. *American Ethnologist* 4: 102–16.

——. 1982. Here there is life: The social and cultural dynamics of successful resistance to resettlement in postdisaster Peru. In *Involuntary migration and resettlement: The problems and responses of dislocated people*, ed. A. Hansen and A. Oliver-Smith, 85–103. Boulder: Westview Press.

——. 1986. *The martyred city: Death and rebirth in the Andes*. Albuquerque: University of New Mexico Press.

Orlove, B. 1993. Putting race in its place: Order in colonial and postcolonial Peruvian geography. *Social Research* 60: 301–36.

——. 1998. Down to earth: Race and substance in the Andes. *Bulletin of Latin American Research* 17: 207–22.

Ortega, J. 1970. Enviado especial comprobó desaparición de siete pueblos. *El Comercio*, June 20.

Portocarrero, C. 1995. Retroceso de glaciares en el Perú: Consecuencias sobre los recursos hídricos y los riesgos geodinámicos. *Bulletin de l'Institut Français d'Études Andines* 24: 697–706.

Prakash, G. 1999. *Another reason: Science and the imagination of modern India*. Princeton: Princeton University Press.

Ramírez Gamarra, H. 1971. *Ancash: Vida y pasión*. Lima: Editorial Universo.

República del Perú. 1968. *Censos nacionales de población, vivienda y agropecuario, 1961*. Vol. 2. *Departamento de Ancash*. Lima: Dirección Nacional de Estadística y Censos.

Rubin, J. W. 2004. Meanings and mobilizations: A cultural politics approach to social movements and states. *Latin American Research Review* 39: 106–42.

Sarewitz, D., R. Pielke Jr., and K. Mojdeh. 2003. Vulnerability and risk: Some thoughts from a political and policy perspective. *Risk Analysis* 23: 805–10.

Scott, J. 1998. *Seeing like a state: How certain schemes to improve the human condition have failed*. New Haven, CT: Yale University Press.

Stein, W. W. 1974. *Countrymen and townsmen in the Callejón de Huaylas, Peru: Two views of Andean social structure*. Buffalo: Council on International Studies, State University of New York at Buffalo.

Steinberg, T. 2000. *Acts of God: The unnatural history of natural disaster in America*. New York: Oxford University Press.

Walter, D. 2003. *La domestication de la nature dans les Andes péruviennes: L'alpiniste, le paysan et le parc national du Huascarán*. Paris: L'Harmattan.

Walton, N. K. 1974. Human spatial organization in an Andean valley: The Callejón de Huaylas. Ph.D. diss., University of Georgia, Athens.

Wisner, B., B. Piers, T. Cannon, and I. Davis. 2004. *At risk: Natural hazards, people's vulnerability, and disasters*. New York: Routledge.

Zamora Cobos, M. 1973. Informe sobre la ascención al pico norte del nevado "Huascarán." Huaraz, octubre. Biblioteca, Unidad de Glaciología y Recursos Hídricos, Huaraz, Doc #. I-GLACIO-010.

Zapata Luyo, M. 2002. La dinámica glaciar en lagunas de la Cordillera Blanca. *Acta Montana* (Czech Republic) 19(123): 37–60.

Zavaleta Figueroa, I. 1970. *El Callejón de Huaylas antes y después del terremoto del 31 de mayo de 1970*. Caraz: Ediciones Parón.

Co-production of Knowledge in Climatic and Social Histories

19

Melting Glaciers and Emerging Histories in the Saint Elias Mountains

Julie Cruikshank

Concepts travel, Andre Beteille (1998) reminds us, carrying and accumulating baggage that may gain unexpected ideological weight. He targets the casual use of "indigenous," an idea with layered historical meanings that accrued initially during expanding European colonialism and proliferated more recently in postcolonial discourses. Indigenous is a concept now lodged worldwide, Beteille suggests, applied indiscriminately to peoples anthropologists formerly called "tribes." He addresses the ironies of exporting this term from classic settler societies (North America, Australia, New Zealand, etc.) to geographical settings where complex populations movements defy such shorthand.

The concept "indigenous," though, has rhetorically expanded from an exogenous category to one of self-designation signifying cultural recognition, defense of human rights, and protection under international law (Stavenhagen 1996).[1] In a world now abstractly universalized as postcolonial, ethnographic investigation of emerging indigenous identities may contribute to social analysis in settings where land rights and the legal, economic and social status of minority rights remain controversial. Anthropological investigations of indigenism's twin traveler, the equally essentialized term *nationalism*, have demonstrated that nationalism is not one thing, and that what may appear superficially to be a European category is frequently embedded in radically differing ideologies.

In everyday practice, the term indigenous is now more often used in relational contexts than as a primary category of self-ascription. In Canada, for instance, the terms *Aboriginal* and *First Nation* emerged as preferred forms for self-reference during the 1980s and 1990s. *First Nation* replaces the administrative term *Indian band* long used by the Government of Canada, whereas *Aboriginal*, the more inclusive term, encompasses Inuit and Métis. In northern Canada, self-designations in local languages are in frequent use. *Indigenous*, however, is becoming the term of choice for some young urban activists who describe Aboriginal and First Nation as too enmeshed with official state discourses. Invoked with reference to emerging international alliances, indigeneity highlights similarities with other populations who share histories of

The Anthropology of Climate Change: An Historical Reader, First Edition. Edited by Michael R. Dove.

dispossession, impoverishment, and enforced schooling. Increasingly authorized by UN working committees, indigenism has been identified as a new kind of global entity currently gaining momentum (Niezen 2000: 119). Yet its expanding usage highlights tensions.

If the new indigenism is increasingly a relational and constructed process in Canada, it emerges from a history of distinct encounters and builds on diverse connections. Early treaty-making processes in British North America suggest that colonial powers tacitly acknowledged local sovereignty. Subsequent emplacement of an international border across North America in 1867 partitioned Canada from the United States and arbitrarily allocated aboriginal populations to one nation state or the other. A century on, a new generation of young activists, emboldened by social protest and radical activism that characterized the 1960s began rebuilding cross-border alliances. On the Canada's Pacific west coast, First Nations forged connections with New Zealand Maori during the 1970s. Further north, circumpolar networks have drawn arctic and subarctic peoples into common causes (Minority Rights Group 1994; Smith and McCarter 1997). Ongoing encounters with postcolonial states, with science, and with international organizations are themes explored in this chapter.

Postcolonial theory forces us to look critically at how Enlightenment categories were exported from Europe through expansion of empire to places like northwestern North America once deemed to be "on the verge of the world" and how those categories have become sedimented in contemporary practice. In this chapter, I look at concepts of indigeneity through the lens of local knowledge, in a setting where such knowledge is seemingly gaining ground yet still received and adjudicated largely within scientific norms of universalism. I trace some of the continuities or similarities in the ways that concepts of knowledge—variously deemed to be "local," "indigenous," "Western," or "universal"—are deployed, arguing that the coloniality of indigeneity is sometimes reinforced by hierarchies even in seemingly progressive contexts.

My chapter originates in a puzzle from my ethnographic research—the appearance of glaciers in life stories told by elderly indigenous women who were born in the late 19th century and lived their entire lives just inland from the Saint Elias Mountains. I spent the 1970s and early 1980s in the Yukon Territory working with several women eager to document memories for younger generations. Among the accounts they recorded were some about a desperately cold year during the mid-19th century when summer failed to arrive, and about glaciers that dammed lakes and eventually burst with catastrophic consequences. Their stories also chronicled voyages made by inland Athapaskan and coastal Tlingit ancestors who traded, traveled, and intermarried between the Yukon plateau and the Gulf of Alaska. Sometimes protagonists crossed over glaciers on foot and other times traveled in hand-hewn cottonwood boats, racing under glacier bridges that periodically spanned major rivers draining to the Pacific from the high-country interior. Other narratives recounted how strangers, *k'och'en* (the colorless "cloud people"), first came inland from the coast, traversing glaciers, and the transformations their arrival heralded.

Initially, I was perplexed by references to glaciers in these life stories. By the 1970s, the women I knew were living well inland from the Saint Elias Icefields, yet insisted on including glacier narratives to explain regional human history. Following connections they made initially led me to literature centered on three seemingly distinct themes that I use to frame this discussion: histories of environmental change; analyses of colonial encounters; and debates about local knowledge.

First, glacier stories directed me to accounts about *environmental change* that occurred in northwestern North America during the lifetimes of these women's parents and grandparents. Stories of geophysical risks that dominate these accounts are associated with late stages of a period some scientists call the "Little Ice Age." Jean Grove, the physical geographer who coined this term for an interval of global cooling between the Middle Ages and the early 20th century, traces some of its slipperiness in her classic volume *The Little Ice Age* (1988), and notes that chronologies vary from region to region. Archaeologist Brian Fagan's recent book of the same title (2000) suggests C.E. 1300–C.E.

1850 for western Europe, but also notes that scientists disagree about dates. In the Pacific Northwest, the years between C.E. 1500 and C.E. 1900 are commonly cited.

Second, I argue that glacier stories also depict *human encounters* that coincided with late stages of this Little Ice Age. In the Gulf of Alaska, two fundamental processes that are often discussed independently coincided: geophysical changes (the turf of natural sciences) and European colonial incursions (a sphere of social sciences and humanities). By the late 1700s, when icefields were especially active, so was commerce in furs transported from America's far northwest to London, Paris, and Moscow. Both oral and written accounts depict encounters with changing landscapes, but also with Europeans who were crossing glaciers from the Pacific to reach the interior by the late 1800s. Notably, another kind of encounter follows from those meetings—between stories written and told about such events and their subsequent readers and listeners as they get taken up in different knowledge traditions. Glacier stories move through time, connect with others, and are being reinvigorated as global climate change influences local glacier conditions.

Third, I use the term *local knowledge* to refer to tacit knowledge embodied in life experiences and reproduced in everyday behavior and speech. Variously characterized as "primitive superstition," as "ancestral wisdom," or as "indigenous science," it has long been framed as a foil for concepts of Western rationality. Local knowledge, then, is a concept often used selectively and in ways that reveal more about histories of Western ideology than about ways of apprehending the world. Its late-20th-century incarnation as "indigenous" or as "ecological" knowledge continues to present local knowledge as an object for science—as potential data—rather than as a kind of knowledge that might inform science. I argue that local knowledge is not something waiting to be "discovered" but, rather, is continuously made in situations of human encounter: between coastal and interior neighbors, between colonial visitors and residents, and among contemporary scientists, managers, environmentalists and First Nations.

I begin with a few words about the physical dimensions of the mountains and glaciers in this place. I next outline some old and some new stories that show how themes of environmental change, human encounters, and local knowledge are still central to struggles in places depicted as "remote" despite long entanglements with world markets. I conclude with reference to conflicting stories circulating in these mountains and their ongoing connections to memory, history, and indigenous rights. Not surprisingly, interpretive frameworks seem to be continually recast on all sides to meet contemporary specifications. Insights from current work on memory and forgetting clarify just how profoundly these three strands—changing environment, transformative human encounters, and local knowledge debates—are entangled as they circulate in transnational contexts under new rubrics like environmentalism, postcolonialism, and "traditional ecological knowledge" in its many acronyms.

America's Northwest Glaciers

The Saint Elias Mountains include some of North America's highest peaks and support the world's largest nonpolar icefields (see Figure 19.1). These glaciers were created by ice ages, maintained by climate, and have been in place for thousands of years. This region fascinates me for several reasons. First, in my ethnographic research in the Yukon Territory and Alaska I have heard vivid accounts of glacier travel transmitted in indigenous oral traditions. Second, a sustained record of scientific research here, at high altitude and high latitude, makes it a key site for contemporary climate change studies. Third, this region has recently become the world's largest UNESCO-designated World Heritage Site spanning the Alaska–Canada border and encompassing four parks: Kluane National Park and Tatshenshini-Alsek Provincial Park on the Canadian side, Wrangell-Saint Elias and Glacier Bay National Parks in the United States. The far northwest, once the source of a fabulously lucrative sea otter trade has re-emerged as a site for new global narratives, this time with strong environmental themes.

Significantly, Icefield Ranges includes glaciers that surge—of great interest to geophysical sciences.

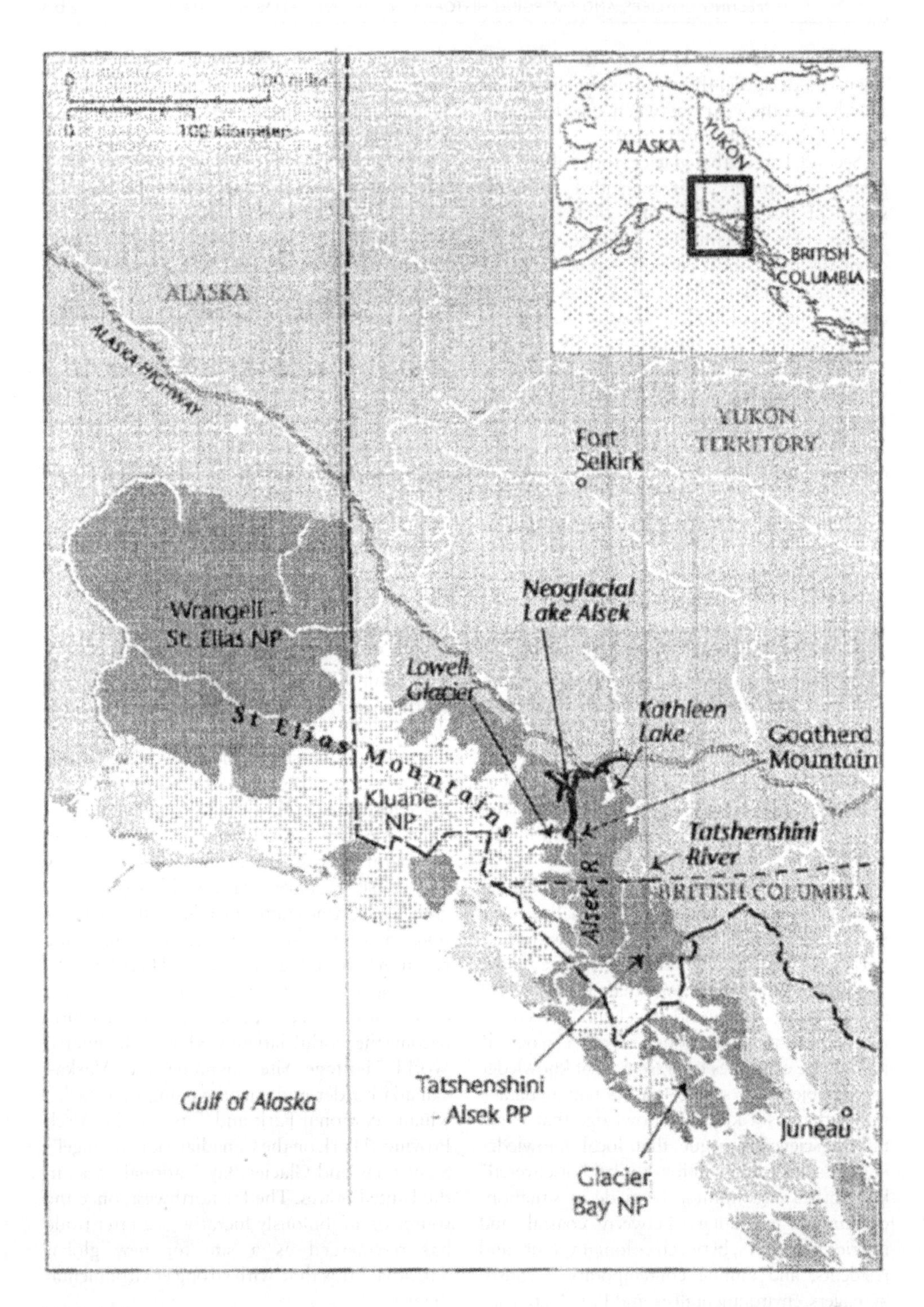
0
100 miles
0
100 kilometers
ALASKA
YUKON
BRITISH
COLUMBIA
ALASKA
ALASKA HIGHWAY
YUKON
TERRITORY
Fort
Selkirk
Wrangell -
St. Elias NP
Neoglacial
Lake Alsek
Lowell
Glacier
St Elias Mountains
Kathleen
Lake
Goatherd
Mountain
Kluane
NP
Tatshenshini
River
Alsek R.
BRITISH COLUMBIA
Gulf of Alaska
Tatshenshini
- Alsek PP
Juneau
Glacier
Bay NP

Figure 19.1

Surging glaciers may advance without warning after years of stability, sometimes several kilometers, and they frequently create ice-dammed lakes that build up and burst out when the ice thins and the dam breaks. Surging glaciers also occur in Greenland and the Antarctic, but issues of scale and accessibility make them easier to study in the Saint Elias Mountains. Of some 4,000 glaciers on these icefields, a relatively large number—at least 200—have this characteristic. In this place, we now see nature represented in many ways—as "primordial wilderness," as a "climate change laboratory," or as a giant "jungle gym" for ecotourists. In northwestern Canada, such depictions now compete with views by local indigenous residents who lived and hunted in these lands until 1943, when they were relocated east of the Alaska Highway after the Kluane Game Sanctuary (and, later, Kluane National Park and Reserve) was set aside as the Yukon's first "protected area."

I begin with one story about a glacier now officially named Lowell Glacier that I first heard in November 1978 when Mrs. Kitty Smith, almost 90 at the time, asked me to record it. Born approximately 1890, she grew up in the Tatshenshini River valley. As a child, she became well acquainted with unpredictable glacial surges and the interpretive challenges of living with glaciers. Lowell Glacier, for instance, has crossed the Alsek River more than once. Mrs. Kitty Smith identified it by the name Nàlùdi, or "Fish Stop" because it interrupted salmon migrations up that river to the interior, leaving land-locked salmon in Kathleen Lake.

Her narrative demonstrates consequences of hubris, a classic theme in stories told here. Nàlùdi, she says, was provoked to surge when a reckless child traveling to the interior with coastal Tlingit traders joked about a balding Athapaskan shaman. "Ah, that old man," he reportedly said, "the top of his head is just like the place where gophers play, a bare stump!" To punish this transgression, the shaman withdrew to the top of a high bluff facing the glacier and began to dream, summoning the glacier to advance across the Alsek River valley. It surged, reached this bluff, and built an immense wall of ice that dammed the river and created an upstream lake a hundred kilometers long. When that ice dam eventually burst, the resulting flood scoured the landscape, drowning Tlingit families camped at the junction of the Alsek and Tatshenshini rivers. Nàlùdi surged again, she says, shortly after her own birth. That summer, the glacier blocked the river and flooded the valley basin for just a few days before draining (Cruikshank with Sidney, Smith, and Ned 1990: 205–208, 332–333; McClellan 2001 [1975]: 71–72).

These events are preserved in the geoscience record, although scientists provide different causal explanations for surges (Clague and Rampton 1982). They estimate that the advancing Lowell Glacier created a two hundred meter high ice dam after it came to rest against Goatherd Mountain and impounded Neoglacial Lake Alsek in the mid-19th century, as it had several times during the previous 2800 years. When the dam broke, it discharged water through the Alsek valley in an enormous flow, emptying the lake in one or two days. Giant ripple marks left in its wake are still visible from the air and on the ground. Scientists now refer to oral histories, tied to genealogies of named persons that suggest 1852 as a possible date for the last major outburst flood (De Laguna 1972: 276).

At issue here are diverging notions of agency and interpretation. One key difference between Athapaskan oral traditions and scientific discourse is that elders' narratives merge natural and social history, whereas scientists assessing environmental change describe one of their objectives as disentangling natural from cultural factors. Elders, for instance, cite the folly of "cooking with grease" near glaciers, lest this excite either the glacier or the being living within such a glacier den. Food should be boiled, never fried, in the presence of glaciers and no grease should ever be allowed to escape from the cooking vessel. Inevitably, such explanations fall out of most contemporary studies of "local knowledge," because they neither fit easily with contemporary scientific understandings of causality nor contribute to databases.

Other stories about glaciers are harder to understand, like those depicting glaciers emitting heat so intense that people were driven to submerge in glacial rivers to avoid being consumed. We know the terrible consequences of epidemics, particularly smallpox, that raged north up the North Pacific coast during the 19th century, and that alternating sweating and submersion in water was a strategy used by coastal victims seeking relief (Gibson 1983). But few details are known about epidemics that traveled up the Alsek. Public health physician Robert Fortuine (1989) identifies the coastal smallpox epidemic that occurred between 1835 and 1840 as one of the most significant events in Alaska history. Starting at Sitka, it spread northward to Lynn Canal. Anthropologist Catharine McClellan estimates that it swept up the Alsek in 1838, and that another smallpox epidemic followed in 1862 (McClellan 2001 [1975]: 24, 223). Two epidemics and an outburst flood must have coincided within one generation but the extent of losses seems to have prevented transmission of firsthand accounts to the present.

Historian Mike Davis has documented how imperial expansion through Asia was enabled when it coincided with El Niño induced droughts and famines in the late 19th century (Davis 2001). In northwestern North America's far northwest, Tlingit traders encountered similar intrusions during late stages of the Little Ice Age. They traveled inland in 1852 and destroyed Fort Selkirk, a trading post that Hudson Bay Company trader Robert Campbell had established three years earlier to divert trade away from long-established Aboriginal networks and into British hands.

Anthropologists, geographers, and historians have demonstrated the enduring power of landscape features to act as points of reference anchoring memories, values, and tacit knowledge. A growing body of research about social memory suggests that landscapes are places of remembrance and sites of transmission, and that culturally significant landforms often provide a kind of archive in which memories can be mentally stored (Boyarin 1994). In the Saint Elias Mountains, though, we can see also how changing landscape features like fluctuating glaciers have also provided imaginative grist for comprehending *changing* social circumstances affecting human affairs. Orally narrated stories indeed provide empirical observations about geophysical changes and their consequences, but also demonstrate how glaciers provide material for evaluating changes wrought by colonial histories.

Such overlapping and conflicting interpretations of glaciers have 21st-century consequences. They seem to typify or even model classic and continuing struggles over cultural meanings replicated in contemporary debates. Implications of what Bruno Latour calls this "Great Divide" differentiating nature from culture continue to cascade internationally through debates about environmentalism, biodiversity, global climate change, and indigenous rights (Latour 1993; see also Franklin 2002; Hornberg and Pálsson 2000; Macnaghten and Urry 1998).

I now turn to some contemporary narratives that provide further points of overlap and contrast.

New Stories from Melting Glaciers

In the 1990s, the Saint Elias Icefields began to reveal fresh surprises. Three recent "discoveries" show just how slippery our views of nature and society can be. In 1991, biologist David Hik spotted a rectangular piece of hide roughly one meter in length and half a meter wide melting from a glacier near the center of what is now Kluane National Park. It had been modified by humans—with slits around the edge and a possible fragment of thong—and looked old. As required by Canadian law, Hik informed Kluane Park staff, who took charge of the hide and then sent it to park headquarters in Winnipeg where it has remained in a freezer. It appears to have been left in the central icefields by a traveler approximately 1,000 years ago (1,110 BP +/– 50, calibrated to account for fluctuations of carbon in the atmosphere). After it was identified as bear hide, parks scientists hoped that it might reveal information about genetic relationships and diversity in the Kluane grizzly bear population over time; however, no genetic information has been salvageable because of repeated thawing and freezing (personal communications from David Hik, University of Alberta, 1992; and David Arthurs, Parks Canada, 2003).

Six years later, in 1997, a wildlife biologist hiking near a north facing alpine basin some 1,830 meters above sea level stumbled on a square-kilometer concentration of caribou droppings melting from an alpine snow patch. Artifacts were literally pouring out of melting ice. Subsequent research revealed evidence of ancient caribou harvesting on the mountain named Thandlät in Southern Tutchone language.

Scientists describe this as a rare opportunity to explore questions about the prehistoric ecology of large caribou populations, implications of climate change for caribou, and human use of high-elevation hunting sites. The droppings contain mitochondrial and nuclear DNA that biologists intend to compare with that of living caribou populations as well as ancient pollen that may help to reconstruct past climates (Kuzyk et al. 1999). By 2003, 72 ancient alpine ice patches, usually no more than a square kilometer and 50 meters deep, had been identified in southwest Yukon. Eighteen of these have yielded tools and the work continues. Radiocarbon dates for rare wooden tools and other organic material melting from these patches demonstrate an enduring relationship between caribou and human hunters at high altitude and latitude for at least 8,000 years, right up until the late 19th century (Farnell et al. 2004; Hare et al. 2004; Krajick 2002).

Archaeologists working in Athapaskan territories are accustomed to making their inferences from a sparse material record. Subarctic hunters made their tools largely from perishable materials—skin, wood, sinew—used them, left them behind and remade them as needed, a technology constructed from ingenious principles and carried in the head rather than on the back. Understandably, then, perfectly preserved tools made from organic materials melting from glaciers are spectacularly interesting to scientists as well as to local people, if for different reasons. In their subsequent investigations of Yukon ice patches, archaeologists and Yukon First Nations have formed partnerships and community members have participated in archaeological field research. Culture camps organized by Champagne-Aishihik First Nation around the theme of ice patches have included visiting scientists who meet with local students. In Alaska, archaeologists are now using global positioning system (GPS) models to "prospect" for potential melting sites that may be releasing similar evidence on the other side of the Saint Elias Range.

More widely publicized was the accidental discovery, by sheep hunters in August 1999, of a young hunter's remains, a man in his late teens or early twenties. He was melting from a glacier in the traditional territories of Champagne-Aishihik First Nation (also within Alsek–Tatshenshini Provincial Park). His death was probably accidental. His woven spruce root hat, part of his squirrel fur robe, some tools (including a bone knife with iron stains), and a piece of fish he was carrying were preserved with him. His robe, radiocarbon dated at 550 BP, carried traces of spruce and pine pollen and some fish scales. Local elders named him Kwäday Dän Ts'ínchi or "Long Ago Person Found" in Southern Tutchone language.

From the beginning, there was close cooperation among scientists and members of the relevant Champagne-Aishihik First Nation, without which the scientific research could not have proceeded. If the Native American Graves and Repatriation Act structures such relationships in the United States (see Brown 2003; Starn 2004; Thomas 2000), partnerships are being carefully negotiated as part of implementing recent land claims agreements in Canada. Members of this First Nation were interested in learning more about this potential ancestor and how his travels overlap with their oral histories. They agreed to allow scientific investigations that included First Nation representatives on the management team and to let the materials travel for scientific analysis. Scientists are especially interested in how this rare evidence—flesh and hair as well as bones—might contribute to understandings of health, nutrition, and disease, but also what his perfectly preserved hat, tools, and fragments of robe may reveal about everyday life from that time (Beattie et al. 2000). As agreed, his remains were returned to the community within a specified time frame. The First Nation held a funeral potlatch for him on July 21, 2001, and his cremated remains were returned to the location where he was found. Radiocarbon and DNA testing continue, but so far DNA indicates only that the young man had closer connections with other Native Americans than with populations in Asia or Greenland (Monsalve et al. 2002). Botanists have determined more about his early diet from bone samples (Dickson et al. 2004). Cross-border connections between coastal Tlingit and inland Champagne-Aishihik First Nations have been strengthened during community negotiations surrounding this research and the funerary arrangements for Kwäday Dän Ts'ínchi.

These events all received wide publicity for a short time—locally, nationally, and internationally, and in sources ranging from news media to scientific journals. In such circumstances, scientists and Aboriginal people are encountering concrete, material evidence of the past but they are also encountering each other at close range. Contemporary encounters fall into new transnational contexts—global climate change, environmentalism, social justice, and scientific studies of human remains—and we see how the same evidence produces different interpretations. The piece of bearskin presents scientists with possibilities of learning about ancient grizzly populations and provides local residents with possible evidence of an early ancestor who traveled in the central icefields. Tools pouring from glaciers suggest that high-latitude, high-altitude landscapes were intensely shared by humans and caribou for thousands of years. Elders born before the turn of the last century, with whom I worked, still remembered large herds at the beginning of the 20th century before they disappeared, and biologists cannot pinpoint the cause of their disappearance. Tellers do not separate the tools from the tool-makers: women, for instance, speculate about the coastal woman who may have made Kwäday Dän Ts'ínchi's coastal hat or the grandmother who might have knitted his inland robe. So his arrival both confirms oral traditions in the minds of local people and authenticates the antiquity of such travel by ancestors. Melting glaciers are revealing material evidence of interest to scientists but they are also reinvigorating longstanding oral histories about travel and trade near the Saint Elias Mountains. Again, questions arise about where stories told by scientists and elders connect and where they slide apart.

Crucially, Athapaskan and Tlingit oral traditions attribute to glaciers characteristics rather different from those discovered through science. Glaciers long provided travel routes or "highways" that enabled human connections between coast and interior. Glaciers are described in many narratives as characterized by sentience. They listen, pay attention, and they are quick to take offense when humans demonstrate hubris or behave indiscreetly. I have been struck by how people who speak knowledgeably about glaciers refer to listening, observing, and participating in ritualized respect relations (see also Anderson 2004). Such visions originate in intense engagement with environment maintained through millennia, creating what anthropologist Tim Ingold calls a "dwelling perspective" so profoundly relational that everyone understands how humans and nature coproduce the world they share (Ingold 2000: esp. 153–156; see also Basso 1996). Glacial landscapes described in oral traditions, then, are intensely *social spaces* that include relationships with nonhuman beings (like glaciers and features of landscape) sharing characteristics of personhood.

In this place, memories of the Little Ice Age are sedimented both in physical processes studied by scientists (bands of grit, layers of ice and rock, etc.) and in memories of long-term residents. Both kinds of knowledge are acquired through close engagement with a physical environment. During the last century, one discourse (science) has gained authority and park managers, ecotourists, and the general public have adopted conceptions of glaciers as places of "raw nature." Once again, glaciers seem to be playing an active role in negotiating the modern terrain of science, history, and politics in these mountains.

Narratives about melting glaciers echo the three interpretive frameworks identified earlier, but also incorporate contemporary concerns. New narratives of *environmental change* associated with melting glaciers now address global climate change rather than Little Ice Age advances. Stories about *human encounters* that once depicted Euro-American incursions across glaciers now speak to implementation of land claims in the context of social justice. *Local knowledge* is again being produced in new contexts and is assuming an expanding role in the rhetorics of comanagement policies. In the remainder of this chapter, then, I offer snapshots of these rapidly changing contexts.

Environmental change

From the Andes to the Arctic, the environmental change of concern at the beginning of the 20th century is global climate, symbolized by visibly melting glaciers. Evidence mounts that warming will be extreme at Arctic and Subarctic latitudes. Scientists may disagree about the magnitude of globally averaged temperature changes or about the role of humans in the process, but they agree that extreme values are being experienced in arctic regions and that that this will continue. Climate change is a global process, but has profoundly local consequences. Yukon First Nations, now completing land claims negotiations and engaged in economic planning as part of self-governance agreements, are raising questions about regional consequences for water levels, forest yield, permafrost, wildlife, and human activities. Their past experience with climate variability evokes risky times and territories (Cruikshank 2001).

There is growing interest in how Aboriginal people and policymakers can work together on questions surrounding climate change and acknowledgement that some solutions must come from local levels. Up close, though, consultations can lead to awkward exchanges. Scientists, for instance, make a distinction between weather and climate. By definition, they tell us, climate is the statistics of weather, including measurement of means (mean temperature, or mean precipitation) and variance. Climate scientists talk about precise and measurable data—temperature, air pressure, precipitation, and wind speed (Weaver 2003). As anthropologists report from one collaborative study in northern Finland, oral traditions are unlikely to provide transferable "data" to climate change scientists, partly because local people are often referring to weather when they talk about environmental change. Memories passed on in the Yukon, for instance, attend to summer warmth from sun-drenched days or biting cold of deep river valleys in winter. People recall chilling boreal winds and nasty hailstorms. "Climate is recorded," Ingold and Kurttila note succinctly, "weather is experienced" (2000: 187). Knowledge about weather, they point out, cannot be transmitted as a set of customary prescriptions or formulae; it accumulates from a lifetime of experience traversing and inhabiting well-known places and is embodied in tacit knowledge.

Climate science presents a more comprehensive picture than weather. But similarly, oral traditions convey understandings that are more comprehensive than data. The two cannot always be conflated, but both reveal a great deal about the human experience of environmental change. One primary value of local traditions about weather is to deploy authoritative local traditions in problem solving during unexpected weather events (McIntosh et al. 2000). A dominant theme in the Yukon concerns living with uncertainty surrounding behavior of glaciers—unexpected advances, violent surges, catastrophic floods, and accompanying weather variations. Another concerns travel: glacier-filled passes between coast and interior provided ancient travel routes. In one well-known story about two trading partners crossing glaciers to reach the coast, one coastal Tlingit and the other interior Athapaskan, the dramatic consequences center on an accident as the coastal Tlingit man slides into a crevasse and his partner, knowing that he will be held responsible, must make choices and construct a rescue plan. Narratives that are useful in times of crisis concern proper relations with land and crystallize quick and timely social responses. Kwäday Dän Ts'ínchi probably lost his way in an unexpected storm, archaeologists speculate. Stories now associated with his appearance, his contribution to science, his ceremonial cremation, and his return to the glacier where he was found also point to practices crucial to maintaining balance in a moral world. Scientists may necessarily distinguish environmental data from social history, but Aboriginal storytellers are just as likely to equate disastrous effects of environmental change (specifically pollution) with history of colonialism and its imbalances, rather than viewing these as isolatable physical problems that science might help "fix."

Human encounters

My larger project traces the role glaciers play in social imagination—in Aboriginal traditions but also in accounts left by Russian, Spanish, French, British, and U.S. and Canadian visitors to the far northwest in the 18th and 19th centuries (Cruikshank 2005). The idea of encounter seems especially useful because of what it reveals about scale and subjectivity. Initially, actors in this region were relatively few and their motives, intentions, and imaginings can be traced partially in diaries and reports, but also in orally narrated stories still told. As national dreams flared in this corner of northwest North America, power shifted decisively.

More than half a century ago, Canadian economic historian Harold Innis (1950) identified arctic and subarctic regions as furnishing a classic illustration of the modernist tendency to conceptualize time as spatially laid out, mechanically segmented, and linear. Colonial projects, he observed, move forward by devising and reinforcing categories—such as objectivity, subjectivity, space, and time. Once normalized as "common sense," these classifications provide a visual template for the annexation of territories and the subjugation of former inhabitants. Innis's analysis is especially apt here. Ironically, a location that gathers much of its imaginative force as a place where boundaries were always being negotiated (between trading partners, between coast and interior, between glaciers and humans, and among residents and strangers) has now become a place festooned with boundaries rather than stories. The international boundary was almost a century in the making, mapping, adjudication, and production (from 1825–1915). Like layers of an onion, successive tribunals and commissions struck to resolve the thorny issue of why, how, and where this boundary would be drawn, demonstrate how an imagined "Nature" can become swept into the formation of nations, and how dreams of nationhood become embedded in borders. As others have noted, these national boundaries were more formidable to negotiate than glaciers.

Boundaries propagate. Fundamental to recent human–environmental histories in this region is the transformation of hunting territories to a nature preserve in linked steps that disenfranchised indigenous hunters at the stroke of a pen. During WWII, the United States became concerned about a possible invasion by Japan through Alaska. They conceived the idea of a military highway connecting their distant northwesterly territories with the national core, an enormous operation that brought 34,000 workers north between April 1942 and December 1943. Following overhunting attributed to U.S. military personnel and Canadian civilians during this Alaska Highway's construction, the Kluane Game Sanctuary was established in 1943. New regulations prohibiting hunting within the sanctuary, including subsistence hunting, meant that a hunter who killed a sheep just east of the boundary was deemed a good provider whereas someone who took a sheep a few feet west was subject to prosecution. When those boundaries were modified to create Kluane National Park in 1979, the region joined a national administrative parks network. The subsequent layering of a UNESCO-designated World Heritage Site across the U.S.–Canadian border drew the region into an international agenda. An international boundary severs people in Alaska from relatives in Canada; provincial and territorial boundaries separate families in British Columbia from those in the Yukon; and boundaries placed around Protected Areas in 1943, and a National Park in 1979, locked ancestral territories *in* behind a boundary, or locked people *out* beyond a boundary, depending on point of view. A crucial problem for those separated by boundaries is how to pass on knowledge about places to those who never experienced them—hence, their appearance in life histories.

Local knowledge

Melting glaciers have generated partnerships among scientists and First Nations that inevitably enliven local debates about knowledge. Internationally, there has been an explosion of interest in indigenous knowledge or "traditional ecological knowledge" during the last decade. Yet the "locality" of such knowledge sometimes disappears as prescriptive methodologies become enshrined. "Local knowledge" has become a commonsense term in early-21st-century rhetoric. Acronyms like TEK [Technical Environmental Knowledge] are ubiquitous in research management plans on topics as diverse as fisheries, wildlife management, and forestry, yet often depict local knowledge as static, timeless, and hermetically sealed within categories like "indigenous" in ways that reinforce the coloniality of that concept.

The implication is that oral sources are somehow stable, like archival documents, and that once spoken and recorded, they are simply there, waiting for interpretation. Yet ethnographic research clearly demonstrates that the content of oral sources depends largely on what goes into the discussions, the dialogue, and the personal relationship in which it is communicated. Oral testimony is never the same twice, even when the same words are used, because, as Alessandro Portelli (1997: 54–55) reminds us, the relationship—the dialogue—is always shifting. Oral traditions are not "natural products." They have social histories and *acquire* meanings in the situations where they are used, in interactions between narrators and listeners. Meanings shift depending on the extent to which cultural understandings are shared by teller and listener. If we think of oral tradition as a social activity rather than as some reified product, we come to view it as part of the equipment for living rather than a set of meanings embedded within texts and waiting to be discovered. One of anthropology's most trenchant observations is that meaning is not fixed, but must be studied in practice—in the small interactions of everyday life.

A growing critique of the uses and abuses of traditional knowledge identifies several problems associated with uses of TEK. One is the underlying premise that different cultural perspectives are bridgeable by concepts in English language (like "sustainable development" or "comanagement") and within scientific discourse (Morrow and Hensel 1992). Another is the idea that statements by knowledgeable people can somehow be "captured," codified, labeled, and recorded in databases. Third is the growing evidence that concepts of "local knowledge" or "tradition" most likely to be selected in management-science studies usually reflect ideas compatible with state administration rather than those understood by local people (Ingold and Kurttila 2000; Nadasdy 1999). Scientists working on TEK projects have not been shy about naturalizing culture as an endangered object, then selecting data that effectively conflates environmental and social agendas (see Raffles 2002: 152).

Gender also plays a significant role in local narratives, a theme I have explored elsewhere at greater length (Cruikshank with Sidney, Smith, and Ned 1990). I heard these glaciers stories from elderly local women who incorporated them into accounts of life experience and made them reference points for interpreting and explaining life transitions and gendered family histories. Their interests and perspectives contrasted sharply with masculine narratives of science and empire that characterize so much of the historical literature from this part of the world. Kitty Smith and her peers paid attention to the specificities of everyday life and made glaciers central images in stories that ground social histories in well-known landscapes. Nineteenth-century narratives of imperial science, by contrast, rarely mention residents of inhabited landscapes other than fellow scientists and sponsors, and they virtually ignore local history. These female storytellers claim agency and embed their knowledge in narrative rather than presenting it as locatable "data." In arctic global cultural flows, though, such storied knowledge tends to be displaced and to slide out of grander narratives (see Bravo and Sörlin 2002).

Crucially, stories transmitted in northern oral traditions make no sharp distinction between environmental and social change and indeed take as axiomatic the connections between biophysical and social worlds. The modernist wedge partitioning nature from culture (reflected in TEK studies and databases) severs important connections that Athapaskan narratives explicitly affirm. The overall effect of segregating environmental snippets from their social context inevitably submerges some memories and recasts others to fit dominant transnational narratives. Once again "indigenous data" is subsumed within universalizing hierarchies. What is included and what is left out is not random (see Cooke and Kothari 2001; Cruikshank 2004; Fienup-Riordan 1990: 167–191; Nadasdy 2003: 114–146; Scott 1996).

Modern science, as Sheila Jasanoff explains so concisely, achieves its accomplishments by abstraction. Scientific observations gain authority by being removed from local contexts and recombined in larger wholes—framed as "universal"—that both travel and frequently transgress boundaries of custom and tradition (Jasanoff 2004). Walter Benjamin's famous essay "The Storyteller" eloquently captures this same distinction between knowledge embedded in stories and disembodied information: "Information," he says, "lays claim to prompt verifiability. The prime requirement is that it appear 'understandable in itself.' ... A story is different. It does not expend itself. It preserves and concentrates its strength and is capable of releasing it after a long time" (Benjamin 1969: 89–90; see also Cruikshank 1998).

A historical approach to memory reveals how socially situated but also how porous knowledge practices are. The field of science studies demonstrates that all knowledge is ultimately local knowledge and has a history. The idea that a measurable world can be pried from its cultural moorings also originated in local knowledge traditions that expanded within Enlightenment Europe. In the space now called Kluane Park, science and oral tradition are both kinds of local knowledge that share a common history. That history includes authoritative gains for one kind of formulation—science—at the expense of another. Since 1960, when the Icefield Ranges Research Project was established under the institutional sponsorship of the Arctic Institute of North America and the American Geophysical Society, the Saint Elias mountain ranges have provided research sites for natural and physical sciences, and now for climate change studies. One ironic consequence is that as part of current comanagement agreements, mandated by land claims and self-governance agreements in Canada, indigenous people living near park boundaries are now being asked to document their "traditional knowledge" about places from which they were evicted 60 years ago. In such instances, it would seem, "our flattery of 'primal peoples' and their knowledge, inevitably deemed to be timeless and ahistorical, can be viewed as an act of immense condescension" (White 1998: 218).

Entangled Narratives

Stories about glaciers in the Saint Elias Mountains, in constant and uneasy interplay, contribute to a two-century debate about humanity's relationship with the natural world. Regional political and economic practices involved in setting aside protected areas (e.g., parks) now intersect with global practices (e.g., scientific research) that make claims from these spaces. The mountains and glaciers are again being reinvented for new purposes—this time as a hybrid comprising management techniques and measurement practices, sometimes circulating on Internet spaces. Just as narratives of Euro-American national dreams once normalized practices of mapping and measuring, global environmental narratives now fill that role. They help to bring primordial wilderness under the human protection of international committees, like Geneva-based UNESCO, in which morally tinged stories of rational use and protection allow global committees to soberly adjudicate local concerns (see Anderson 2004). The United Nations, centrally positioned within international debates about indigeneity, appears in many guises.

The UNESCO World Heritage List categorizes sites given this status into one of three categories: "natural," "cultural," or "mixed properties," reasserting modernist opposition between nature and lived experience. As of April 6, 2007, the 830 properties inscribed on the World Heritage List included 644 sites deemed cultural, 162 classified as natural, and 24 as having mixed properties. The awkwardly named "Kluane/Wrangell-Saint Elias/Glacier Bay/Tatshenshini-Alsek" World Heritage Site, the first to cross an international boundary, has been allocated to the "natural" World Heritage Site category, to the consternation of local First Nations. Breaking the bond between people and place along lines so arbitrary as one imagined between cultural heritage and natural environment marks a decisive rift (Giles-Vernick 2002; Ingold 2000).

The nature we are most likely to hear about in the early 21st century is increasingly represented as marvelous but endangered, pristine or biodiverse. Such depictions tend to exclude other ways of seeing and make it more difficult to hear or appreciate unfamiliar points of view (Franklin 2002; Slater 2002). Environmental politics have so normalized our understandings of what "nature" means that we can no longer imagine how other stories might be significant. As claims and counterclaims made in nature's name proliferate, areas deemed to be primordial wilderness are reimagined as uncontaminated by humans. Such views largely exclude other practices, memories, conceptions, and beliefs of people that do not match this vision. Ever-narrowing subsets of nature push humans out except as subjects for management-science to regulate.

Indigenous visions passed on in narratives about glaciers (like those about caribou, forests, or rivers) seem uniquely important because they position nature and culture in a single social field and graft colonial and environmental histories onto older stories. They draw connections between relationships and activities on the land and proper social comportment. They provide rich, complex alternatives to normalized values that now conventionally frame nature as a redeemable object to be "saved." Always in motion even when they appear static, surging glaciers encompass both the materiality of the biophysical world and the agency of the nonhuman, and draw on traditions of thought quite different from those of academic materialism (see Raffles 2002: 38, 181 for a differently situated discussion of this). They *are* grounded in material circumstances but also carry a multitude of historical, cultural, and social values that slide away when they are relegated uncritically to "nature."

Narrative recollections about history, tradition, and life experience represent distinct and powerful bodies of local knowledge that have to be appreciated in their totality, rather than fragmented into data, if we are to learn anything from them. Rarely do either management-driven studies of TEK or environmentalist parables tap into the range of human engagements with nature—diverse beliefs, practices, knowledge and everyday histories of nature that might expand the often crisis-ridden focus of environmental politics. What looks similar on the surface often turns out to have different meanings and different aims. Codified as TEK, and engulfed by frameworks of North American management science, local knowledge shifts its shape, with sentient and social spaces transformed to measurable commodities called "lands" and "resources." Indigenous peoples then continue to face double exclusion, initially by colonial processes that expropriate the land and ultimately from neocolonial discourses that appropriate and reformulate their ideas. Environmentalist values may now shape our understandings of nature (much as science or survey did in the past) but they, too, become entangled with questions of justice working their way through local debates in northwestern North America.

Successive visions of Saint Elias Mountains, continuously recast to serve present purposes, become entangled with those of contemporary First Nations whose visions deserve more space in such schema. In the Gulf of Alaska where European and indigenous forms of internationalism have been enmeshed for two centuries, physical places and people have always been entangled, and in the future they are likely to be more entangled than ever before. Local knowledge in northern narratives is *about* unique entanglements of culture and nature, humans and landscapes, objects and their makers. Material evidence of human history—from Kwäday Dän Ts'ínchi, caribou pellets, or from the modified bear hide—may be naturalized as probable genetic evidence of natural history. But memories covered by appliquéd layers of sanctuary, park, and World Heritage Site are also being reenergized as human history emerges. The glacier stories I began hearing more than two decades ago may originate in the past but they continue to resonate with current struggles surrounding environmentalism, indigenous rights, land claims, nationhood, and national parks. Such narratives will undoubtedly continue to lead entangled social lives.

NOTE

1 The *Oxford English Dictionary* defines *indigenous* as "born or produced naturally in a land or region; native to [the soil or region]" and traces its earliest use to 1646.

REFERENCES

Anderson, David G. 2004. Reindeer, caribou and "fairy stories" of state power. In *Cultivating arctic landscapes: Knowing and managing animals in the circumpolar North*, edited by David G. Anderson and Mark Nuttall, 1–16. Oxford: Berghahn.

Basso, Keith. 1996. *Wisdom sits in places: Landscape and language among the Western Apache*. Albuquerque: University of New Mexico Press.

Beattie, Owen, Brian Apland, Eric W. Blake, James A. Cosgrove, Sarah Gaunt, Sheila Greer, Alexander P. Mackie, Kjerstin E. Mackie, Dan Straathof, Valerie Thorp, and Peter M. Troffe. 2000. The Kwäday Dän Ts'ínchi discovery from a glacier in British Columbia. *Canadian Journal of Archaeology* 24: 129–147.

Benjamin, Walter. 1969. The Storyteller. In *Illuminations*, edited by Hannah Arendt, 83–109. New York: Schocken.

Beteille, Andre. 1998. The idea of indigenous people. *Current Anthropology* 39 (2): 187–91.

Boyarin, Jonathan. 1994. *Remapping memory: The politics of timespace*, edited by Jonathan Boyarin, 1–37. Minneapolis: University of Minnesota Press.

Bravo, Michael, and Sverker Sörlin, eds. 2002. *Narrating the arctic: cultural history of Nordic scientific practices*. Canton, MA: Watson Publishing International.

Brown, Michael F. 2003. *Who owns Native culture?* Cambridge, MA: Harvard University Press.

Clague, John J., and V. N. Rampton. 1982. Neoglacial Lake Alsek. *Canadian Journal of Earth Sciences* 19: 94–117.

Cooke, Bill, and Uma Kothari, eds. 2001. *Participation: The new tyranny?* London: Zed Books.

Cruikshank, Julie. 1998. *The social life of stories: Narrative and knowledge in the Yukon territory*. Lincoln: University of Nebraska Press.

——. 2001. Glaciers and climate change: Perspectives from oral tradition. *Arctic* 54 (4): 377–393.

——. 2004. Uses and abuses of "traditional" knowledge: Perspectives from the Yukon Territory. In *Cultivating arctic landscapes: Knowing and managing animals in the circumpolar North*, edited by David G. Anderson and Mark Nuttall, 1–16. Oxford: Berghahn.

——. 2005. *Do glaciers listen? Local knowledge, colonial encounters and social imagination*. Vancouver: UBC Press.

Cruikshank, Julie, with Angela Sidney, Kitty Smith, and Annie Ned. 1990. *Life lived like a story: Life stories of three Yukon elders*. Lincoln: University of Nebraska Press.

Davis, Mike. 2001. *Late Victorian holocausts: El Niño Famines and the making of the Third World*. London: Verso Press.

De Laguna, Frederica. 1972. *Under Mount Saint Elias: The history and culture of the Yakutat Tlingit*. 3 vols. Smithsonian Contributions to Anthropology, 7. Washington, DC: Smithsonian Institution Press.

Dickson, James H., Michael P. Richards, Richard J. Hebda, Petra J. Mudie, Owen Beattie, Susan Ramsay, Nancy J. Turner, Bruce J. Leighton, John M. Webster, Niki R. Hobischak, Gail S. Anderson, Peter M. Troffe, and Rebecca J. Wigen. 2004. Kwäday Dän Ts'ìnchí. *The Holocene* 14 (4): 481–486.

Fagan, Brian. 2000. *The little ice age: How climate made history, 1300–1850*. New York: Basic Books.

Farnell, Richard, P. Gregory Hare, Erik Blake, Vandy Bower, Charles Schweger, Sheila Greer, and Ruth Gotthardt. 2004. Multidisciplinary investigations of alpine ice patches in Southwest Yukon, Canada: Paleoenviromental and paleobiological investigations. *Arctic* 57 (3): 247–259.

Fienup-Riordan, Ann. 1990. *Eskimo essays: Yu'pik lives and how we see them*. New Brunswick, NJ: Rutgers University Press.

Fortuine, Robert. 1989. *Chills and fevers: Health and disease in the early history of Alaska*. Fairbanks: University of Alaska Press.

Franklin, Adrian. 2002. *Nature and social theory*. London: Sage.

Gibson, James R. 1983. Smallpox on the Northwest Coast, 1835–38. *BC Studies* 56: 61–81.

Giles-Vernick, Tamara. 2002. *Cutting the vines of the past: Environmental histories of the Central African rain forest*. Charlottesville: University Press of Virginia.

Grove, Jean. 1988. *The little ice age*. London: Methuen.

Hare, P. Gregory, Sheila Greer, Ruth Gotthardt, Richard Farnell, Vandy Bower, Charley Schweger, and Diane Strand. 2004. Ethnographic and archaeological investigations of alpine ice patches in Southwest Yukon, Canada. *Arctic* 57 (3): 260–272.

Hornborg, Alf, and Gísli Pálsson, eds. 2000. *Negotiating nature: Culture, power and environmental argument*. Lund, Sweden: Lund University Press.

Ingold, Tim. 2000. *The perception of the environment*. London: Routledge.

Ingold, Tim, and Terhi Kurttila. 2000. Perceiving the environment in Finnish Lapland. *Body and Society* 6 (3–4): 183–196.

Innis, Harold. 1950. *Empire and communications*. Oxford: Clarendon.

Jasanoff, Sheila, ed. 2004. *States of knowledge: The co-production of science and social order*. London: Routledge.

Krajick, Kevin. 2002. Melting glaciers release ancient relics. *Science* 296 (April 19): 454–456.

Kuzyk, Gerald W., Donald E. Russell, Richard S. Farnell, Ruth M. Gotthardt, P. Gregory Hare, and Erik Blake. 1999. In pursuit of prehistoric caribou on Thandlät, Southern Yukon. *Arctic* 52 (2): 214–219.

Latour, Bruno. 1993. *We have never been modern*. Cambridge, MA: Harvard University Press.

McClellan, Catharine. 2001 [1975]. *My old people say: An ethnographic survey of southern Yukon Territory*. 2 vols. Mercury Series, Canadian Ethnology Service Paper, 137. Ottawa: Canadian Museum of Civilization.

McIntosh, Roderick J., Joseph A. Tainter, and Susan Keech McIntosh, eds. 2000. *The way the wind blows: Climate, history and human action*. New York: Columbia University Press.

Macnaghten, Phil, and John Urry. 1998. *Contested natures*. London: Sage.

Minority Rights Group. 1994. *Polar peoples: Self-determination and development*. London: Minority Rights Publications.

Monsalve, M. Victoria, Anne C. Stone, Cecil M. Lewis, Allan Rempel, Michael Richards, Dan Straathof, and Dana V. Devine. 2002. Brief communication: Molecular analysis of the Kwäday Dän Ts'ìnchi ancient remains found in a glacier in Canada. *American Journal of Physical Anthropology* 119: 288–291.

Morrow, Phyllis, and Chase Hensel. 1992. Hidden dissension: Minority-majority relationships and the uses of contested terminology. *Arctic Anthropology* 29 (1): 38–53.

Nadasdy, Paul. 1999. The politics of TEK: Power and the "integration" of knowledge. *Arctic Anthropology* 36 (1–2): 1–18.

——. 2003. *Hunters and bureaucrats: Power, knowledge, and Aboriginal-state relations in the Southwest Yukon*. Vancouver: University of British Columbia Press.

Niezen, Ronald. 2000. Recognizing indigenism: Canadian unity and the international movement of indigenous peoples. *Comparative Studies in Society and History* 42 (1): 119–148.

Portelli, Alessandro. 1997. *The battle of Valle Giulia: Oral history and the art of dialogue*. Madison: University of Wisconsin Press.

Raffles, Hugh. 2002. *In Amazonia: A natural history*. Princeton: Princeton University Press.

Scott, Colin. 1996. Science for the West, myth for the rest? The case of James Bay Cree knowledge construction. In *Naked science: Anthropological inquiry into boundaries, power and knowledge*, edited by Laura Nader, 69–86. London: Routledge.

Slater, Candace. 2002. *Entangled Edens: Visions of the Amazon*. Berkeley: University of California Press.

Smith, Eric Alden, and Joan McCarter, eds. 1997. *Contested arctic: Indigenous peoples, industrial states, and the circumpolar environment*. Seattle: University of Washington Press.

Starn, Orin. 2004. *Ishi's brain: In search of America's last "wild" Indian*. New York: W. W. Norton.

Stavenhagen, Rudolfo. 1996. Indigenous rights: Some conceptual problems. In *Constructing democracy: Human rights, citizenship, and society in Latin America*, edited by Elizabeth Jelin and Eric Hershberg, 141–160. Westview Press, CO: Boulder.

Thomas, David Hurst. 2000. *Skull wars: Kennewick Man, archaeology and the battle for Native American identity*. New York: Basic Books.

Weaver, Andrew J. 2003. The science of climate change. *Geoscience Canada* 30 (3): 91–109.

White, Richard. 1998. Using the past: History and Native American studies. In *Studying Native America*, edited by Russell Thornton, 217–243. Madison: University of Wisconsin Press.

20

The Making and Unmaking of Rains and Reigns

Todd Sanders

With little food and no more European goods to give to (and hopefully impress) local leaders, the tired German explorer C.W. Werther, followed by a nearly endless queue of even more tired trunk-toting porters, reached the Wembere swamps just north-west of Ihanzu. The year was 1893. It was the rainy season, and rain there was. Lots of it.

As is common even in dry years, all the more so in wet ones, the Wembere swamps had flooded, making any passage a potentially perilous one. Rather than turn back, the party toiled determinedly for hours to construct a bridge across the Sibiti River, a feat they accomplished in spite of a hippo's savage attack on their temporary structure. After all had safely reached the other side, the bridge promptly collapsed and was swept away by the floodwaters.

Several hours later, Werther entered central Ihanzu, thus becoming the first European to have done so. He sent some of his messengers ahead to the male rainmaker's homestead ('the Sultan,' Werther calls him) to make him aware of their arrival. And to demand foodstuffs. The full party arrived shortly thereafter.

Perplexed by such an oddity, some Ihanzu men looked on curiously as Werther's party set up camp. Recognizing one particularly exquisite bovine with Werther's party as belonging to a man in Sukumaland, one local came forward and asked where Werther had acquired the beast. It was true, the man was assured – it had belonged to the Mwanangwa from Miatu, but had been purchased from him. Curiosity aroused, the man persisted with his questioning.

How had Werther and his party managed to cross the river, given that it was completely flooded? Again, one of Werther's assistants answered: 'The Great Lord here' – gesturing to Werther – 'made a powerful *dawa* [medicine] that made the river drop, we crossed and the waters swelled once again.' This must have been astonishing news, as such medicines were the purview of powerful persons alone, like their own royal rainmakers. Werther, so it appeared, possessed medicines – potent ones at that – that could influence the elements, causing rivers to rise and fall and who knows what else.

Soon thereafter, fierce fighting erupted between the two sides, which eventually led to Werther's and his party's hasty retreat to the east (Werther 1894: 221ff). Thus began the Ihanzu's first contact – and conflict – with Europeans. It would not be the last.

The Anthropology of Climate Change: An Historical Reader, First Edition. Edited by Michael R. Dove.

In this initial, tentative Ihanzu encounter with a European explorer, Ihanzu rainmakers and their medicines featured prominently. This would be so thereafter in one way, shape, or form throughout Ihanzu history and into the present. For this reason, it is impossible to speak of Ihanzu history without speaking of Ihanzu rainmakers and the medicinal powers they command. Ihanzu rainmakers played a pivotal role in integrating and regulating certain types of regional, precolonial trade; they also provided the basis for an Ihanzu moral community. Years later, they featured crucially in anticolonial resistance movements. Beginning in the 1920s, they gave specific form to the newly invented Ihanzu 'chiefdom.' And when, in the 1960s, Tanzanian chiefdoms were abolished across the land, Ihanzu rainmakers continued to hold sway. They still do today. In short, throughout a tumultuous history, a period that witnessed sweeping political, economic, and social changes of every imaginable sort, Ihanzu rainmaking and rainmakers have remained remarkably important (cf. Packard 1981; Feierman 1990).

Much of this importance has turned on Ihanzu's two royal rainmakers – one male, the other female – and the powers they jointly command. In Ihanzu eyes, when it comes to rainmaking, one gender without the other is pointless and impotent. Neither the male rainmaker nor the female one can bring rain alone. Thus to speak of rainmaking is to speak of gender. Curiously enough, this fact, so apparent to the Ihanzu themselves, was repeatedly missed or ignored by successive waves of outsiders: by German and British colonials and later by postcolonial administrators. No colonial or postcolonial administration ever formally recognized Ihanzu's female rainmaker, even though she has been pivotal for as far back as oral and written histories take us (Adam 1963b; Kohl-Larsen 1943: 290). As we shall see, this fundamental neglect or oversight has had profound consequences, not least in framing how anticolonial resistance was dealt with; how the Ihanzu chiefship was eventually given its contours; and how postcolonial administrators view Ihanzu rainmaking today.

Early History, Rainmaking and Identity

For the women and men of Ihanzu, rainmaking is inextricably linked to their earliest history, migration, and identity. I have heard only one Ihanzu origin story, the one all Ihanzu know, the one many have told over the years to non-Ihanzu with evident zeal (Kohl-Larsen 1943: 194–5; Adam 1963b: 14–15). The story comes with minor variations, though all versions tell of an ancient migration from Ukerewe Island in Lake Victoria to the Ihanzu's present location. Different clans, driven by famine and drought, made this lengthy journey. During the migration, people say, each clan or clan-section rested or settled temporarily at various sites along the way. These places are remembered by name, and are today used as ancestral offering sites, particularly those within Ihanzu proper.

After some warring between clans and with others en route, the migrants, people say, took refuge in Ihanzu's central mountainous region, an area called *Ihanzu la ng'wa Kingwele* (Kingwele's Ihanzu), named after one of the first *Anyansuli* clan members. The village of Kirumi is the focal point here, Ihanzu's 'sacred centre.' This is a rocky, highland area surrounded on all sides by vast open plains. There, the story goes, they built a large, fenced-in enclosure in which to live in safety. When men or beasts threatened, they could seal up the entrance with a large door made of sturdy poles called *mahanzu* (sing. *ihanzu*). The people thus became known as Anyĩhanzu: 'the people of the byre door pole.'

During the migration, each clan allegedly brought particular things with it. Some came with seeds, others with cattle. Few can name all the clans and clan-sections or spell out what, exactly, each supposedly brought with it. People never fail to mention, though, that the first Ihanzu rainmakers also came from Ukerewe, bringing with them their rainmaking knowledge and ritual paraphernalia. For many, this seems to be the point of telling the migration story in the first place – to say, in so many words, 'We came from Ukerewe with our rainmakers, rain medicines, and rain.'

Whether this original migration ever took place we may never know. But fact or fiction, the idea that it did occur has informed Ihanzu notions of their history and identity since at least the turn of the last century.[1] Everyone I asked about what makes an Ihanzu an Ihanzu – male or female, young or old, 'rich' or poor, religious or not – explicitly noted as much, often pointing proudly in the northerly direction of Ukerewe for added emphasis. It is no doubt true, as Lambek (1996: 239; also 2002) reminds us, that such expressed visions and versions of the past are inextricably linked to their consequences for relations in the present. But it must equally be said, at least for the Ihanzu, that this present has been recognizable for more than a century. The women and men of Ihanzu have long identified themselves as a 'rainmaking people' who, from the very beginning of time, have owned and controlled their own rain.

Precolonial Rainmaking Economies

In the 1880s and 1890s, just before the arrival of German colonial forces, the men and women of Ihanzu lived over a relatively small highland area centred on Kirumi. They were also embedded in an expansive regional economy, warring and exchanging, raiding and trading with their neighbours (see Figure 20.1).

The pastoral Maasai and Tatog made periodic cattle raids into Ihanzu (Adam 1961: 3, 5; Kidamala and Danielson 1961: 74–5; Reche 1914: 69),[2] and many elders tell of the heroic exploits of Ihanzu men who, hidden among the many boulders and caves in the area, shot and killed Maasai with arrows as they passed through with their stolen Iramba, Iambi, and Turu cattle. From time to time, Ihanzu returned the stolen bovine goods to their rightful owners in the south, sometimes for a profit.[3]

During this period, Arab and Nyamwezi caravan traders passed to the south and north of Ihanzu (Alpers 1969; Roberts 1970). Some Ihanzu men obtained beads from southern caravan traders, and iron for hoes from Nyamwezi traders (Kidamala and Danielson 1961: 77; Obst 1923: 218n, 222; Stuhlmann 1894: 759, 763; Werther 1894: 238). From Sukuma traders to the north, Ihanzu traders acquired iron in return for salt gathered at Lake Eyasi (Obst 1912a: 112; Reche 1914: 84; 1915: 261). The hunting and gathering Hadza, also to the north, provided the Ihanzu with ivory, rhino horns, honey, and arrow poison in return for calabashes, beads, cloth, knives, axes, and metal arrowheads (Obst 1912a: 112; 1912b: 24; Reche 1914: 19, 71; Woodburn 1988a: 51; 1988b).[4] To the north-east, the Ihanzu exchanged goods with Iraqw traders: articles made by their own smiths – arrowheads, knives, hoes, and axes – in return for livestock, red earthen body paint, tobacco, and arrow poison (Obst 1912a: 112; Reche 1914: 69, 71). And as far east as Mbugwe, some Ihanzu traders obtained brass and copper jewellery, since Ihanzu smiths were adept at working with iron but unskilled at working with softer metals (Reche 1914: 84).

Thus, far from being isolated in their mountainous homeland in precolonial times, the Ihanzu – or rather, some men from Ihanzu – ranged far and wide, maintaining extensive trading networks across the region. They were always involved in a broader regional political-economy.

The people of Ihanzu and their royal *Anyampanda* clan rainmakers were renowned for their powerful rain medicines well beyond the boulder-strewn confines of their own land. In the 1890s, and likely earlier, the Ihanzu played a key role in what might be called the regional rainmaking economy. People from Turu, Iambi, Iramba, Sukumaland, and Hadza country made annual pilgrimages to the royal Ihanzu village of Kirumi, bringing tribute of black sheep, among other things, to Ihanzu rainmakers. In return, they were given rain medicines so that the rains would be plentiful in their own lands (Adam 1963b). Royal *Anyampanda* rainmaking powers were considered so potent, in fact, that peoples from Turu and Iramba to the south – who had their own rainmaking traditions, medicines, and shrines – routinely visited the Kirumi rainshrine in harsh years to obtain stronger medicines (Adam 1961: 2; Jellicoe 1969: 3). (During my stay in Kirumi I met several young and middle-aged men who had come from Iramba and

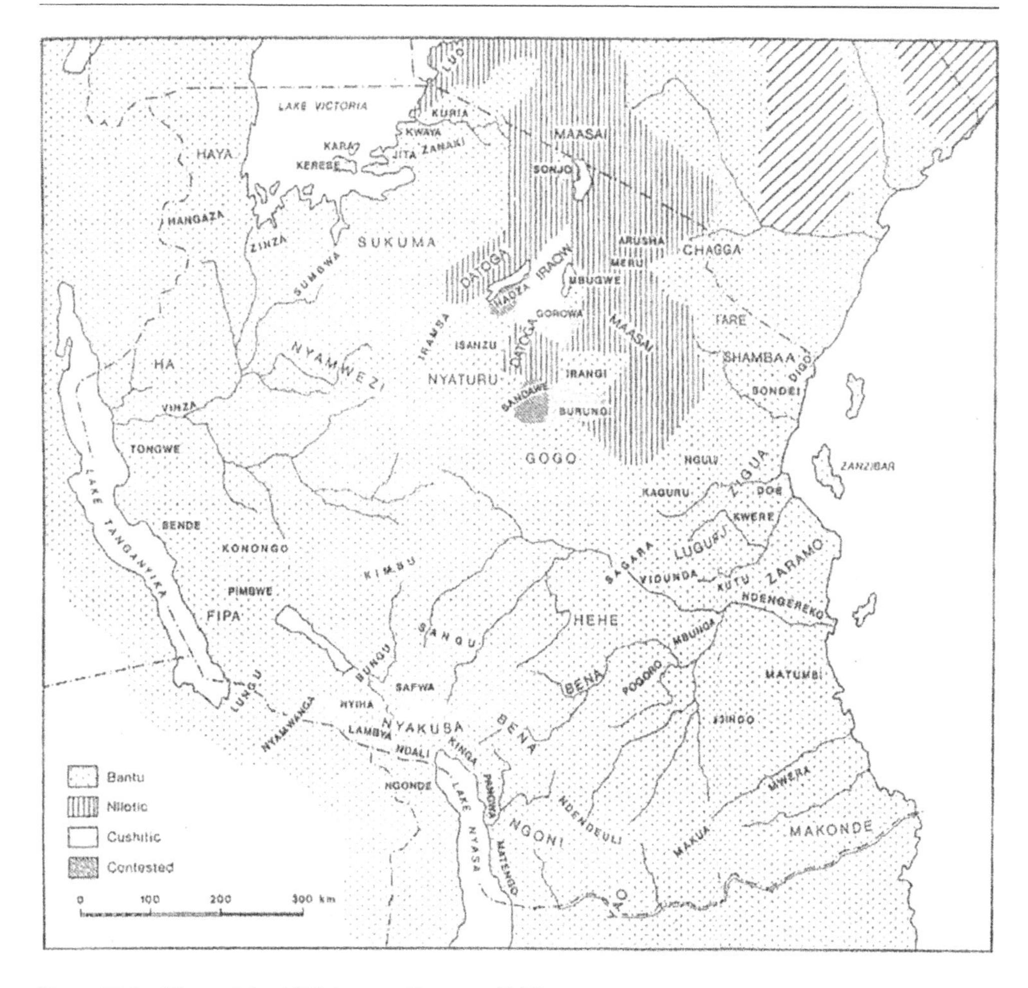

Figure 20.1 *'Tanzania' c. 1890 (source: Koponen 1988).*

Sukumaland to visit the royal rainmakers for this purpose) (see Figure 20.1).

Others in the region who made no regular pilgrimages to Kirumi still recognized the medicinal superiority of Ihanzu's royal rainmaking clan. Today rainmakers in faraway places derive their authority from their Ihanzu origins. In Mbulu, the Iraqw's *Manda* clan regulates the rain, and people there claim that *Manda* powers originated with the Ihanzu *Anyampanda* rainmaking clan (Thornton 1980: 203–4; Winter 1955: 11; Snyder 2005: 126–7). Similarly, the Wambugwe claim that their royal rainmaking clan had its origins in the Ihanzu *Anyampanda* clan (Gray 1955: 42; 1963: 145; Kesby 1981: 41; Thornton 1980: 216).[5]

Precolonial Ihanzu rainmakers did more than prepare rain and fertility medicines for themselves and others; they also prepared war medicines. Royal *Anyampanda* medicines were so potent, my informants commonly claimed, that Maasai and Tatog were defeated time and time again and eventually repelled from Ihanzu never to return.

For various reasons, then – namely, because they provided rainmaking, fertility, and war medicines – the Kirumi rainshrine, Ihanzu royal rainmakers, and the powers they controlled served as a precolonial focal point for many peoples across an expansive area.[6] If this was true across the region, it was equally the case within Ihanzu itself.

In the 1890s, the people of Ihanzu numbered no more than a few thousand. Villagers farmed sorghum, millet, groundnuts, manioc, sweet potatoes, beans, and tobacco. Domestic livestock included cattle, sheep, goats, and donkeys (*Deutsches Kolonialblatt* 1901: 903; Obst 1923: 218; Reche 1914: 69–70; 1915: 260; Werther 1894: 238; 1898: 72).[7] Ihanzu men had a reputation as keen hunters as well as competent smiths (Obst 1923: 218–22; Reche 1914: 84; Werther 1898: 72). In this mountainous area, which was fairly densely populated (*Deutsches Kolonialblatt* 1901: 903; Obst 1912a: 114), men and women worked their fields together. All lands were owned and allocated by particular matriclans. When a man cleared the bush in order to farm or live, that land became his matriclan's property (Reche 1914: 74).[8]

Men and women alike greatly valued ornamental beads, which they obtained from caravans passing through Turu and Iramba (Stuhlmann 1894: 759, 763; Werther 1894: 238). Blue and white beads were the most sought after (Obst 1923: 223; Reche 1915: 260), being ritually auspicious colours and associated with rainmaking (cf. Tanner 1957: 199). The fact that Ihanzu men wore bead necklaces, and that women wore beads on their arms and legs, around their waists, and in their hair, prompted one early German observer to dub them 'the Bead People' (*Perlenvolk*) (Obst 1912a: 115).

Ihanzu villages were fairly independent and largely autonomous with regard to their own internal affairs. All villages were governed informally by groups of male elders, who were the final arbiters in village matters. When inter- or intra-village conflicts developed over murders or adultery, fines were negotiated between the parties involved (Reche 1914: 85). Movement between villages was sometimes dangerous and often required certain medicinal protections and precautions (Adam 1963b: 17).

Yet villages and villagers, in spite of some tensions between them, were connected in practice, in that all recognized the supreme authority of the royal rainmaking clan section: the *Anyampanda wa Kirumi* (Adam 1963b).[9] There were on any given occasion two royal rainmakers – one male, the other female – known as the 'owners of the land' (*akola ihĩ*). In the 1890s the male 'owner' was Semu Malekela (Werther 1898: 72–3); Semu's female counterpart was his sister's daughter, Nya Matalũ.

These two royal owners of the land lived in and rarely ventured beyond the royal village of Kirumi. They were jointly responsible for the general welfare and prosperity of all Ihanzu villagers. Semu initiated male circumcision ceremonies (*kidamu*), hunting parties, and salt-fetching caravans to the northern salt flats. He could also offer sanctuary to murderers, regardless of the Ihanzu village from which they came (Adam 1963b: 16; cf., S.F. Moore 1986: 57; Rigby 1971: 397). Nya Matalũ, for her part, initiated and coordinated women's rain dances (*masĩmpũlya*) in dry years and played an indispensable role in various other rain rites.

Together, this royal duo regulated the Ihanzu agricultural cycle from the first cutting-of-the-sod ceremony (*kũtema ilĩma*), which initiated each new agricultural season, to giving the order to begin the harvest (*kũpegwa lupyu*). They also made war medicines together. In a variety of ways, the fertility of the land and the well-being of its inhabitants were in the hands of these two royal owners. This, in turn, gave them a degree of political authority within Ihanzu (Adam 1963b).[10]

Semu and Nya Matalũ were assisted in their rainmaking tasks by male and female rainmaking assistants known as *ataata* (sing. *mũtaata*). Each Ihanzu village had at least one such assistant. These assistants exercised no particular power or authority over fellow villagers on the basis of their positions; they functioned more as intermediaries between the two royal *Anyampanda* rainmakers in Kirumi and villagers from other parts of Ihanzu.[11]

In August or September of each year, rainmaking assistants collected grain from their respective villages, which they then brought as tribute to the owners of the land in Kirumi (Reche 1914: 85; 1915: 261; Werther 1898: 99). Some of this grain, informants say, was divided between the two royal rainmakers, blessed, and later returned by the assistants to the villages from which it came to ensure all-round fertility. The remainder was later used by Nya Matalũ and her female assistants to brew sorghum beer for the annual rainmaking or *kũtema ilĩma* ceremonies.

Semu and Nya Matalũ jointly initiated these rites each year in Kirumi, usually in October. Ihanzu rainmaking assistants attended, bringing with them a number of children from their home villages. Rainmaking assistants from outside Ihanzu – those from farther afield who received rainmaking medicine from Ihanzu royals – are also said to have attended these rites. After a 'grandson' and 'granddaughter' cut the sod with special hoes (a long-handled 'male' hoe and short-handled 'female' one), the children, who probably numbered in the hundreds, began hoeing the royal *Anyampanda* fields adjacent to the Kirumi rainshrine. These annual cutting-of-the-sod ceremonies normally lasted a day and a half (cf. Adam 1963b).

The pre-colonial picture of Ihanzu in the early 1890s that emerges is one of a small number of decentralised and largely autonomous villages, clustered around the boulder-strewn centre of the country. Each village was responsible for its own internal political, legal, and economic affairs. Village elders – any men of advanced age – governed their own daily activities. From this highland 'fortress,' some members of the community occasionally trekked over vast distances and were deeply involved in long-distance trading of various goods in neighbouring areas.

There was limited cooperation between villages, occasionally there was fighting. People did, however, share a common purpose in both warfare and rainmaking. And in both instances, villagers looked to the two *Anyampanda* owners of the land in Kirumi for protection and to cure their ills. Indeed, the very flow of day-to-day life throughout Ihanzu – the farming cycle, hunting, circumcision, the rains, and the state of the land itself – hinged inexorably on the powers of the two rainmaking royals.

Unmaking Reigns and Rains

It was into this world of cross-regional trading, raiding, and rainmaking that German military forces made their first forays in the late 1800s. A principal objective of these forces was to 'pacify the natives,' as it were, not least because much 'tribal' hostility was being directed against the Germans themselves. These pacification efforts and local resistance to them are best understood from a regional perspective, since colonial forts and forces in one area often made their influence felt in places farther afield.

For some years, Gogo people had been extracting tariffs from passing caravans (including German ones) – a practice that German colonialists, for obvious reasons, deplored (Peters 1891: 521). Thus, in 1894 they erected a massive fortress in Ugogo (Obst 1923: 303–4; Prince 1895; Sick 1915: 59). Kilimatinde, as the fortress was called, was about thirteen days south of Ihanzu country by caravan, or five days for runners (Admiralty 1916: 326). From Fort Kilimatinde colonial forces tentatively extended their influence to the west and north, into Turu, Iramba, Iambi, and Ihanzu (Sick 1915: 59–60).[12] But not without grave difficulties. They often encountered people who, as one early observer meekly and misleadingly put it, 'did not like the protection of the Germans' (Obst 1923: 304). More to the point, anticolonial resistance and revolts were rife right across the region for decades.

The German military response was often a simple one: to kill as many as was deemed necessary to quell the situation. One popular strategy was to capture and hang local leaders, diviners, and rainmakers, who were seen as roguish and responsible for inciting anticolonial violence. In many cases, German suspicions were patently correct: traditional leaders, rainmakers, and the like were often uniquely situated to unite people against aggressive colonial forces, both locally and across vast expanses (cf. Lan 1985; Jellicoe, Sima, and Sombi 1968; Maddox 1988: 759; Sick 1915: 45; Swantz 1974: 77).[13]

From Kilimatinde, German military patrols made their first large-scale expedition into Iramba and Ihanzu areas in 1899. In Iramba, they swiftly captured some local leaders, some of whom were hanged (Lindström 1987: 38). In Ihanzu, the story was different. The Ihanzu male rainmaker, Semu, who had repelled Werther's party seven years before, had by this time died; his younger brother, Kitentemi, had succeeded him as male rainmaker sometime in 1897 or 1898. Nya Matalũ was still female rainmaker (see Figure 20.2). The Germans were either unaware of or unconcerned about

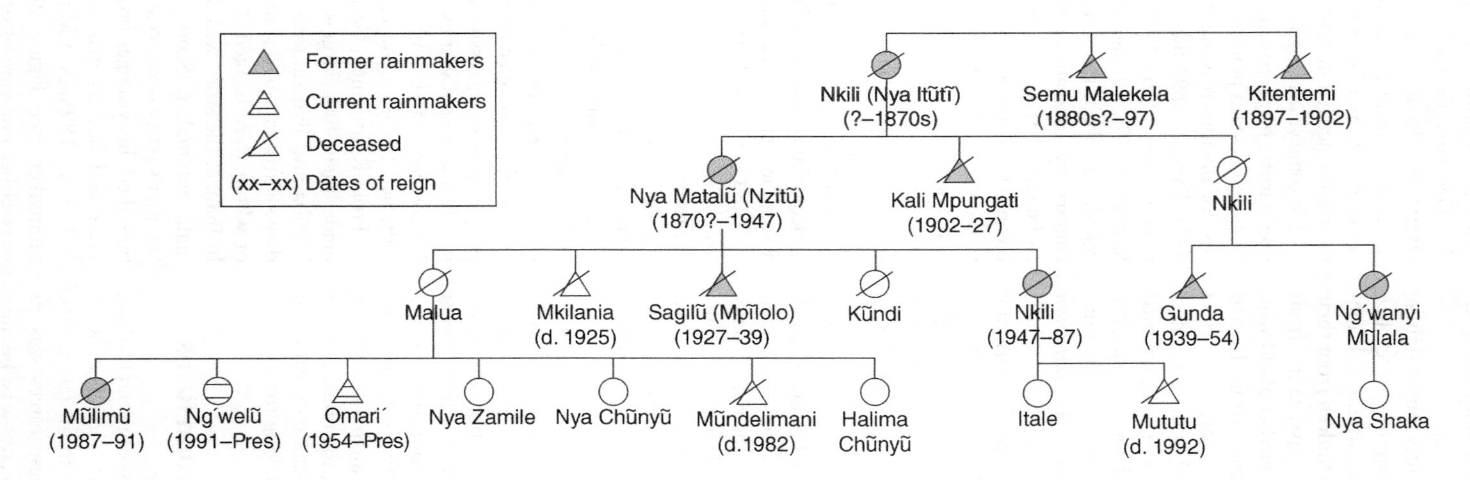

Figure 20.2 *Anyampanda wa Kirumi royal matrilineage.*

these 'leaders.' For instead of capturing and/or hanging them, they simply appointed a *jumbe*,[14] a man by the name of Mũgunda s/o (son of) Nzega, who was to rule over Ihanzu and who would periodically report to them at Kilimatinde.[15]

Somewhat predictably, this proved an unsatisfactory arrangement. People in Ihanzu continued to cause problems. As a result, colonial officials decided to establish an outpost of Fort Kilimatinde, a new fort in Ihanzu proper (Obst 1923: 304; Sick 1915: 60). They chose Mkalama village, just southwest of central Ihanzu. On 26 May 1901, one Sergeant Künster and his troops arrived and built a temporary 'thorn fortress' as well as necessary living quarters for the soldiers. The German-appointed *jumbe* assured Künster that the men and women of Ihanzu, now reputedly under his control, had surrendered completely to the colonial government. Indeed, in certain respects, things were looking up: around a thousand locals showed up at the site, supposedly to make peace with the new administration, and they celebrated well into the night. Not insignificantly, Kitentemi was not among them (*Deutsches Kolonialblatt* 1901: 903).

German forces soon began erecting a large stone fort at the same location. This immense structure, part of which still stands, is perched high atop a hill, surrounded by massive stone walls. The location, wind-swept though it is, affords a view of much of western Ihanzu and, to the south and west, well into Iramba and Sukumaland. The fortress took eight years to complete, largely by forced Ihanzu and Iramba labour.

Peace dances aside, the Germans' problems did not end in 1901. Not even close. For just after beginning their Mkalama fort – only weeks after the grand 'peace dance,' in fact – there was a district-wide uprising directed against colonial forces and their instruments, the German-appointed *jumbe* (*Deutsches Kolonialblatt* 1902: 587).[16] Revolts barely subsided.

In June 1902, Ihanzu, Iramba, and Iambi forces attacked and annihilated a number of colonial caravans (*Deutsches Kolonialblatt* 1903: 1). Following several such successful campaigns, 'the natives were so emboldened as to send a formal declaration of war to the military post at Mkalama.'[17] Zahn, the German sergeant in charge at Mkalama, dispatched troops to Iambi. They suffered heavy losses at the hands of locals, who had joined forces with men from Ihanzu and Iramba. In the end, many African colonial troops and two European officers were killed in the battle.[18] 'The position was now so serious that Zahn had to retire to Mkalama and call in all his patrols, a move which appeared very necessary when a series of attacks on Mkalama ensued which were repulsed only with great difficulty.'[19] Days later, reinforcements arrived at Mkalama from the German fort at Mpwapwa, and the revamped colonial forces mounted another assault. This time they were victorious (*Deutsches Kolonialblatt* 1902: 587; Thurnwald 1935: 18).

From oral and written sources, it is clear that Ihanzu's then rainmakers, Kitentemi and Nya Matalũ, played an instrumental role in these anticolonial battles (Adam 1961: 9; 1963b: 17; Jellicoe 1969: 3). Little wonder, really, since they had already proved themselves and their medicinal powers the previous decade by successfully repelling invasive Maasai and Tatog cattle thieves. Nor was this fact lost on the German administration – or at least half this fact. For this time the rainmaker Kitentemi was captured, hauled to Fort Kilimatinde and hanged.[20] Nya Matalũ, for her part, seems to have benefited greatly from colonialists' lack of knowledge about differently gendered rainmaking pairs: she survived the encounter unscathed and, according to informants, continued to prepare rain secretly in Kirumi with Kitentemi's sister's son, the new rainmaker, Kali (Figure 20.2). Kitentemi's capture and removal signalled the end of violent anticolonial resistance in Ihanzu.

In Ihanzu today, Kitentemi is a household name. He is remembered not so much for what he did during his brief period as male rainmaker, but for what became of him: he was captured and taken away, never to return to Ihanzu again. During my fieldwork, women and men seemed to enjoy speculating, when asked, about Kitentemi's fate. Some suggested that he might have been taken to Germany to start a rainmaking clan there. Most, however, proffered that Kitentemi's fate is simply a mystery and that we shall never know for certain – this, even though most are well aware that German forces captured and hanged rainmakers, diviners, and leaders from all across the region. It is as if the many men and women who spoke to me about Kitentemi and German colonial history refuse to see the defeat of royal rainmakers or their powers, as if they prefer to cast their own history (at least when it comes to rainmaking) as an open question. What is more, people claim that when German forces stole their male rainmaker, Kitentemi, in so doing they stole the rain. And life turned bad. Very bad.

On balance, the first decade of the twentieth century was not a kind one to the men and women of Ihanzu. The decade saw drought after drought, famine after famine, particularly towards its end. Faced on and off with starvation, many desperate villagers bartered away the few cattle they had for grain in Iramba, Turu, and Sukumaland (Sick 1915: 18).[21] To complicate matters further, the German administration imposed taxes during this period (Admiralty 1916: 19–20; Jellicoe 1969: 10). This was one reason why, during the famine years of 1907–10, many Ihanzu and Iramba villagers began trekking to the northern part of the territory to work on the northern railway and plantations in Moshi and Arusha (Adam 1963a; Iliffe 1979: 161). At home in the villages and beyond, forced labour under German rule became a deplorable fact of life. This included bush clearing, road construction, and fort building at Mkalama. For the latter task, Ihanzu men had to walk to Mbulu District, where they picked up (quite literally) beams and carried them back to Ihanzu. Each trip took a few days.

Everyday resistance, to borrow Scott's (1985) term – foot-dragging, apparent sluggishness, and extreme reluctance to carry out colonial labour of any sort – became part of Ihanzu's everyday political landscape (Admiralty 1916: 80–1). Some men refused to carry out such work at all. Others worked at such an exaggeratedly lethargic pace that some German observers remained in a constant state of anxiety, annoyance, and incomprehension.[22]

But not all stayed to resist the new colonial regime. Many simply left. One elderly man born in Kirumi village said this:

> I was born during the period (I don't know the date) when they started building that fort at Mkalama [1901]. After they began building that fort, the Germans gathered people to do the work there. My parents both saw that the work was too hard and wanted to leave so we moved to Sukumaland. I was only small at the time and was carried on my mother's back …
>
> When we arrived in Sukumaland, we lived there until I grew up a bit, and it was then that my father was forced to go and fight in the [First World] War. His name was Kea. When my father was taken, I was fully aware of what was going on. I saw my father was going to war …
>
> After my father was taken off to fight, I went to live with my grandfather who had come with us to Sukumaland. He also wanted to get far away from those Germans and their fort.

I collected many life histories like this one that attest to the fact that countless men, women, and children left Ihanzu during this first decade of German occupation. They did so not just to get food and return, as happens under other circumstances, but rather to escape the clutches of the governmental menace at their doorstep. In all directions they scattered: into Hadza country, Sukumaland, and Mbulu.

It was during these trying, turbulent times, my informants claim, that the two royal *Anyampanda* rainmakers, Kali and his sister Nya Matalũ, fled Ihanzu for Hadza country to the north. One of my most reliable informants, an elderly woman from Ibaga village, gave the year as 1910, the same year the fort at Mkalama was completed. Their relocation notwithstanding, the evidence suggests that men and women continued to see their rainmakers, and not the German-appointed *jumbe*, as Ihanzu's only legitimate figureheads. A British assistant political officer based at Mkalama in the early 1920s wrote that during the transition period, after Kitentemi had been hanged and *jumbe* had been appointed, 'the tribal chiefs still controlled their people and from them permission was still sought for planting, harvesting, marriage, circumcision, for the cure of ills, the detection of thieves, for the punishment of murderers and the rectification of tribal ills, while at the same time they were looked to to produce favourable conditions through the Sun-God for the year's harvest.

The Germans tried hard to stamp out this system, and the measure of their failure is in the degree of fidelity with which the people still cling to their hereditary rulers.'[23]

Thus, while German forces may have removed and hanged Ihanzu's male rainmaker, Kitentemi, in an effort to quell the violence directed against them – and in this they may have in part succeeded – their actions also contributed in significant ways to their eventual undoing. What they failed to see was that, in local eyes, reigns and rains were inextricably linked; to attack a royal rainmaker was to attack the rain and hence all Ihanzu women and men. The British administration, whatever else it did, would not repeat this mistake. Quite the opposite.

Making Reigns and Rains

The First World War devastated the whole of German East Africa. Countless Africans fought and died on both sides; and severe drought and famine once again plagued the region. In 1919, F.J. Bagshawe, a British District Political Officer, painted a haunting picture of German East Africa's central region: 'On my arrival in February an appalling state of affairs existed. Excepting for the Masai and Tatoga who are not dependant upon cereal food, and for the Wafiomi, all tribes were practically starving. On every road were to be met thousands of natives with their families travelling – too often unable to travel further – in search of food, ready to barter their last remaining possessions, their children, even their wives, for food ... I do not know how many died, and I have never tried to find out. To look back upon that period is like recalling a nightmare.'[24]

Ihanzu was among the worst affected areas. Villagers starved owing to lack of rain, but equally because German forces throughout the war had shipped vast quantities of grain out of Ihanzu by carrier safaris, and by transport wagons of the Arusha Dutch, to the front in Moshi.[25] Local labour shortages exacerbated an already precarious situation.

Life histories reveal the extent of the tragedy. Many able-bodied Ihanzu men who might otherwise have farmed were, during the war, recruited by both sides – some were enlisted and paid, others were abducted at night – to fight a foreign war they scarcely understood. Some were marched to distant locations to fight or work as porters; others did so closer to home. Towards the end of the war, as the Germans retreated south into Iramba from advancing South African forces, they abandoned their fort at Mkalama and abducted many Ihanzu and Iramba men – any who had enough strength to carry their loads – to hasten their retreat. Many died in the process. German forces pillaged and plundered, taking with them any and all foodstuffs and livestock (cf. Brooke 1967: 15; Ten Raa 1968; Maddox 1988: ch. 3).

Many remember this period as *nzala ndege* or 'the aeroplane famine': a phrase evoking wartime Allied forces' aeroplanes that flew overhead.[26] William Kali, now an elderly man, vividly recalls this monumentally chaotic period. At the time, he was a young boy living in Matongo village: 'Many people moved out of Ihanzu. Many people. They went in all directions in search of food. Many of those died. Many who stayed died too. We stayed. My father went into the bush, into Hadza country, and learned how to dig roots in the bush. It was hard. We picked some wild fruits and ate roots from the bush. There was no food. None. I even ate baboon meat sometimes! There was nothing. We survived.'

Ali Gimbi, who was born in Ikũlũngu la Mpepo village, was the son of a wealthy cattle owner and farmer. He, too, remembers the aeroplane famine: 'We sold all our cattle. We – father, mother, my maternal uncle, the children, everyone – went to Mbulu in search of food, and we stayed there for three years [until the war's end]. It was hard there too.'

At the war's end, dealing with the immediate disaster was among British administrators' top priorities. As part of this process they sought to re-establish 'chiefdoms' across the land, much as they had existed prior to the arrival of the Germans. Or so they believed. (Re)establishing chiefdoms meant, among other things, finding 'chiefs' and making sure they were doing what they were intended to do: ruling over their 'tribes' (Graham 1976; Iliffe 1979: ch. 10). War, forced labour, drought, and famine had sent a flood of 'chiefs' and 'tribesmen' in all directions. Early British administrators took it as part of their task to correct this, to redraw and underscore tribal boundaries and to ensure that people stayed within them: a massive geo-political exercise, as it were, in reducing to a minimum 'tribal matter out of place.'

British administrators had set up shop at Mkalama in the old German fort. From their earliest days in Ihanzu, they paid close attention to the German-appointed *jumbe*. One thing that quickly became clear was that these 'leaders' were nothing of the sort: they were seen as illegitimate, were highly unpopular, and had little or no influence among ordinary villagers. Many, it seems, had used their positions for personal gain: 'Careful inquiries were made into the methods of each jumbe; as to the number of under-strappers employed by him, and as to their and his actual tribal status. Following this, numbers of jumbes have been charged and convicted of crimes (varying from the concealment of murder to cases of mere petty theft) and five jumbes are undergoing detention at the present moment.'[27]

With most of the *jumbe* in prison, and the need to establish a functioning government, it fell to the new British administration at Mkalama to find and empower local leaders in the district. This meant finding male 'chiefs.' Because the administrators' guiding principle was 'to ignore those jumbes and others who have no status tribally, and to enlist the support of the chiefs and jumbes who, by reason of their hereditary position, enjoy the actual respect of their people,'[28] it took little time to locate the exiled rainmaker Kali Mpungati in Hadza country and return him to Ihanzu as the 'rightful, hereditary chief.' He was one among three chiefs installed in Mkalama District in 1920:

> During this year the hereditary chiefs of the Mkalama tribes were restored to power. The Germans recognised none of them and their influence, which remained never the less at full strength amongst the tribesmen, was a potential source of mischief, instead of being, as now, a useful instrument in the hands of the Government. A change of administrative officers took place during this re-organization, whilst at the same time the Police officer was removed and a native clerk of many years local standing was sent to prison. This combination of events gave an excellent opportunity for trouble amongst the backward tribes of the sub-district: that none at all occurred is, I consider, satisfactory proof that the tribal changes were in the right direction.[29]

The right direction? Perhaps. But Ihanzu royal rainmakers were never 'chiefs' in the sense that administrators imagined: not during German times, not before. For one thing, Ihanzu royals never 'ruled' over 'their people.' Rather, they attracted a certain amount of respect and authority because of the ancestrally sanctioned rainmaking powers they controlled for the benefit of all. For another, the only reason this was possible at all was because royal rainmakers came in differently gendered pairs. Yet British administrators never seemed to realize this; or if they did, their concerns remained elsewhere. While Kali Mpungati was returned from exile and installed as the first 'chief' of Ihanzu, his sister, Nya Matalũ, who is said to have returned with him, was given neither governmental office nor official recognition.

When Kali and Nya Matalũ returned to Ihanzu, so did the rains. The aeroplane famine, at long last, was over. And with rainmakers and rains, Ihanzu itself held new possibilities and promise that had been absent since the Germans first entered the area decades earlier. As a result, by early 1920, 'many hundreds of Anisanzu [Ihanzu], who, during the past years, had fled to neighbouring districts to avoid the impositions inflicted upon them, have now returned.'[30]

One elderly man who, along with many of his relatives, had spent the aeroplane famine far from Ihanzu in Mbulu District, had this to say: 'We got the news that Mpungati and Nya Matalũ had lots of sorghum at home so we returned ... That is how he got his name, Mpungati. It means "someone who brings [people] together." His real name is Kali. Our owners of the land had returned and so did the rains and the food. When we returned, I didn't leave the country again.'

Many female and male elders, like this one, delight in telling dramatic tales of the triumphant return of their exiled royal rainmakers. The stories often involve a motorcar of some sort, in which colonials and rainmakers rode, and a parade of followers on foot. In these stories, it is not just the leaders' physical return that is at issue. Of equal importance is the return of the rain. Two young women told me that as the two rainmakers were driven triumphantly back into Kirumi by British officials, the rains followed immediately behind them as they proceeded, drenching the crowds that followed. Thus, in the Ihanzu popular imagination, early British administrators not only returned the rightful Ihanzu owners of the land to Kirumi, but in so doing – and unbeknownst to themselves, I suspect – returned the rain in the bargain. If the Germans destroyed the rain, the British returned it.

Plus ça Change ...

But British administrators' attitudes towards Ihanzu rainmaking nonetheless remained mixed throughout the colonial era, a promising start notwithstanding. On the one hand, I have found no evidence that the British administration ever actively discouraged Ihanzu rainmaking practices. It is equally clear, though, that administrators lived in perpetual hope that Ihanzu rainmaking beliefs and practices would one day vanish. Hopefully soon. Yet in spite of the many comings and goings of 'chiefs,' administrators, education officers, and missionaries throughout the British colonial era, rainmaking institutions and practices did more than survive. They thrived.

Under Chief Kali, throughout the 1920s, large, public annual rainmaking rites once again took place at the Kirumi rainshrine, as they had before Kitentemi's capture and hanging by German forces. Yet even while his newly created position as 'chief' allowed him, with Nya Matalũ, to bring rain, Kali routinely refused to take on administrative responsibilities with any degree of enthusiasm, and commonly opposed government policies. On Kali's death in 1927, the Singida District Officer reported:

> Chief Kali, who died during the year and who was an autocratic and obstinate old man, was the principal stumbling block to amalgamation [of Ihanzu with Iramba]. It may be possible to induce amalgamation or at least a federation with Iramba in the coming year as the present chief, Asmani, does not appear to have the same tribal pride as his uncle. The matter, however, rests to a large extent with the elders of the tribe now that old Kali has gone and their prejudices are not easily broken down. It would, of course, be easy to direct an amalgamation but such a course would be most indiscreet and would probably lead to trouble.
>
> In the Mkalama area the mass of the people and their leaders are steeped in superstition and they have to be handled carefully ... The question of rainmaking in this area is one which must be approached with the greatest caution.[31]

From earlier German experiences, British administrators understood perfectly well the extent to which resistance, reigns, and rains were linked, even if they failed to recognize the central conceptual and practical position the female rainmaker filled for the men and women of Ihanzu.

Sagilũ Asmani succeeded Kali in 1927 as 'chief of Ihanzu' and male rainmaker. Nya Matalũ, Sagilũ's mother, continued as female rainmaker. From the start, Sagilũ held great promise for British administrators. He had travelled far and wide and had lived in Tanga and Dar es Salaam while serving in the army. While away, he had converted to Islam (he even took the name Athmani) and had acquired a taste for European fashion and for conspicuous European consumption habits.[32] Ihanzu-based colonial officials thus held high hopes that Sagilũ, unlike Kali, would be able to wrench 'his people' from a 'primitive past,' as some other colonial chiefs in neighbouring areas had ostensibly managed: 'The authority, Asmani, comes of a line of rainmakers, but having spent the last seven years in the K.A.R. [King's African Rifles] prior to succeeding to the chiefship this year on the death of his uncle, Kali, it is more than probable that he does not know much about the art which was always jealously guarded by old Kali. As far as can be ascertained, Asmani has not been initiated in accordance with tribal custom. It is anticipated therefore that rainmaking will not play such a prominent part in the tribal life as hitherto.'[33]

Such hopes were not realized. Rainmaking remained central (misguided colonial assumptions about 'tribal initiation' notwithstanding). What is more, Sagilũ eventually proved to be one of the most loathed chiefs in the territory, at least among administrators. With his extraordinary annual income of £35,[34] Chief Sagilũ was (nearly) able to satisfy his seemingly insatiable desire for foreign goods: my informants claimed that he owned, among other things, a bicycle, a motorcycle, and a rifle, and that he routinely drank beer and smoked tobacco and marijuana. Years later he even purchased a Land Rover. In spite of his dazzling display of material wealth, however, Sagilũ never became the 'progressive' chief that administrators had hoped for. Not even close. Only a year after Sagilũ took office, C. Lyons, an administrator based at Mkalama, was already longing for bygone days: 'During the lifetime of Kali the old mtemi of Isanzu the court of Isanzu was the best controlled court in Mkalama [District], but with the advent of his successor I regret to say it is the worst. Sagilu (or Asmani as he prefers to call himself) is still young and very inexperienced. It is possible that he is a man of good character and has many latent good qualities, but up to date he has only succeeded in making a bad impression on all who have come into contact with him.'[35]

Elders who can remember, and younger women and men who cannot, relish telling tales of Sagilũ's and his mother's defiance of British officials. One popular story tells that when administrators drove from Mkalama to call on Sagilũ in Kirumi, instead of humbly greeting them, he remained inside his house with his friends drinking beer and smoking a number of substances, forcing colonial officials to wait inordinate amounts of time, or in some versions, to return on another occasion.

Other commonplace stories tell of the royal rainmakers' impressive displays of ancestrally sanctioned powers in administrators' presence. As Nya Chũnyũ, an elderly woman from Kirumi, told me:

> Sagilũ was around during British rule and, after asking around at Kiomboi they [some high-up colonial officials] arrived at Sagilũ's house by car, saying they had heard he could make rain. He went to his mother [Nya Matalũ] and told her the white men wanted some rain and she told him to bring it. He donned his black clothes, his lionskin loincloth, and went to the rainshrine. The British were afraid to enter. There he did his stuff with those pots and the rain began to pour! They said goodbye and went on their way, leaving him there as chief.

A similar story revolves around colonial administrators' desires to establish chiefly legitimacy – to see, among the three chiefs of the Mkalama Federation, who the 'real' one was. Though I have recorded several versions of this story myself (e.g., Sanders 2001b), I take one recorded by an anthropologist in 1963 to underscore its historical durability:

> The 3 chiefs used to meet at Mkalama, with the D[istrict] O[fficer], and each had his separate house. The DO decided to hold a competition to see who was the greatest rainmaker. He ordered Kingu [of Iramba] to bring rain on the first day. The hours passed, and no rain came. He ordered Jima [of Iambi] to bring rain on the 2nd day. Hours passed, no rain came. Sagilu realised that his turn would come next, borrowed a bike, and rode all the way to Kirumi. He told his mother that rain must fall during the lunch hour break on the following day. They prepared their medicine, and he returned immediately to Mkalama, so that his absence was not even missed. When the DO asked him if he could bring rain, he just said 'We will see.' The rainclouds gathered in the morning over Mkalama, and it grew darker. At noon it began to pour down, and it rained heavily … It was so heavy that many of the tembe roofs caved in. The DO said that indeed Sagilu was the greatest of them all.[36]

These well-known, often-told tales of Ihanzu in the late 1920s and 1930s are interesting for several reasons, not least for what they tell us about submerged subaltern histories of the colonial enterprise and local notions of power and authority. For present purposes I wish only to highlight the fact that the women and men of Ihanzu have long imagined their own rainmakers as more powerful than those of their neighbours; and that the powers these royals wield do what they do because they come in differently gendered productive pairs. Only when male and female rainmakers join forces, when they cooperate, will the rains come.

If early British administrators were publicly supportive and privately ambivalent about Ihanzu rainmaking, the same cannot be said for the Augustana Lutheran missionaries, who first opened their doors in Ihanzu in 1931. As with colonial administrators, both early and later missionaries cast the Ihanzu and their rainmaking beliefs and practices as 'superstitious' and 'traditional,' assuming they would eventually vanish under the weight of modernity. Unlike colonial administrators, however, these self-designated 'messengers of love' (Ward 1999) positively loathed such things and aimed explicitly at the 'breaking down of their primitive tribal religion before the advance of civilization' (Johnson 1934: 23). From this pious perspective, not only was rainmaking seen as 'superstitious,' 'primitive,' and 'traditional'; it was also understood as irrevocably evil, something that had to be eradicated at all cost.

Ihanzu Lutheran views have changed little in recent years. The local reverend, himself an Ihanzu man, continues to preach on the perils of tradition, including rainmaking, and on the salvation Jesus offers in the form of moral and material betterment. But today, seventy-some years after missionaries' entrance into the area, this missionary message falls mostly on deaf ears: around 80 per cent of Ihanzu men and women call themselves 'pagans' (*wapagani*), and do so unabashedly.[37] Monumental missionary efforts from the 1930s onwards aside, few Ihanzu women and men today have any enthusiasm for hearing The Word of a distant demigod if this means the wholesale abandonment of rainmaking practices and royals.

Chief Sagilũ died in 1939. When he did, British officials successfully steered the royal rainmaking line of succession from a teenaged (and potentially troublesome) rainmaking 'chief' named Omari, to a more congenial character named Gunda.

Gunda was Sagilũ's mother's sister's son and was thus a member of the royal rainmaking lineage (Figure 20.2). He remained in office for fifteen years, from 1939 to 1954. In addition to being a greatly respected and gentle man who is said, with his female counterpart, to have controlled the rains admirably and with great skill, he was an efficient bureaucrat and administrator, much as colonial administrators had hoped. Villagers themselves, it appears, could not have cared less about his aptitude for administration, so long as he cooperated with Nya Matalũ and brought rain, which he did each year at the royal rainshrine.

It was during Gunda's chiefly reign, in 1947, that the elderly female rainmaker, Nya Matalũ, passed away, ending possibly the longest reign of any Ihanzu female rainmaker. This ancient woman had managed to outlive all of her own daughters, save one, who immediately succeeded her. Nkili, or Nzĩtũ as she was sometimes known, became the next female rainmaker, and a powerful rainmaker she proved to be over the next forty years.

In 1954, Gunda the regent chief stepped down, and Sagilũ's sister's son, Omari Nkinto, now about thirty years of age, became 'chief.' At the same time, he became Ihanzu's male rainmaker, who with his mother Nkili now conducted rain rites each year.

As far as government functionaries were concerned, Omari was always a difficult man. As with his chiefly predecessors (with the exception of Gunda), his central concern was not with policy or politics, but rainmaking. If Ihanzu men and women appreciated this aspect of his personality, colonial officials by and large did not. Marguerite Jellicoe, a Tanganyika government sociologist, wrote of Chief Omari in the early 1960s:

> An apparently young man of very difficult personality. Is variously called [by colonial officials?] lazy, weak, ineffectual, shy. Avoids whenever possible meeting strangers, especially Government officials and Europeans. Is said to be practically illiterate, which fact helps to undermine his present-day position, literacy now being admired. Clearly has an immense 'inferiority complex' coupled with a firm belief in his powers as a rainmaker and his rights as traditional chief. Very touchy and ready to take offence at any imagined slight, but still has great power over the more traditionally-minded of his people, and if he is offended it is difficult to make any headway ... He should on no account be asked about his rain-making rites, until perhaps he is known very well. The best approach is through the history of his clan and of the Masai wars ... Neither should any initial curiosity whatever be shown about the

rainmaking house (the *mpilimo*) which can be seen close to the old tumbleddown resthouse. Nor should Kirumi Ridge [Ng'waũngu] be climbed (the rainmaking hill behind the Resthouse), nor any drum cave visited, without the permission of the Mtemi. The Mtemi speaks Swahili.[38]

Jellicoe's comments make plain the extent to which many 'traditionally minded' men and women in the 1960s saw Chief Omari much as he saw himself: first and foremost as a rainmaker. Even if literacy was admired – which it likely was, among a small minority – this had little bearing on whether or not Omari, together with his mother, was respected or brought rain. For at the end of the day, 'In the eyes of most of the Isanzu, the chief is not regarded as a government servant but as the giver of rain' (Adam 1963b: 10; cf. Packard 1981: 169).

Even if administrators failed to notice, the same was true for Nkili, Omari's mother. One anthropologist noted in the 1960s: 'She is a woman who is feared and respected throughout the kingdom … She succeeded her mother [Nya Matalũ], a woman who lived to a great age and outlived all of her daughters except this one. Nkili is said to possess her own rainstones … People are afraid to thwart her wishes, and she wields great influence over the chief and all the royal family' (Adam 1963b: 17–18).

Omari, who lives in Kirumi today, speaks fondly of his years as government chief: he owned a Land Rover; lived in a palatial, modern government house in Kirumi; afforded imported brandy and whisky; supported five wives and his children; and conducted rainmaking ceremonies with his mother at the Kirumi rainshrine without governmental interference. In fact, Omari is not the only one who tends to remember the 1950s and 1960s as 'the good old days.' So does the female rainmaker, and ordinary villages, too. Those were decades when royal powers were strong, times when Ihanzu rainmaking practices in manifold forms flourished, times when white people – the British social anthropologist Virginia Adam, to be precise – came from faraway places and took special notice (Adam 1963b). Although the two royal rainmakers had conducted rain rites publicly since the 1920s, it is clear with hindsight why these particular years in the twilight of British rule are sometimes singled out as the heyday of Ihanzu rainmaking. This has everything to do with what happened immediately after independence.

Gaining Independence, Losing Rain

When Tanganyika gained independence on 9 December 1961, Julius Nyerere and his Tanganyika African National Union party (TANU for short) ruled the nascent nation. Although some Ihanzu men and women I know played active roles in the process, on balance, independence to many Ihanzu villagers seems to have meant very little. Many with whom I have spoken on the topic, in fact, do not remember with any clarity those particular years. What did make an indelible mark on people's imaginations was the new government's decision the year after independence to abolish chiefdoms across the land. It was not so much that ordinary Ihanzu villagers had come to love their 'chief' or the 'chiefdom' that the British administration had, over many decades, created. Rather, men and women feared that with the destruction of the chiefdom, Chief Omari and his mother Nkili would be prohibited from bringing rain (Adam 1963b: 15). These fears were not entirely unfounded.

Former chief Omari recalled a meeting that he and the two other former chiefs of the Mkalama Federation had with the new District Commissioner, after being relieved of their chiefships in December 1962:

> The DC was really angry, crazy you could say. He told us, 'So you think you are chiefs? Real big men? Well, let me tell you you're nothing. TANU is in charge now. You are just rubbish, savages [*washenzi*]. All of you. There's no place for any of you lot.'
>
> You know, *they* are the savages. Do you know what happened then? We left Kiomboi and returned home, all of us. And the rains did not fall over the whole of Iramba District that year, and the government had to start importing food aid, dishing it out by the cup, to help the people.

Coincidence is a strange beast. With Tanganyika's independence and the demise of her chiefships came the total failure of the rains across the country. The central and northern parts of the nation were particular hard hit, 'causing a famine which was the north's worst since the 1890s' (Iliffe 1979: 576). Many villagers spoke then, as Omari and others do today, in a single breath, of the demise of the Ihanzu chiefdom and the massive drought that ensued (e.g., Adam 1963b). And to the extent they did, and still do, Chief Omari's deposition harks back to a bygone but not entirely forgotten era many decades earlier when German forces captured, hauled off and hanged the male rainmaker, Kitentemi. In both cases the demise of male rainmakers signalled the demise of the rain. Time had seemingly come full circle, the end mirrored its own beginning.

With Omari's removal in the early 1960s, a young and relatively well educated TANU man filled the newly created position of Division Clerk (*Katibu Tarafa*). He was himself an Ihanzu man and a Lutheran, and he had no connection with the royal *Anyampanda* rainmaking clan-section – something that no doubt underscored the desired break between new and older orders (cf. Abrahams 1981: 37). The clerk soon moved to the administrative centre of Kirumi village. There, to make the transition from chiefs to TANU complete, he moved into the former chief's house, a massive and positively ostentatious structure built towards the end of the British colonial era to convey an unmistakable aura of chiefly authority. With the chief now living in a modest mud-brick home near the rainshrine, the grand State House and its single occupant now embodied all the power and hopes of a new but as yet unproved post-independence order.

Some of my informants claim that certain local TANU administrators tried to pressure the Ihanzu royals to stop making rain and, on occasion, tried to stop villagers from giving grain tribute to their rainmakers; such instances, though, seen to have been more the exception than the rule. More often than not, it appears that administrators turned a blind eye to royals and their rainmaking rituals. For some, this was because they had better things to do; they saw rainmaking rites as relics of the past, and they supposed, as did the missionaries and colonial administrators before them, that they would fall by the wayside as the nation modernized and developed. Others ignored such things because they were themselves of Ihanzu origin and understood perfectly well the importance that rainmaking had both for themselves and for all others with whom they lived. It was thus that Omari and Nkili continued to carry out rainmaking rites throughout the 1960s, albeit no longer on the grand public scale that Adam had witnessed in 1961 and 1962 (Adam 1963b).

By the mid-1960s, the devastating drought and famine that had plagued the nation had ended, and the rains of the early 1970s were sufficient. This state of affairs was once again threatened in the mid-1970s by *ujamaa*, or African socialism. This was due to an attack, not on Ihanzu's rainmakers, but rather on their royal home, Kirumi village.

Ujamaa was the most elaborate social experiment ever attempted in Tanzania (save for colonialism, of course). This complex policy turned on the interrelated notions of independence, freedom, and self-reliance. One of *ujamaa*'s central pillars – and it affected millions of rural Tanzanians – was villagization. Its premise was simple: if people lived in close proximity, all would benefit from easy access to roads, transportation, shops, schools, dispensaries, water pumps, and other modern conveniences. The government would provide the raw materials for these ventures; villagers, in the spirit of cooperation, would provide the labour necessary to assemble the greater good. This thinking also extended to large, village-owned collective farms, on which people were urged to work together for the common good. The logistics of such a massive, nationwide relocation programme – millions of Tanzanians were moved into large, centralized villages – were daunting, the final results mixed.

In 1975, district-level officials in Kiomboi decided that Ihanzu would be divided into sixteen *ujamaa* villages (*vijiji*), each comprising several smaller subvillages (*vitongoji*). The criteria for such decisions were straightforward: those villages with more 'modern' amenities that were easily accessible were favoured, while those that were remote, with fewer amenities, were not. (Modern markers here included water pumps, road access, dispensaries, courthouses, schools, shops, and churches.) On these grounds, dozens of Ihanzu villages were slated for closure, their residents to be relocated to more 'modern' *ujamaa* villages. Some areas of Tanzania saw confrontations, sometimes violent, between overzealous government officials and reluctant villagers, but there were few signs of open protest in Ihanzu. With one notable exception, that is: Kirumi.

The thought of closing Kirumi disturbed villagers. As far back as anyone could remember, Kirumi had been the rightful home of their royal rainmakers, the place where their ancestral migrants from Ukerewe first settled. It was here that former rainmakers, male and female (except Kitentemi), were buried. And it was here that rainmaking rites had been conducted since at least the 1800s. Kirumi meaningfully connected Ihanzu women and men with their past, and sketched a tentative path into an uncertain future. Having survived Maasai and Tatog raids in the nineteenth century, and having withstood the onslaught of two different colonial regimes in the twentieth, now it appeared that Nyerere's villagization programme was going to destroy Kirumi – and with it, the rain – forever. As one elderly woman from Isene village pointedly put it, 'Closing Kirumi is like closing Ihanzu.'

Zugika Maua, a now-elderly man who was in 1975 the head village representative of Kirumi (*shina wa Kirumi*), explained that there were lots of troubled public meetings that year during which villagers expressed profound concerns about Kirumi's closure. Following one meeting, he said, he wrote a letter of protest on villagers' behalf, pleading that Kirumi be incorporated as a subvillage under a larger *ujamaa* umbrella. The letter went to the Division Secretary, an Iramba man named Welia who lived in Kirumi and who was a staunch supporter of the closure. He paid no attention.

In his letter, Maua suggested that there was no good reason to close Kirumi. It had, after all, some modern facilities: a dispensary, a courthouse, and an almost reasonable road. He did not mention rain in the letter, though he told me that this, really, was what at the time concerned most people. 'To convince the government,' he said, 'they did not want to hear about our problems of rain. We had to use a different approach and show them how modern we were. We told them about roads and dispensaries, not chiefs and rains.' A subtle irony, it would seem, deploying British-built 'modernity' to save Ihanzu 'tradition.' For his efforts, Maua was reprimanded in court and fined 500 shillings (about $1.50 today). Other villagers were ignored.

In the middle of it all, the unpopular Division Secretary, Welia, was unexpectedly transferred from Kirumi and a new secretary, Obedi Lange, took his place. Like his predecessor, Lange was an Iramba man and a Lutheran. Unlike Welia, however, he was sympathetic to villagers' concerns. In fact, little time passed before Lange sided with villagers, arguing at the district level in favour of Kirumi's incorporation under Matongo village.

Eventually, after nearly a year of commotion by villagers, district-level officials abandoned their original position and issued a revised and updated statement: Kirumi would *not* be closed. It would be incorporated as a subvillage of Matongo village. Men and women were elated. In former chief Omari's words: 'That time when the government wanted us to move out of Kirumi to Matongo was a real disaster. They said all of us would go, everyone. But there was one thing that stopped them: that house [pointing to the rainshrine]. In the end they realized that if they moved us from here, they would absolutely destroy our rainmaking traditions and so, in the end, they let us stay. What do you think would have happened if we had moved? All this would be bush, and there would be no rain.'

It is unlikely that rainmaking was the reason district administrators eventually changed their minds. Even so, Omari speaks in tune with most villagers when he points to the significance that rainmaking practices had for most in the 1970s, and indeed have today. That Kirumi village be closed was the only proposal in Ihanzu that raised a public outcry, which lasted for nearly a year. It was also the only village in the whole of Ihanzu that was, once wiped clean from the map, re-drawn by government officials in ink.

Rainmaking in the New Millennium

Today, Ihanzu rainmaking continues to shape encounters with postcolonial administrators, though only, I think, when anthropologists occasion such reflection. For the most part, Ihanzu's royal rainmakers have continued to do what they do in Kirumi, and representatives of Tanzania's modern nation-state show little interest in them.

Iramba's District Commissioner made a brief visit to Ihanzu in 1995 to hold a public meeting about adopting new fast-grow maize seeds, and to persuade villagers that they should 'develop' using 'the free market.' Before the meeting, he asked me how my research was going on Ihanzu 'traditions and customs' (*mila na desturi*), those things generally 'known' to preoccupy anthropologists. He also wanted to know, among other things, why 'the chief' had not seen fit to show up at this particular government meeting, as all villagers are expected to. I replied that I had not seen him that morning, that he might be in Kirumi or, most probably, was on the way as we spoke. Somewhat irritated, the DC made clear his view on chiefs and rainmaking: 'For a long time the chiefs here in Tanzania stopped modernity. They kept people from becoming modern (*watu wa kisasa*). They were against education, against good roads, against business and against change. They only wanted old customs [*mila za zamani*] ... Perhaps it's better that the chief stays there in Kirumi with his rainshrine. He had many years to send modernity backwards [*kurudisha nyuma maendeleo*]. Now they're gone and the government's here. We will develop these people!'

The DC's comments raise a few points. First, postcolonial officials today consider Ihanzu rainmaking and rainmakers (when they consider them at all) to be outmoded relics that have no place in our 'modern' world. These ideas – no, fervent hopes – are informed by a particular unilinear vision of the world, a world where modernity destroys tradition, where the present inevitably overwrites the past. Such ideas or hopes are hardly new. As we have seen, similar notions have underpinned administrative thinking since the advent of colonialism more than a century ago. Second, the DC's comments suggest that postcolonial administrators do not recognize female rainmakers and/or 'chiefs' and the central role they play for locals any more than their colonial predecessors did. Rainmaking chiefs are just men. (This of course is not an issue that concerns them, given rainmakers' alleged irrelevance in and to the modern world.)

These points notwithstanding, Ihanzu villagers remain adamant that rainmaking is not just a thing of the past, but equally something highly desirable in the present, something they actually want, something they actually *need*, and in the particular gendered form it takes. But why? What is at stake?

Claims about 'tradition,' 'culture,' and 'identity,' anthropologists have long known and frequently shown, can serve particular class or clan, generation or gender interests. This is commonly the case, as for instance in northern Tanzania, where there are resources like land, livestock, and labour worth fighting over (S.F. Moore 1986). In such cases and places, what counts as 'tradition,' or what makes one more 'traditional' than someone else, is crucially linked to managing one's livelihood successfully. Ihanzu 'traditions,' however, are not linked to identity politics in the same way, or to the same extent: being more 'traditional' or insisting on maintaining rainmaking 'traditions' provides no obvious material benefits, no privileged access to scarce resources, to specific socially situated actors. Such claims are these days more likely to attract derision from administrators. Truth be told, within Ihanzu, there are no 'traditional' resources worth struggling for: the long-farmed matriclan lands are largely exhausted, and people cut new plots from the bush in preference to them; and there are no 'traditional' corporate herd holdings that villagers might tap into. The point here is that associating oneself with 'traditional' rainmaking is no means to secure scarce resources for oneself, or to deny others.

So why bother? First, by linking rainmaking, 'tradition,' and Ihanzuness, villagers attempt to forge a solid conceptual mooring in an ever-changing world. In this sense, rainmaking provides Ihanzu women and men with the means by which to establish meaningful historical connections with their past, and to concretize their own place-in-the-world at present. 'Rainmaking,' villagers frequently told me, 'is our tradition' (*jadi yetu*). Yet as important as such moorings are – and here is the point to underscore – for the women and men of Ihanzu, rainmaking is first and foremost crucial because it brings rain: a tautology, to be sure, but one that nonetheless captures the essence of what is at stake. For without their two differently gendered royals and the powers and rains they jointly control, all would cease to exist. There would be no rain. There would be no harvest, food, or beer. There would be no animals. There would be no villagers.

NOTES

1 Hichens, 'Mkalama Annual Report 1919/1920 (16 April 1920),' p. 4, TNA 1733/1.

2 Wyatt, n.d., p. 6, 'Mkalama District Book,' School of Oriental and African Studies (hereafter SOAS); Hichens, 'Mkalama Annual Report 1919/1920 (16 April 1920),' TNA 1733/1; V. Adam 1963. 'Draft Report on Isanzu for Community Development Department of Tanzania,' p. 9, BLPES.

3 M. Jellicoe 1961, 'Interim Report on Isanzu,' p. 45, RH MSS. Afr. s. 2038 (4); Wyatt, n.d., p. 6, 'Mkalama District Book,' SOAS.

4 M. Jellicoe 1961, 'Interim Report on Isanzu,' p. 45, RH MSS. Afr. s. 2038 (4); Wyatt, n.d., p. 6, 'Mkalama District Book,' SOAS.

5 These claims are well-established for the Turu, Iambi, and Iramba. The claim that Sukuma visited Kirumi in the 1890s is based on what informants told me and the nature of Sukuma society at the time. The fact that multiple, small-scale Sukuma communities had by the late nineteenth century developed across Sukumaland in which people relied heavily on ritual leaders (*ntemi*) to provide rain, coupled with the fact that these same leaders were increasingly being deposed, their ritual powers increasingly being called into question, makes it plausible that some eastern Sukuma would have visited Kirumi in an effort to seek out legitimate rainmakers (see Holmes and Austen 1972: 386; Tanner 1957: 201; Sanders 2001b). Informants also say that Hadza visited the Kirumi rainshrine long ago, and it is certain that two Hadza men named Majui and Tawashi regularly visited Chief Omari in the 1950s (Woodburn 1979: 262n). Intriguingly, the Ihanzu claim that *all* Hadza are members of the royal *Anyampanda wa Kirumi* clan section, and James Woodburn (pers. comm.) informs me that, when asked by Ihanzu, they admit this is true, though relevant only for their occasional dealings with Ihanzu. Woodburn also claims that Hadza have no rainmaking institutions. Ihanzu men and women see the Hadza, like their own rainmakers, as 'owners of the land,' a status that would seem to derive from their hunting and gathering way of life.

6 The role of the Ihanzu rainshrine and rainmakers is similar to the role played by other rainshrines and rainmakers in other parts of precolonial Africa (see Iliffe 1979: 28–30; Kimambo 1991: 30, 34; Lan 1985; O'Brien 1983; Swantz 1974: 75–82; Young and Fosbrooke 1960: 41–2).

7 Hichens, 'Mkalama Annual Report 1919/1920 (16 April 1920),' pp. 13–16, TNA 1733/1.

8 Virginia Adam 1961, 'Land Ownership and Local Descent Groups in Kirumi,' BLPES 1/15.

9 Wyatt, n.d., p. 9, 'Mkalama District Book,' SOAS; Hichens, 'Mkalama Annual Report 1919/1920 (16 April 1920),' p. 7, TNA 1733/1; Bagshawe, 'Kondoa-Irangi Annual Report 1920/1921,' pp. 14–15, TNA 1733: 5; V. Adam 1963. 'Draft Report,' p. 9, BLPES.

10 Wyatt wrote in the 1920s of precolonial Ihanzu 'chiefs': 'They appear to have derived their authority solely from their supposed powers as "rain makers."' Wyatt, n.d., p. 7, 'Mkalama District Book,' SOAS. While Wyatt is referring here to male 'chiefs,' the same was so for the female ones, too.

11 V. Adam 1963. 'Draft Report,' p. 9, BLPES.

12 Dundas, 1914, 'History of Germans in East Africa, 1884–1910' (m.s.), pp. 34–5, RH MSS. Af. s. 948.

13 The most famous of these movements was the Maji-Maji rebellion (see Gwassa 1970; Iliffe 1967, 1969: 9–29).

14 *Jumbe*, a term that is variously translated as 'chief' or 'headman,' were political agents employed under the German regime all across German East Africa.

15 Wyatt, n.d., p. 6, 'Mkalama District Book,' SOAS.
16 Wyatt, n.d., p. 6, 'Mkalama District Book,' SOAS.
17 Dundas, 1914, 'History,' p. 34, RH MSS. Af. s. 948.
18 Ibid., Hichens, 'Mkalama Annual Report 1919/1920,' p. 4, TNA 1733/1.
19 See note 17.
20 Lyons, n.d., Iramba (Kiomboi) 'District Book,' vol. I, RH Micro. Afr. 472, reel 22; Wyatt, n.d., pp. 6–7, 'Mkalama District Book,' SOAS.
21 Oberleutnant Ruff, 24 March 1910, 'Einsiedler Adolf Siedentopf,' TNA G55/27.
22 Ibid.
23 Hichens,'Mkalama Annual Report 1919/1920 (16 April 1920),' p. 7, TNA 1733/1. See also Wyatt, n.d., p. 6, 'Mkalama District Book,' SOAS.
24 Bagshawe, 'Kondoa-Irangi Annual Report 1919/1920 (14 April 1920),' pp. 15–16, TNA 1733/1.
25 Bagshawe, 'Kondoa-Irangi Annual Report 1919/1920 (14 April 1920),' pp. 13–14, TNA 1733/1.
26 *Ndege* means 'bird' in Swahili, but not in the Ihanzu language.
27 Hichens,'Mkalama Annual Report 1919/1920 (16 April 1920),' p. 8, TNA 1733/1.
28 Ibid., pp. 8–9, TNA 1733/1.
29 F.J. Bagshawe, 'Annual Report, Kondoa Irangi, 1920/1921,' pp. 14–15, TNA 1733, 5.
30 Hichens,'Mkalama Annual Report 1919/1920 (16 April 1920),' p. 10, TNA 1733/1.
31 Singida District Officer, 'Singida Annual Report 1927,' pp. 8–10, TNA 967, 823.
32 Ibid., p. 12. Chief Sagilũ is also occasionally referred to as Mpĩlolo.
33 Ibid.
34 Ibid.
35 Lyons, March 1928, 'Iramba (Kiomboi) District Book,' vol. I, RH Micro. Afr. 472, reel 22.
36 Virginia Adam, 'Field notes' (July 1963), pp. 19–21, BLPES 1/10.
37 As for the rest, around 18 per cent self-identify as Christian (1 per cent Catholic and 17 per cent Lutheran), and 2 per cent as Muslim.
38 M. Jellicoe, 'Field notes' (April/May 1961). Archived with Virginia Adam's documents at BLPES.

REFERENCES

Abrahams, Ray G. 1981. The Nyamwezi Today: A Tanzanian People in the 1970s. Cambridge: Cambridge University Press.

Adam, Virgina. 1961. Preliminary Report on Fieldwork in Isanzu. Conference Proceedings from the East African Institute of Social Research. Makerere College.

Adam, Virginia. 1963a. Migrant Labour from Ihanzu. Conference Proceedings from the East African Institute of Social Research. Makerere College.

Adam, Virginia. 1963b. Rain Making Rites in Ihanzu. Conference Proceedings from the East African Institute of Social Research. Makerere College.

Admiralty War Staff (Intelligence Division). 1916. A Handbook of German East Africa. London: H.M.S.O.

Alpers, E. A. 1969. The Coast and the Development of the Caravan Trade. *In*: A History of Tanzania, I. N. Kimambo and A. J. Temu, eds. Nairobi: East African Publishing House.

Brooke, C. 1967. The Heritage of Famine in Central Tanzania. Tanzania Notes and Records 67:15–22.

Deutches Kolonialblatt. 1901. Vol. 12, no. 24, 15 December.

Deutches Kolonialblatt. 1902. Vol. 13, no. 23, 1 December.

Deutches Kolonialblatt. 1903. Vol. 14, no.18, 15 September.

Feierman, Steven. 1990. Peasant Intellectuals: Anthropology and History in Tanzania. Madison: University of Wisconsin Press.

Graham, J.D. 1976. Indirect Rule: The Establishment of 'Chiefs' and 'Tribes' in Cameron's Tanganyika. Tanzania Notes and Records 77–78:1–9.

Gray, Robert F. 1955. The Mbugwe Tribe: Origin and Development. Tanganyika Notes and Records 38:39–50.

Gray, Robert F. 1963. Some Structural Aspects of Mbugwe Witchcraft. *In*: Witchcraft and Sorcery in East Africa, J. Middleton and E. H. Winter, eds., pp. 143–73. London: Routledge and Kegan Paul.

Gwassa, G. C. K. 1970. The Outbreak and Development of the Maji Maji War, 1905–1907. PhD thesis, University of Dar es Salaaam.

Holmes, C.F. and R.A. Austen. 1972. The Precolonial Sukuma. Journal of World History 2:377–405.

Iliffe, John. 1979. A Modern History of Tanganyika. Cambridge: Cambridge University Press.

Jellicoe, Marguerite. 1969. The Turu Resistance Movement. Tanganyika Notes and Records 70: 1–12.

Jellicoe, Marguerite, Vincent Sima, and Jeremiah Sombi. 1968. The Shrine in the Desert. Transition 34(7):43–9.

Johnson, V. Eugene. 1934. The Augustana Lutheran Mission of Tanganyika Territory, East Africa. Rock Island: Board of Foreign Missions of the Augustana Synod.

Kesby, John D. 1981. The Rangi of Tanzania: An Introduction to Their Culture. New Haven, CT: HRAF.

Kidamala, D. and E. R. Danielson. 1961. A Brief History of the Waniramba People up to the Time of the German Occupation. Tanganyika Notes and Records 56:67–78.

Kimambo, I. N. 1991. Penetration and Protest in Tanzania: The Impact of the World Economy on the Pare, 1860–1960. London: James Currey.

Kohl-Larsen, Ludwig L. 1943. Auf den Spuren des Vormenschen (Deutsche Africa-Expedition 1934–1939). Stuttgart: Strecker und Schröder.

Koponen, J. 1988. People and Production in Late Precolonial Tanzania: History and Structures. Jyväskylä: Scandinavian Institute of African Studies.

Lambek, Michael. 1996. The Past Imperfect: Remembering as Moral Practice. *In*: Tense Past: Cultural Essays in Trauma and Memory, Paul Antze and Michael Lambek, eds., pp. 235–54. New York: Routledge.

Lambek, Michael. 2002. The Weight of the Past: Living with History in Mahajanga, Madagascar. New York: Palgrave-Macmillan.

Lan, David. 1985. Guns and Rain: Guerrillas and Spirit Mediums in Zimbabwe. Berkeley and Los Angeles: University of California Press.

Lindström, Jan. 1987. Iramba Pleases Us: Agro-Pastoralism among the Plateau Iramba of Central Tanzania. PhD. thesis, Univerity of Göteborg.

Maddox, Gregory H. 1986. Njaa: Food Shortages and Famines in Tanzania Between the Wars. International Journal of African Historical Studies 19:17–34.

Moore, Sally Falk. 1986. Social Facts and Fabrications: 'Customary' Law on Kilimanjaro, 1880–1980. Cambridge: Cambridge University Press.

O'Brien, Dan. 1983. Chiefs of Rain – Chiefs of Ruling: A Reinterpretation of Pre-Colonial Tonga (Zambia) Social and Political Structure. Africa 53:23–42.

Obst, Erich. 1912a. Die Landschaften Issansu und Iramba (Deutsch-Ostafrika). Mitteilungen der geographischen Gesellschaft in Hamburg 26:108–32.

Obst, Erich. 1912b. Von Mkalama ins Land der Wakindiga. Mitteilungen der geographischen Gesellschaft in Hamburg 26:1–45.

Obst, Erich. 1923. Das abflußlose Rumpfschollenland im nordöstlichen Deutsch-Ostafricka (Teil II). Mitteilungen der geographischen Gesellschaft in Hamburg 35:1–330.

Packard, Randall M. 1981. Chiefship and Cosmology: An Historical Study of Political Competition. Bloomington: Indiana University Press.

Peters, C. 1891. New Light on Dark Africa: The German Emin Pasha Expedition. London: Ward, Lock.

Prince, von T. 1895. Deutsch-Ostafrika, von der Station Kilimatinde. Deutsches Kolonialblatt 6:243–344.

Reche, O. 1914. Zur Ethnographie des abflußlosen Gebietes Deutsch-Ostafrikas. Hamburg: L. Friederichesen.

Reche, O. 1915. Dr Obst's ethnographische Sammlung aus dem abflußlosen Rumpfschollenland des nordöstlichen Deutsch-Ostafrika. Mitteilungen der geographischen Gesellschaft in Hamburg 29:251–65.

Rigby, Peter. 1971. Politics and Modern Leadership Roles in Ugogo. *In*: Colonialism in Africa, 1870–1960: Vol. 3, Profiles of Change: African Society and Colonial Rule, V. Turner, ed. Cambridge: Cambridge University Press.

Roberts, A. 1970. Nyamwezi Trade. *In*: Pre-Colonial African Trade: Essays on Trade in Central and Eastern Africa before 1900, Robert Gray and D. Birmingham, eds. London: Oxford University Press.

Sanders, Todd. 2001. Territorial and Magical Migration in Tanzania. *In*: Mobile Africa: Changing Patterns of Movement in Africa and Beyond, M. de Bruin, Rijk van Dijk, and D. Foeken, eds. Leiden: Brill.

Scott, James C. 1985. Weapons of the Weak: Everyday Forms of Peasant Resistance. New Haven, CT: Yale University Press.

Sick, E. von. 1915. Die Waniaturu (Walimi): Ethnographische Skizze eines Bantu-Stammes. Baessler-Archiv 5:1–62.

Snyder, Katherine. 2005. The Iraqw of Tanzania: Negotiating Rural Development. Cambridge, MA: Westview.

Stuhlmann, F. 1894. Mit Emin Pascha ins Herz von Afrika. Berlin: Dietrich Reimer.

Swantz, L. W. 1974. The Role of the Medicine Man among the Zaramo of Dar es Salaam. PhD. thesis, University of Dar es Salaam.

Tanner, R. E. S. 1957. The Installation of Sukuma Chiefs in Mwanza District, Tanganyika. African Studies 16(4):197–209.

Ten Raa, Eric. 1968. Bush Foraging and Agricultural Development: A History of Sandawe Famines. Tanzania Notes and Records 69:33–40.

Thornton, Robert J. 1980. Space, Time, and Culture among the Iraqw of Tanzania. New York: Academic.

Thurnwald, R. C. 1935. Black and White in East Africa: The Fabric of a New Civilization. London: George Routledge and Sons.

Ward, Robert E. 1999. Messengers of Love. Kearney, NE: Morris.

Werther, C. W. 1894. Zum Victoria Nyanza: eine Antisklaverei-Expedition und Forschungsreise. Berlin: Gergonne.

Werther, C. W. 1898. Die mittleren Hochländer des nördlichen Deutsch-Ost-Afrika. Berlin: Hermann Paetel.

Winter, Edward H. 1955. Some Aspects of Political Organization and Land Tenure among the Iraqw. Conference Proceedings from the East African Institute of Social Research. Makerere College.

Woodburn, James. 1979. Minimal Politics: The Political Organisation of the Hadza of North Tanzania. *In*: Politics in Leadership: A Comparative Perspective, W.A. Shack and P.S. Cohen, eds. Oxford: Clarendon.

Woodburn, James. 1988a. African Hunter-Gatherer Social Organisation: Is It Best Understood as a Product of Encapsulation? *In*: Hunters and Gatherers: History, Evolution, and Social Change, Tim Ingold, David Riches, and James Woodburn, eds., pp. 31–64. Oxford: Berg.

Woodburn, James. 1988b. Hunter-Gatherer 'Silent Trade' with Outsiders and the History of Anthropology. Fifth International Conference on Hunting and Gathering Societies, Darwin, Australia.

Young, R. and H.A. Fosbrooke. 1960. Land and Politics among the Luguru of Tanganyika. London: Routledge and Kegan Paul.

"Friction" in the Global Circulation of Climate Knowledge

21

Transnational Locals: Brazilian Experiences of the Climate Regime

Myanna Lahsen

The Politics of Universalizing Discourses

Popular environmental discourses often evoke impressions of planetary beauty, fragility, and interdependence. They merge with other universalizing discourses that similarly offer images of globalization as liberations from present and past limitations and the promise of a unified humanity no longer divided by East and West, North and South, the rich and the poor. These discourses have broad appeal, and they may be beneficial to the extent that they create a common ground through shared cognitive and normative understandings that transcend local, parochial interests and perspectives. Global environmental indicators suggest the need for some level of convergence in the form of ecological thinking that transcends narrowly nationalistic frames of reference. However, it is important to understand what is elided by universalizing discourses. Actors, including social scientists, easily fall for the "charisma" of the global (Tsing 2000), oversimplifying globalizing processes by highlighting primarily their virtues and promises. Universalizing discourses can serve problematic political programs by hiding contradictions, ambiguities, and complexities of socio-political reality. In the environmental arena, universalizing discourses—for instance, the "life-boat ethic" (Enzensberger 1974) or "sustainability"—avoid the need to analyze concrete inequalities, such as the distribution of power, costs, profits, and responsibility.[1]

Scientific knowledge is commonly associated with universality. Science and scientists are often looked to as models and stimulators of global citizenship and ecological thinking that transcends narrow, self-serving concerns and frames of reference. The sociologist of science Robert K. Merton (1973 [1942]) formulated an early sociological framework that identified a scientific "ethos," a set of cultural values and norms governing knowledge production and knowledge sharing. Merton argued that these values and norms are internalized and transmitted by scientists, guiding their thoughts and actions regardless of where they are situated in terms of politics or culture. More recently, theorists in the field of international relations (IR) have advanced a

The Anthropology of Climate Change: An Historical Reader, First Edition. Edited by Michael R. Dove.

similar understanding of the role of science and scientists in global environmental politics. For instance, Peter Haas has promoted the notion of transnational "epistemic communities" (1992), transnational networks of professionals from a variety of disciplines and backgrounds with shared ways of knowing, shared patterns of reasoning, shared concern about transboundary environmental problems, and shared environmental policy agendas. These networks of actors shape national and international environmental policies. Haas distinguishes epistemic communities from scientific communities in general by the former's adherence to a shared normative framework and an associated policy agenda. Members of an epistemic community are said to identify with those who share their normative commitments, and to succeed occasionally in efforts to persuade states to transcend rational utilitarian considerations in favor of international environmental cooperation.

Powerful critiques of the epistemic community framework have emerged from the field of science and technology studies. These analyses note that support for a shared perception of environmental problems may be the result of diverse social and political influences, for example, shared disciplinary orientations, economic interests, political ideologies or discursive framings (Jasanoff 1996; see also Miller 1998). This chapter builds on and extends these observations through an ethnographic study of Brazilian scientists and policy makers engaged with international climate science and politics. Empirical data are used to chart some of the "uncharted ideological and political minefields"[2] encountered in efforts to find common ground on a planetary scale concerning the global environment. The data suggest that members of the "climate epistemic community" in Brazil have far more complex identities, more agency, more ambivalence, and more selective allegiances to the climate "episteme" than assumed by IR scholars.

The research reported here was conducted for a total of nine months spread out over a period of three and a half years,[3] ending with the end of the Fernando Henrique Cardoso government (December 2002).[4] By contrast to the standard IR depiction of epistemic communities, this study reveals the circumscribed, internally fragmented and unstable nature of Brazilian actors' adherence to the so-called epistemic community that has formed around climate science and policy. The network studied involves professional collaboration—and, often, personal friendships—that stretch across continents, yet involve frequent interaction among some members around issues of shared interest and concern. To the extent that this social formation congeals, the "glue" holding it together is the research and policy-process related to human-induced climate change, overseen in part by the Intergovernmental Panel on Climate Change (IPCC). The IPCC is a panel of hundreds of scientists from around the world charged with assessing the state of scientific knowledge related to climate change and identifying potential remedial action. Created jointly by the World Meteorological Society and the United Nations Environmental program in 1988, the IPCC provides technical assistance to international negotiations under the Framework Convention on Climate Change (FCCC) to reduce global emissions of greenhouse gases.

While identifying important transnational cultural and political dimensions of Brazilian scientists' and policy makers' life worlds, this chapter challenges identifications of this "community" with cognitive and normative homogeneity. The identities of Brazilian actors participating in IPCC-related arenas are complex and hybrid, with important consequences for their interpretations of international science, the global environment, and their own agency. Some of their frameworks of understanding reflect the transnational nature of their professional networks. Other levels of understanding reflect the continued impact of history, geography and socio-economic realities. Among other things, Brazilian climate scientists and policy makers are keenly aware of structures of difference within international and transnational climate-related forums in which they participate, including conditions of inequality, reflecting the dominance of actors from richer countries ("the North").

Temptation to gloss over these persistent differences and unresolved tensions is considerable, not the least in the environmental arena where various actors—whether politicians, environmental activists, and academic scholars studying international regimes—seek to consolidate support for their visions of social reality and social change. Giving in to those temptations, however, may undermine rather than strengthen efforts to create a more just, sustainable world; universalizing discourses of science and the environment risk aggravating North–South relations around global environmental problems.[5] Institutions related to global environmental governance need to integrate more sophisticated understandings of the role of science in world affairs, including how power can operate through scientific processes. The interview data presented here may serve to nurture these more multidimensional and complex understandings.

Transnational Normative Convergence

The international community of climate scientists and policy makers exhibits some of the shared norms and beliefs identified by Haas as a key component of epistemic communities. Interviews with Brazilian scientists and policy makers produced expressions of concern about human-induced climate change, supporting the notion of cognitive convergence at the transnational level within the climate epistemic community. Brazilian climate-related affairs have been directed by a few key persons affiliated with the Brazilian Ministry of Science and Technology (MCT) and the Brazilian Ministry of External Relations ("Itamaraty"). I will refer to this group as the "MCT group," which has centrally shaped Brazil's official position on climate change in international negotiations.

In his first conversation with me, one of the two longest-standing and most central actors in the MCT group stated that he is "very concerned with climate change."[6] In numerous subsequent interviews he expressed frustration with national politicians' reluctance to engage with the issue and include it among more traditional policy issues.

A second, equally central and long-standing actor in the MCT group also accepted the idea that humans are altering the global climate and endorsed the policy framework under the FCCC. He strongly identified with the IPCC and its conclusions. For instance, he characterized as "insulting" (to himself and the IPCC as a whole) the fact that President George W. Bush chose to look to the US National Academy of Sciences (NAS) rather than the IPCC in assessing the threat of human-induced climate change, and even asked the NAS to reassess the conclusions and objectivity of the IPCC. This Brazilian found consolation in the fact that most of the scientists on the NAS panel evaluating the IPCC themselves are IPCC participants, implying that privileging US scientists over the IPCC failed to take into account the interpenetration of the two communities. Himself a distinguished scientist and IPCC participant, as well as a politically appointed overseer of climate research programs and policy making in Brazil, this actor epitomizes the difficulty of establishing any neat separation between "scientists" and "policy makers."

In interviews, Brazilian scientists involved with climate science and policy expressed concern about human-induced climate change. Arguing for ecological interdependence and the need to transcend narrowly nationalistic concerns, these actors expressed frustration with Brazilian politicians. Some described the latter as stuck in mindsets characterized by zero-sum thinking and national economic interests. "All the talk is about economics: 'What can we get out of this?' As if getting money out of it was the priority," one of them said about Brazilian political meetings convened to discuss climate change. Another regretted the fact that national politicians often "have the attitude that the developed countries got rich by polluting and therefore conclude 'So why can't we?!'" The Brazilian members of the climate epistemic community also wished that the Brazilian public as a whole would get more informed about the issue. One suggested that science courses in primary education do not sufficiently include environmental issues.[7]

In this way, these Brazilian actors appear to share a set of concerns and a policy agenda with their Northern counterparts (Lahsen 1998a,b; see also Boehmer-Christiansen 1994a,b). Part of the reason for convergence is socio-cultural; Brazilian scientists' life worlds are intimately interlinked with those of scientists abroad, especially Europeans and Americans. During my interviews with Brazilian scientists, I sometimes found it difficult to steer the discussion away from US scientists and politicians and toward the Brazilian context. Many of these interviews conveyed a level of knowledge of central personalities in US science and politics surpassing that of many mid- and lower-level US scientists. The discussions thus reveal the globalizing trend whereby segments of societies—and especially "symbolic analysts" (Reich 1992) involved in problem solving, problem identifying, and other knowledge-brokering activities—increasingly are connected to their counterparts throughout the world. Through face-to-face interaction as they travel, and through communications networks, Brazilian climate scientists come to collaborate and interact more with fellow scientists abroad than with most of their domestic colleagues and fellow citizens.[8]

As posited by the epistemic community framework, Brazilian scientists have been important in pushing Brazilian political leaders to focus on the issue of climate change. Brazil's diplomatic position in international discussions about the global environment reflects a partial and on-going transformation from resistance to international environmental regimes to more openness and interest in the development of environmental science. Brazil led the opposition of less-developed countries (LDCs) to the first international environmental initiative, the 1972 United Nations Conference on the Human Environment in Stockholm, which helped to identify the biosphere as a legitimate object of national and international policy (Caldwell 1996: 68, 101). An important subset of Brazil's political leaders changed course in the 1980s and the 1990s, gradually becoming more receptive to international negotiations about global environmental problems. This change in receptivity was both reflected in and reinforced by the country's hosting the first Earth Summit in 1992, the meeting that gave rise to the Framework Convention on Climate Change. Scientists I interviewed claimed to have stimulated this shift.[9] They initially alerted the Brazilian government to the importance of human-induced climate change and "pushed hard" for government attention to the issue and for the appointment of technically competent persons within the government to deal with the issue.

Brazilian Cohesion and International Fragmentation along the North–South Axis

Despite some level of cognitive and normative convergence within the transnational network of climate scientists and policy makers, profound differences also persist. This reality was belatedly impressed upon a US administrator of global change science. A key promoter of the focus on global environmental change in the United States and abroad, he was also an important force behind the creation of the Inter-American Institute (IAI). Based in Brazil, and so far funded predominantly by Northern institutions (especially the US National Science Foundation),[10] the IAI funds global change science projects involving Latin American scientists. In an interview, this science administrator (a Ph.D. environmental scientist) described how his assumption of mutual understanding and transnational fraternity was exposed as an illusion after years of collaboration with Latin American colleagues:

> Ever since 1989, we have been building an Inter-American Institute. President Bush called for it at a meeting in 1990. There are now eighteen or so countries of the Americas that are members of the Inter-American Institute. And its purpose is global change research. The other day I heard something really interesting [*slight laugh*]. I do not even remember what precipitated it but somehow something came along and a person from one of the countries of the Americas—from Chile—after nine years of [being involved with] this, said "There it is!

> There is the US motive for IAI. I knew they were up to something, I knew there was a larger political motive. It took eight years, but now it has been revealed." It was actually a group of people from several countries, joined by Chile, who said that IAI was an American rip-off. Now, I was there from the word "go." I know the motives, I know every iota of thinking behind it. There is no conspiracy. There is no hidden purpose. There is no political agenda. [*Laughs*] It is basically altruistic as hell. But it was never ever perceived that way by the other players. These are friends of mine, people I have known for years, and I suddenly realized: oh my God, they have been sitting there in their respective countries, these pals of mine, wondering what devious thing I was up to.

This science administrator calls for greater attention to the "gulf" that exists among scientists from the North and the South, a gulf involving suspicions he described as "really startling—especially to scientists."

Cognitive differences among climate scientists and policy makers at the international level are rooted in national and political identities, cultural memories, and other aspects of personal experience. For Brazilian climate scientists and policy makers, experiences of colonialism and perceptions of continuities between the colonial past and present conditions in international science constitute one line of division. Brazilian scientists and politicians share certain framings of the climate problem and its attendant science and policy frameworks, and they are attuned to attempts to promote alternative framings of responsibility. Following other members of "the South" (Agarwal and Narain 1990; Mwandosya 2000), they typically frame the issue of responsibility in terms of historic contributions to the problem, and they prefer to discuss responsibility in terms of per capita emissions rather than national totals that penalize population size. Interviews with Brazilian scientists and environmental activists alike revealed resounding support for the Brazilian government's position, indicating strong national cohesion around this issue.

Nationalist sentiment is an important element in Brazilian positions on issues of responsibility related to climate change. Brazil has been identified by some as "Latin America's most nationalist country," even during "sell-out periods" (Adler 1987: 201). It is "one of those countries which, in spite of its liberal rhetoric and its rhetoric in favor of foreign capital, has systematically used its bargaining power, i.e. the bargaining power of its dominant classes, of its government technocracy, and of its national entrepreneurs, to resist.... Even if the national technocrats consider themselves transatlantic, consider themselves liberals, in practice they have increased State intervention, have increased the strength of State enterprises augmenting Brazil's political control" (Maria da Conceicao Taveres, in Adler 1987: 201).

As Brazilian scientists move around international and national arenas, they shift back and forth between different realities. As Brazilians and "Southerners" in the international climate regime, they tend to be suspicious not only of Northern-created science institutions such as the IAI but also of the science that those institutions support. A nationalist ethos, memories of colonialism, and awareness of continued inequity—in the world, generally, and in science more specifically—limit the extent to which Brazilian scientists and policy makers involved in the climate regime accept at least some Northern scientific interpretations of the global environment, along with the associated policy implications. These Brazilian actors portray international science—science legitimized by transnational expert networks—as biased by Northern framings and interests, and therefore not to be accepted at face value. They have witnessed how important framings associated with the Framework Convention were put in place by primarily Northern actors with little, if any, input from developing countries, yet with important unfavorable consequences for the latter. These framings and their implications for notions of responsibility in less developed nations have been widely discussed (see for instance Agarwal and Narain 1990; Biermann 2000; Jasanoff and Wynne 1998; Kandlikar and Sagar 1999; Miller 1998; Yearley 1996).[11]

Factors Reducing (Overt) Dissent

As evident in the US science administrator's comments on the IAI, mistrust can persist despite apparent agreement on scientific and normative matters. Material and socio-cognitive factors can contribute to this state of affairs that can, erroneously, be interpreted as an intersubjective consensus.

Scarcity of resources

Scarce resources—of funding, time and expertise—limit the ability of poorer nations to contest points they may disagree with in international climate science and policy. One policy maker noted that he and his colleagues in Brazil "have no time" to contest issues in these forums ("if we direct too much time in that direction then we can't do our business in Brazil. We are very few. So we have to forget about this").

Poorer nations such as Brazil also have difficulty producing authoritative knowledge in the realm of climate modeling. Financial limitations translate into technical limitations, which in turn translate into limited ability to contest dominant constructions in science and policy. Up to the present, Brazil has been neither able nor inclined—given more pressing domestic problems and priorities—to invest the capital necessary to produce its own projections of future global climate changes and their socio-environmental impacts. The resources required to run the most complex computer-based numerical models that simulate the interactions between the atmosphere, oceans, and land-masses are so expensive that only relatively few countries and institutions can afford them. To the extent that poor nations and other actors resort to simpler models, they are disadvantaged by a value hierarchy within the climate sciences that privileges as more reliable output produced by complex models—a hierarchy some scholars have identified as possibly arbitrary and culturally biased (Lahsen 1998a; Shackley et al. 1998). The MCT group encountered this bias when presenting a policy proposal under the FCCC on the basis of what they acknowledged as a "rather radical simplification" of a model US scientists already considered too simple to qualify as a reliable policy tool.[12]

Interviews with Brazilian scientists repeatedly revealed a vague sense that, given sufficient resources, they and other LDC scientists might possibly develop knowledge of a different kind that could serve as a critical alternative to what is now known as "international science."[13]

Financial interests and needs

The political economy of climate science in Brazil may also serve to muffle dissent. Brazilian climate scientists depend on foreign institutions for an important part of the funding for their science. Moreover, they may fear forgoing invitations to participate in forums such as the IPCC if they become known as troublemakers. Participation in the IPCC confers prestige at the national levels and visibility at the international levels that can result in lucrative consulting assignments with both national and international governmental and non-governmental entities (e.g., the national oil company Petrobras, the World Bank, the US Pew Foundation, the US National Environmental Trust, and United Nations, among many others). Some Brazilian institutions allow funds from consulting services to be paid as supplements to university scientists' personal salaries, a particular attraction given the national freeze on civil servants' salaries since 1995 and the present weakness of the national currency.

Intimidation, perceptions of prejudice, and lack of voice

A third reason for superficial appearances of agreement is that LDC scientists do not always feel free to voice their differences with the preferred social order. In interviews, Brazilian scientists expressed perceptions of prejudice, which they sometimes related to cultural experiences of colonialism in Brazil. In the words of a policy maker associated with the MCT group:

> When the Portuguese came to Brazil, in the 1500s, they gave mirrors to the Indians. They obliged them to cut the trees and took the

> wood back to sell in Portugal. And they gave mirrors in payment for cutting down the trees. They worked for the whole year cutting trees and then the Portuguese loaded up all the timber and gave them a few mirrors and that was it! And some metal axes to cut the trees that were more efficient. It is something like that [with the IPCC]! They always look to the third world like that. We are Indians, we don't know how to talk, we are not.... We don't know what we are doing. "Poor guys, forget about them." That's how people look at us.

This same policy maker repeatedly returned to difficulties rooted in language and to associated experiences of feeling intimidated or judged. He also contrasted my credibility with his own as follows: "[Y]ou are from the developed countries. So in this sense you have more credibility than the Indians. If I talk, I have no credibility!" To be sure, this policy maker was tall, and though his hair was dark and his skin olive in tone, he looked more like a "white" Southern European than an indigenous "Indian." Nevertheless, his remarks suggest the ways in which he experiences the role of race, ethnicity, and differential levels of national development in international science. He thus points to the existence of credibility economies (Shapin 1995) in which factors other than scientific merit are at play, affecting whose voices are heard and respected in international science.

In interviews and personal conversations with me, a number of Brazilian climate scientists expressed similar feelings of tokenism, prejudice, and bias in international science:

> I think that there is a very big North American bias. I think these international programs are extremely elitist, you know, English-speaking scientists getting together. But now, there is this awareness that you have to include the developing world. You can't make a global network of observations if you don't have Brazil chipping in money, Africa chipping in money, or whatever; you have to have these countries participate. But I feel that the participation is at a minimum. You're a token scientist. If you speak English, great, but....

This speaker said that she often has the impression that her input as a participant in international forums is not taken seriously, making her feel like a token rather than a true participant. She also commented about international scientific projects with which she had been involved: "You might have science talk, but it is totally political." Such statements contrast with Merton's (1972: 270) account, according to which only knowledge that is international, impersonal, and "virtually anonymous" is sanctioned as true.

Science as Situated Knowledge: Distrust along a North–South Axis

Interviews revealed that Brazilian actors participating in the climate regime do not, as a rule, presume Northern science to be universal and disinterested. Policy makers largely responsible for formulating the official Brazilian position on climate change in international negotiations appeared particularly opposed to such a universalizing and depoliticized understanding of science. They suspect "foreign science" of advancing foreign interests. The existence of geopolitically structured suspicions suggests that norms and values are shared only to a limited extent. Intersubjectivity is strongly circumscribed, at best, in a context marked by such suspicions.

The discourses of Brazilian climate scientists reflect some distrust of international scientific institutions such as the IPCC, which they describe as dominated by Northern framings of the problems and therefore biased against the interpretations and interests of the South. The distrust is greater among politicians, but scientists without official political roles in the domestic context express similar views. For instance, a leading Brazilian climate scientist, who received his Ph.D. from a top US university, argued that scientific arguments benefiting only poor nations probably would not be influential within the United States nor, by extension, within the IPCC, which is dominated by US scientists. "Say, if it could be proven that only tropical forests remove atmospheric carbon dioxide while Boreal forests are a source of CO2, then I think the US would cut funding for carbon cycle studies of that source," he said.

At the time of my first round of field work in Brazil (April 1999), the IPCC was working on a report on the environmental impact of aviation and possible policy responses. A policy maker and two university-based scientists independently brought up this example as reflecting a Northern bias that they portrayed as prevalent within the IPCC. They themselves did not see aviation as a national problem, partly because of other more pressing socio-economic and environmental needs, partly because the average Brazilian flies little compared to the average Northern citizen. To the extent that aviation ought to be subjected to additional environmental regulation, they therefore supported more stringent measures than those adopted by the IPCC. Several interviewees suggested that the IPCC's weak response reflected Northerners' unwillingness to sacrifice the luxury of frequent flying.

After attending another Northern-designed and Northern-dominated international science meeting in May 2002,[14] one of these Brazilian scientists criticized what he perceived as an over-emphasis on sources of "black carbon." Black carbon is soot generated from outdoor fires and household burning of coal and biomass fuels, as well as polluting industrial activities and traffic. Black carbon emissions are particularly large in developing countries where wood, field residue, cow dung and coal are widely used for cooking and heating. At the meeting, the Brazilian scientist observed a large number of overheads used by Northern scientists to demonstrate the inefficiency of cooking stoves used in China and India. He wryly commented to me that "the same scientists go back to the US and drives a SUV just to transport himself" [*sic*], referring to the recent trend in the US towards highly inefficient fossil-fuel Sports Utility Vehicles for personal transport. In this way, this scientist evoked a normative framework that distinguishes between "luxury emissions" and "survival emissions" (Agarwal and Narain 1990).

In short, in many instances, Brazilian scientists and policy makers alike question the impartiality of "international" science and associated framings of responsibility. This distrust extends to international peer review processes as a whole. A policy maker in the MCT group who has participated in the FCCC negotiations, and also in the review of IPCC reports as an energy expert, expressed his preference for the FCCC on the grounds that it, unlike the IPCC, does not cover its politics behind a veil of pretended objectivity:

> POLICY MAKER: I don't like the IPCC. I think it is too biased. Now, I would like to have a lot of new discussion about responsibility and attribution. As it is, all this game is dirty. The science is biased; it reflects their point of view [*voice turns heated*]. In all the working groups, if you have 90 percent of the people coming from the developed countries, it is bound to reflect their perspective. I am not saying that they have prior bad intentions. Both the IPCC and the FCCC are biased by governments.... It is very clear when it comes to the Convention. There are no misunderstandings there, no trying to convince people through the wrong methods, no trying to give certainty to several points of view that are completely biased. The IPCC, they make believe that their point of view is objective.
>
> LAHSEN: So you perceive them as using science disguised as being objective when it is really not?
>
> POLICY MAKER: As a basis for political discussion at the Convention, yes. And using this to help arguments supporting their point of view. That is why we present our proposals directly to the Convention. Because it cannot go through the literature; they will not allow us. If we go through peer review, the peer reviewers will be part of IPCC and they all reject our proposal saying that we are doing junky literature, and so on. They will not approve our work. They will not allow our papers to get published in any scientific magazine. Because you have to go through a peer review process and they will say that this is junky science, that what we are doing is junky science because we are Indians and we do not know what we are talking about. That kind of thing.[15]

Perceptions of bias were also evident in interviews with other persons associated with the MCT group, including the Minister of Science and Technology under Fernando Henrique

Cardoso. This group of policy makers within the Ministry of Science and Technology repeatedly cited a 1995 article in *Scientific American* (Gibbs 1995) in interviews with me. The article's discussion of inequities and prejudices in international science supports their perceptions of widespread Northern prejudice in international peer review processes. Among other things, the article documents how imperfect ability to communicate in English, combined with the fact of having a mailing address in a developing country, tend to undermine third world scientists in peer review processes.[16]

National Divisions

Brazilian climate scientists and policy makers may be united in their tendency to bring an interpretive framework shaped by experiences of bias and shared memories of colonialism to bear on climate science and politics. On other issues, however, they are themselves divided.

Particularly divisive among Brazilian scientists, policy makers and even environmental groups is the issue of deforestation. Brazil's deforestation trend has brought intense criticism on the Brazilian government abroad and at home. Much of the criticism has come from environmental NGOs and scientists. Recently, a coalition of over 25 NGOs in Brazil launched the Climate Observatory, a coalition designed to push the issue of deforestation to the center of climate discussions in Brazil. These groups have looked to climate change as an opportunity to strengthen forest conservation efforts, as deforestation accounts for two thirds of Brazil's national emissions of greenhouse gases.

By contrast, the Brazilian government, including (and sometimes represented by) the MCT group, has sought to minimize attention to deforestation. The Amazon has historically been viewed in Brazil as a resource in the nation's project to modernize and become a superpower and the region is invested with nationalist sentiments, as reflected in the often-evoked slogan by former Brazilian President José Sarney: "The Amazon is ours" (Barbosa 1993: 119; Kuehls 1996:ix; Rohter 2002). Important parts of the Brazilian government and many in the general population suspect foreigners of wanting to lay claim to the Amazon. Strongly opposed to foreign involvement around this issue, factions in the Brazilian government and population frame foreign concern about the deforestation of the Amazon in terms of national sovereignty, and conspiracy theories circulate widely in Brazilian society. Such theories—circulating as fact—posit that Americans students are taught to think of the Amazon region as an "international reserve" administered by the United Nations and that US Army forces are training to seize control of the region (see Rohter 2002).

In interviews, members of the MCT group dismissed such "black helicopter" conspiracy theories. However, the MCT group is keenly aware of the play of geopolitics in science. Given the role of both foreign and domestic scientists in pressuring the government on the issue of forest conservation, the MCT group has been circumspect in their use of scientists' estimates concerning carbon emissions resulting from land-use changes in Brazil. They portray national scientists as variously naïve about how international science perpetuates Northern interests and as strategically playing along with foreign interests in order to obtain research funds and to pressure the Brazilian government. As one person in the MCT group related to me, "Normally [both foreign and Brazilian] scientists are very biased in terms of policy; they use [science] to blame Brazil for misusing the Amazon forest."

To meet its commitment under the FCCC in the form of national inventories of its levels and sources of greenhouse gases, those within the MCT responsible for the effort sought to produce their own independent estimates of Brazilian emissions from deforestation. They resisted using estimates circulating in international science and assessment processes, referring to the latter as extrapolations made by scientists on the basis of limited, available data. They believe that both foreign and national scientists exaggerate the estimates of carbon emissions from deforestation.

The MCT group thus avoided use of estimates emanating from both the IPCC and the Large-Scale Biosphere Atmosphere (LBA) experiment, an international science project seeking to understand how the Amazon functions with respect to natural cycles of water, energy, carbon, trace-gases and nutrients, and how these processes may change due to future land-use and climate changes. In interviews, the MCT representative responsible for coordinating the inventories dismissed estimates emanating from the LBA as unreliable. The same person referred disapprovingly to the IPCC special report on land-use change, pointing to a subset of IPCC lead authors and contributing authors whose estimates of greenhouse gases he considered unreliable. This MCT interviewee considered Brazilian scientists who partook in that report similarly "political," since they had apparently accepted the foreign scientists' views—and in some cases even co-authored articles on the subject with some of these scientists.

Such lack of confidence in national scientists was also reflected in the fact that the Brazilian government under the Cardoso administration did not commonly ask non-MCT Brazilian scientists and IPCC participants—leading scientific experts on the issues—to partake in its review and critique of IPCC reports and summaries for policy makers. By contrast, other countries (e.g., the United States) include a broad base of national—and IPCC participating—scientists in the review process. It appears that the MCT suspects the national scientists in question to be too closely connected to the IPCC, and to international science generally, to perform "independent" review of the IPCC reports.

Faced with this attitude on the part of the MCT, the excluded Brazilian scientists criticize the MCT group for failing to recognize that they (the scientists) can be part of international science teams while remaining "critically independent," as one scientist put it. Some point out that they have influenced foreign and national scientists to lower their estimates of deforestation and to accept the MCT group's estimates of deforestation rates. As we see below, however, several Brazilians did recognize that participating in international science likely shapes their world views.

Personal Divisions—Self-Doubts

With Emanuel Adler, Haas has claimed to "regard learning as a process that has to do more with politics than with science, turning the study of politics process into a question about who learns what, when, to whose benefit, and why" (Adler and Haas 1992: 370). This rare and brief acknowledgment aside, however, Adler and Haas do not sufficiently recognize the influence of power and identity formation in shaping an epistemic community's shared normative and cognitive frameworks. Several Brazilian scientists revealed perceptions of such a knowledge/politics/power nexus, even to the point of revealing doubts about the validity and independence of their own scientific and normative assumptions. A Brazilian scientist's foreign education and continued professional connections even led him to question his own ability to represent the national interest:

> You know, you can't escape the international invasion of science .... Its definition of future research agendas is bound to trickle down. And if you want to do state-of-the-art-stuff, you want to get involved with that—but it changes things. I am a biased subject because my education has been only in the US. So no matter how nationalist I am, I don't know .... Do you understand what I mean? There are a lot of interests that I try to defend in Brazil. But maybe my education doesn't allow me to see that [i.e., the Brazilian] side as well as I should.

It is significant that he retains such a critical perspective despite his US training and strong transnational personal and professional networks. When I asked the above scientist whether he really believes what he asserts, he responded with an emphatic "Yes," noting that national reactions to a new president of Brazil's central bank had prompted him to think about it. That president was educated in an Ivy League school and had worked for many years for the international investor George Soros. This scientist had heard some fellow Brazilians say that Brazil had thereby "given them over to the International Monetary Fund, that they just put this guy there who caters to the Americans." This led him to reflect:

> ... I was thinking about that—that we, scientists, are the same. We [Brazilian scientists educated in the US] get a vision of science in the US and we don't know how to do it any differently. Maybe my criteria are just

> American and I try to apply them to a Brazilian reality; there are these things, like the-way-you-do science paradigm and what you choose to focus on.

He then conjectured that maybe if I interviewed Brazilian scientists who were educated in Brazil, I would find them to have "a different view of how things should be conducted, or about global warming or the IPCC." After some thought, he concluded that Brazilian scientists who have not been educated abroad and who are not part of international science tend to study problems that are more local in nature.

The central member of the MCT group quoted repeatedly above was himself described by a different scientist and political appointee in the Brazilian government as shaped by Northern efforts to influence Brazilian affairs:

> Governments work on bureaucracies. If you don't have some bureaucrats in place, it just doesn't work; they create a momentum. The United States, they are bad but not dumb. They gave money for the national inventories. So [the persons in the Brazilian government responsible for producing these reports], who didn't have much money from the Brazilian government, got one or two million dollars from the US government to do an inventory of emissions. [The person in charge of it] was delighted, you know. It's a career for him, you know!

In a globalizing world, complex subjectivities and the role of contexts in forming them render essentialism difficult; nationality cannot necessarily be equated with ability to represent a national point of view.

Conclusion

Much like the geopolitical constructs of "North" and "South," the epistemic community concept is useful. However, like the former, it must be used with care, with sustained attention to the important divisions they fail to capture and threaten to obscure. Rather than a single, cohesive, transnational "community" united by a shared professional ethos and by scientific and policy-related concerns, this study reveals a complex domain characterized by transnational networks and cognitive convergence, but also important differences. Persistent lines of division exist—sometimes under surface appearances of shared frameworks of understanding and policy action—testimony to the precarious and unstable nature of any apparent consensus on global environmental problems and the associated knowledge (Brosius 1999; Escobar 1999). Divisions continually replicate themselves as one looks from the transnational to the national and even the individual level. On second glance, not only does the transnational epistemic community appear internally fractured along geopolitical lines; important fractures also reveal themselves at the national level and even within the subjectivities of individual scientists, at the most intimate level of personal commitments and understandings of self and the world.

Even so, patterns appear in the fractures, conditioned by history and socio-economic realities. The scientists identify as Brazilians, as citizens of "the South" and of a formerly colonized country. They are transnational locals. They are more local than transnational, perhaps, in that their discourses consistently evoke the "North–South divide," even as they know themselves to have been intimately shaped by their foreign (overwhelmingly US) educations, work experiences, and associated personal relationships. The recurrence of the North–South construct in Brazilian scientists' discourses bears testimony to the reality glossed over by epistemic community theorists, namely, the extent to which climate science and geopolitical conditions co-produce each other. Privy to this insight, and less advantaged by the associated politics, Brazilian scientists and policy makers see the shadow side of universalizing discourses about science. Matters such as race, ethnicity, and level of development shape their understandings and experiences of international science, and they perceive prevailing international climate-related knowledge as typically reflecting developed ("Northern") countries' assumptions, agendas, and interests. Such perceptions, and the differences they reveal, add complexity to the dominant, purified images of science, representations that overlook important questions about the intersection of science, culture, power, and politics.

NOTES

1. See, e.g., Agarwal and Narain 1990; Escobar 1995a; Goldman 2004; Gupta 1998; Jasanoff 1996a; Jasanoff and Wynne 1998; Sachs 1993; Tsing 2000; Yearley 1996.
2. See Jasanoff and Martello 2004.
3. About 70 interviews were conducted with scientists, science administrators, and policy makers in institutions in São Paulo, Rio de Janeiro, and Brasília. A first round of interviews took place in April 1999. The same persons—along with an additional set of new actors—were interviewed in 2002 over a period of eight months. The interviews were complemented by many informal discussions and participant observation when possible. The group of Brazilian policy makers and scientists discussed in this paper is quite small. Anonymity has been preserved to the extent possible, and sources are generally not cited. The gender of interviewees has occasionally been altered, to conceal the identify of Brazilian female climate scientists, of whom there are just a few.
4. At the time of writing, Luiz Inácio Lula da Silva of the Workers' Party has been president for 7 months. Though the Lula government may provoke changes in the governmental structures and policies with a bearing on Brazil's position on climate change, no significant changes have happened at those levels yet.
5. See below. See also Yearley 1996 (especially p. 103) and Lahsen 2001.
6. I had in fact only expressed scholarly interest in understanding Brazilian engagements and concerns related to climate change.
7. This scientist did say that this is changing, however, that younger Brazilians do learn about climate change and ozone depletion and such things now in their text books, beginning in the fifth grade or so (that is, around the age of 11). So she attaches hope to the younger generations.
8. My field work identified members of the latter group who were not similarly transnationally connected nor regular participants in international science forums. They complained about not being able to break into these arenas and associated networks due to protective measures on the part of their Brazilian colleagues serving to retain the network and associated benefits to themselves, to be shared only with a select few students and colleagues.
9. Environmental groups are also widely credited with effecting this change (Torres 1997).
10. The plan is to gradually obtain an increasing level of financial support from developing countries themselves.
11. For instance, the IPCC's decision to include land-use driven changes in carbon stored in above-ground biomass as an anthropogenic source but to exclude similar changes in below ground biomass (soil carbon) as natural source was made without input from LDCs and unfavorable to them. The same was the case with the decision to include methane from rice agriculture and cattle and not from deer and other ruminants which exist in the North (the rationale being that the latter are "wildlife," despite the fact that they are managed populations) and with the lack of distinction between subsistence and luxury emissions and the prevalent method of calculating emissions in terms of national emissions rather than per capita emissions (Miller 1998).
12. I base this statement on years of field work among US climate modelers during which I came to know this simple US model and the associated criticisms.
13. This applies to scientists partaking in Working Groups II and III of the IPCC (involved with impact assessment and mitigation and adaptation measures) rather than in Working Group I, the group devoted to the natural science.
14. The meeting was headed by US scientists and involved five key IPCC members, one of whom was an advisor to President George W. Bush. The meeting involved roughly 80 scientists, roughly 70 of them from "the North."
15. All this is expressed with raised voice, indicating frustration and strong feelings on the subject.
16. Particularly revealing of the kind of bias was the comment by an editor for *Science*. When asked about his editorial practices, this editor noted that articles with imperfect English submitted for publication weren't likely to be published: if they make mistakes in English one would suspect that they are similarly sloppy in their science, he said.

REFERENCES

Adler, Emanuel. 1987. *The Power of Ideology: The Quest for Technological Autonomy in Argentina and Brazil*. University of California Press.

Agarwal, A., and S. Narain. 1990. *Global Warming in an Unequal World: A Case of Environmental Colonialism*. Centre for Science and Development, New Delhi.

Barbosa, Luiz C. 1993. "The 'Greening' of the Ecopolitics of the World-System: Amazonia and Changes in the Ecopolitics of Brazil." *Journal of Political and Military Sociology* 21, no. 1 : 107–334.

Biermann, Frank. 2000. Science as Power in International Environmental Negotiations: Global Environmental Assessments Between North and South. Discussion paper, Environment and Natural Resources Program, Belfer Center for Science and International Affairs, John F. Kennedy School of Government, Harvard University.

Boehmer-Christiansen, Sonja. 1994a. "Global Climate Protection Policy: Part 1." *Global Environmental Change* 4, no. 2: 140–159.

Boehmer-Christiansen, Sonja. 1994b. "Global Climate Protection Policy: Part 2." *Global Environmental Change* 4, no. 3: 185–200.

Brosius, J. Peter. 1999. "Analyses and Interventions: Anthropological Engagements with Environmentalism." *Current Anthropology* 40, no. 3, June: 277–309.

Caldwell, Lynton Keith. 1996. *International Environmental Policy: From the Twentieth to the Twenty-First Century*. Duke University Press.

Enzensberger, Hans-Magnus. 1974. "A Critique of Political Ecology." *New Left Review* 84: 3–31.

Escobar, Arturo. 1995. *Encountering Development: The Making and Unmaking of the Third World*. Princeton University Press.

Escobar, Arturo. 1999. "Comments." *Current Anthropology* 40, no. 3: 291–293.

Fearnside, Philip M. 2001. "Saving Tropical Forests as a Global Warming Countermeasure: An Issue that Divides the Environmental Movement." *Ecological Economics* 39, no. 2, November: 167–184.

Gibbs, W. Wayt. 1995. "Lost Science in the Third World." *Scientific American*, August: 92–99.

Gieryn, Thomas. 1983. "Boundary-Work and the Demarcation of Science from Non-Science: Strains and Interests in Professional Ideologies of Scientists." *American Sociology Review* 48: 781–795.

Gieryn, Thomas. 1995. "Boundaries of Science." In *Handbook of Science and Technology Studies*, ed. S. Jasanoff et al. Sage.

Goldman, Michael. 2004. "Imperial Science, Imperial Nature: Environmental Knowledge for the World (Bank)." In *Earthly Politics: Local and Global in Environmental Governance*, ed. Sheila Jasanoff and Marybeth Long Martello. Cambridge, MA: MIT Press.

Gupta, Akhil. 1998. *Postcolonial Developments: Agriculture in the Making of Modern India*. Duke University Press.

Guston, David H. 1999. "Stabilizing the Boundary between US Politics and Science: The Role of the Office of Technology Transfer as a Boundary Organization." *Social Studies of Science* 29, no. 1: 1–25.

Haas, Peter M. 1992. "Introduction: Epistemic Communities and International Policy Coordination." *International Organization* 46, no. 1: 1–35.

Jasanoff, Sheila. 1987. "Contested Boundaries in Policy-Relevant Science." *Social Studies of Science* 17: 195–230.

Jasanoff, Sheila. 1992. "Review of Ezrahi." *American Political Science Review* 86: 233–234.

Jasanoff, Sheila. 1996. "Science and Norms in Global Environmental Regimes." In *Earthly Goods*, ed. F. Hampson and J. Reppy. Cornell University Press.

Jasanoff, Sheila and Marybeth Long Martello. 2004. "Introduction." In *Earthly Politics: Local and Global in Environmental Governance*. Cambridge, MA: MIT Press.

Jasanoff, Sheila, and Brian Wynne. 1998. "Science and Decisionmaking." In *Human Choice and Climate Change* Volume One, ed. S. Rayner and E. Malone. Batelle.

Kandlikar, Milind, and Ambuj Sagar. 1999. "Climate Change Research and Analysis in India: An Integrated Assessment of a South–North Divide." *Global Environmental Change* 9: 119–138.

Lahsen, Myanna. 1998a. "The Detection and Attribution of Conspiracies: The Controversy over Chapter 8." In *Paranoia within Reason*, ed. G. Marcus. University of Chicago Press.

Lahsen, Myanna. 1998b. Climate Rhetoric: Constructions of Climate Science in the Age of Environmentalism. Ph.D. thesis, Rice University.

Lahsen, Myanna. 2001. Brazilian Epistemers' Multiple Epistemes: An Exploration of Shared Meaning, Diverse Identities, and Geopolitics, in Global Change Science. Discussion paper, Environment and Natural Resources Program, Belfer Center for Science and International Affairs, John F. Kennedy School of Government, Harvard University.

Merton, Robert K. 1973 (1942). "The Normative Structure of Science." In *The Sociology of Sciences*, ed. N. Storer. University of Chicago Press.

Miller, Clark. 1998. Extending Assessment Communities to Developing Countries. Discussion paper, Environment and Natural Resources Program, Belfer Center for Science and International Affairs, John F. Kennedy School of Government, Harvard University.

Mwandosya, Mark J. 2000. *Survival Emissions: A Perspective from the South on Global Climate Change Negotiations*. DUP and The Center for Energy, Environment, Science and Technology.

Reich, Robert B. 1992. *The Work of Nations*. Vintage Books.

Sachs, Wolfgang, ed. 1993. *Global Ecology: A New Arena of Political Conflict*. Zed Books.

Schoijet, Mauricio, and Richard Worthington. 1993. "Globalization of Science and Repression of Scientists in Mexico." *Science, Technology, and Human Values* 18, no. 2: 209–230.

Shackley, Simon, Peter Young, Stuart Parkinson, and Brian Wynne. 1998. "Uncertainty, Complexity and Concepts of Good Science in Climate Change Modelling: Are GCMs the Best Tools?" *Climatic Change* 38: 159–205.

Shapin, Stephen. 1995. "Cordelia's Love: Credibility and the Social Studies of Science." *Perspectives on Science* 3: 255–275.

Torres, Blanca. 1997. "Transnational Environmental NGOs: Linkages and Impact on Policy," in *Latin American Environmental Policy in International Perspective*, ed. G. MacDonald et al. Westview.

Tsing, Anna. 2000. "The Global Situation." *Cultural Anthropology* 15, no. 3: 327–360.

Yearley, Steven. 1996. *Sociology, Environmentalism, Globalization*. Sage.

22

Channeling Globality

The 1997–98 El Niño Climate Event in Peru

Kenneth Broad and Ben Orlove

Among natural phenomena, the 1997–98 El Niño event received an extraordinary amount of print and electronic media attention around the world, surpassed only recently by the coverage of the Indian Ocean tsunami in 2004 and Hurricane Katrina in 2005. In part, this interest was attributable to the planetary scale of the event. The shifts in the oceans and atmosphere affected every continent, leading to increased rains in some areas and droughts in others and, consequently, to floods, epidemics, landslides, forest fires, and crop failures. Equally impressive was the proliferation of the phrase "El Niño," which traveled around the globe accompanied by dramatic imagery and diverse political responses. The social and cultural processes that interacted on multiple scales as the 1997–98 environmental event unfolded allow interpretation, from a range of theoretical perspectives, of how global phenomena acquire meaning for different actors.

To examine this large-scale event and to gain the insights it can provide into globalized environmental incidents more generally, we trace its trajectory in a single country, Peru. By considering one country, we can see more clearly the connections between actors who were engaged with the event. El Niño events have a particularly strong impact in Peru, and Peruvians have been aware of them since the late 19th century, longer than people in most other countries, so this case offers some specific and attractive features to consider.

In our account of the El Niño event in Peru, we note elements that originate from outside the spatial and social boundaries of Peru—currents of warm water that cross the Pacific Ocean, Internet images, scientific models of ocean and atmospheric circulation, foreign experts, funds from international agencies—and link up with other elements that come from within Peru—the national government, national and regional NGOs, media (newspapers, magazines, and television channels), and forms of humor identified with the country. The scale and nature of these connections bring to mind the word *global.*

Many writers describe globalization as a phase that the world has entered, marked by the unimpeded extensive movement of people, objects, ideas, images, and capital; they suggest that elements move, or flow, under their own power. One famous formulation

The Anthropology of Climate Change: An Historical Reader, First Edition. Edited by Michael R. Dove.

(Appadurai 1996) describes the scapes, the fluid, amorphous, interconnected spaces of the contemporary world, that have replaced an earlier, more structured hierarchy of centers and peripheries. This notion of flow has informed studies of many widely varying topics, including migration, media, social movements, and religion (Brenner 1999; Hannerz 2003).

The distinctiveness of the strand of work on which we draw most is suggested by the contrast between different ways of talking about the nature of movement associated with a globalized era. Anna L. Tsing (2005) uses the word *friction* to contrast with the earlier term *flow*, which suggests an ease and spontaneity of movement and a lack of interruption. The term *friction* condenses in a single word four different attributes of connection. Friction enables motion (as Tsing points out, wheels can turn and move because they grip the road; without friction, they only spin on their axles). Friction also directs motion: The friction on a surface can cause an object to move straight ahead or to turn. Friction slows objects down, so that something can move only if it is given additional force or impetus. And the heat created by friction serves as an example of the consequences—often unanticipated—of movement. In Tsing's view, globality is not a preexisting scale, given in the current international order. She writes, "Scale must be brought into being: proposed, practiced and evaded, as well as taken for granted" (Tsing 2005: 58). This active creation of globality is particularly striking in the undertakings that Tsing terms "global projects." She provides a detailed example of a corporation created by Canadian entrepreneurs, who sought to use science, publicity, and discourses of global development to direct international capital to a gold mine in Indonesia. She carefully analyzes the frictions that supported the enterprise and also led to its demise—the linkages between different entrepreneurs and political actors, the use of imagery of untapped nature, the availability of certain forms of labor in specific regions. Her view is quite different from one that presents such enterprises as characteristic of an era of globalization in which capital, expertise, and labor flow to a mining site, balanced by a counterflow of commodities.

Aihwa Ong and Stephen J. Collier (2005) also find an alternative to the discussion of flow with their discussion of "global assemblages." This term stresses the active construction of links between diverse elements on a variety of scales. The globality of the assemblages often derives not only from the long distances that separate the elements they contain but also from the placelessness and universality of the claims to the commonality of these elements. They are linked within discursive frameworks that are worldwide: scientific systems, sets of humanist values and ethics, and neoliberal structures of exchange and governance. Ong and Collier present two clear, if distinct, examples of global assemblages: stem-cell research and the trade in donated human organs. These assemblages require technoscientific expertise and international capital to come into being; they also rest on ethical principles about human needs and human rights that make the specific treatments of human tissue comprehensible and valued and on the systems of regulation that create uneven and complementary distribution of these activities in different parts of the world. The unevenness of elements in an assemblage and the tensions between these elements, like the friction in Tsing's discussion, contrast with a view of ready flow around the globe.

We draw from these discussions the idea of globality as a worldwide or universal scale of connections between disparate elements, and we use the word *channeling* to underscore the way that these connections are actively constructed. Like the word *friction*, *channeling* contrasts with *flow*. Even more than *friction*, it suggests an intentional directing of distant or new elements along certain paths. The term seems well suited to our case, granted the active seeking of linkages to globality on the part of many individuals and organizations in Peru. The term also conveys that Peruvian culture and media attend selectively to certain international images and ideas, only to rework them in distinctively Peruvian fashion, while finding other international images and ideas less engaging. Moreover, some efforts to channel globality succeed in attracting considerable public attention, whereas others fail; we find that certain features of Peruvian public attentiveness help account for this difference. In this instance, attentiveness was fueled by the political *telenovela* (soap opera) starring then president Alberto Fujimori, whose theme revolved around El Niño and whose actions involved intrigue, humor, and manipulation.

El Niño

A brief explanation of the El Niño phenomenon illustrates its global scale. El Niño is a combined oceanic–atmospheric phenomenon. Every two to ten years, the body of warm water normally located in the western Pacific Ocean shifts eastward toward South America. This movement alters the circulation of winds and rainfall patterns around the globe (Glantz 2001). El Niño events vary greatly in their intensity and duration.

Multinational research programs since the 1950s have increased observation and collection of environmental data. This research has led to the establishment of the World Climate Program, run by two branches of the United Nations, the UN Environment Program and the World Meteorological Organization. The World Climate Program organized international programs in the 1980s and 1990s to understand El Niño events. Much of this research resonated with growing public concern about ozone thinning and global warming, problems often perceived by the public to be causally linked (Kempton et al. 1996). Scientists now can predict the onset of an El Niño event and its likely regional impacts several months in advance. This scientific advance has led to major international efforts to disseminate climate forecasts. Climate-forecast institutions link their efforts to those of development agencies. These activities have created what can be called, in Ong and Collier's parlance, a "global assemblage" that connects El Niño researchers, program administrators, applications specialists, and many social groups with each other and with elements of universal discourses of technoscience and development. Through its treatment of information as a resource to be sold and invested, this assemblage connects with the universal discourse of neoliberalism as well. Nonetheless, we note that, despite the efforts of international organizations to study the climate and share weather data, national meteorological services of individual countries retain control over local data, predictions, and dissemination of information.

Although research and prediction efforts occur among a worldwide network of organizations, impacts manifest themselves on a local scale. In Peru, strong El Niño events bring torrential rains and flooding to the normally arid north coast. Ordinarily cool coastal waters turn warmer, sparking mass migration and die-offs of commercially important fish species. Elevated air temperatures and increased moisture create health problems throughout the country. Strong events lead to economic downturns, as agriculture and fisheries are severely affected and roads, bridges, and ports are damaged.

The 1997–98 El Niño was one of the most powerful of the century and the one most extensively forecast as well. The first official announcement that an event was developing was made by a Peruvian governmental agency in June 1997, following earlier rumors. When air and water temperatures in coastal Peru in 1997 did not drop as usual during winter months (June, July, and August), the country as a whole realized that an El Niño was likely. Satellite images and speculative articles appeared in several media, emergency meetings were held, and government committees were created to develop action plans. Debate was intensified by public disagreement among scientific agencies over the potential magnitude of the event, and rumors about the event became a major focus of conversation. In September 1997, President Fujimori presented a mitigation plan to the congress that focused on massive public-works projects; although the opposition strongly disagreed with this plan, its implementation was assured by Fujimori's successful use of global imagery and his authoritarian control of congress and other key institutions. An international meeting held in Lima in October 1997 led to more debate over the nature of the event, but it did not undercut Fujimori's position.

Heavy rains fell in late December and washed out roads and bridges, damaged many buildings, and created standing water that led to disease outbreaks. These rains continued through April 1998. In May, reconstruction efforts began, largely funded by loans from international agencies. This reconstruction coincided with political scandals, which later led to Fujimori's downfall (Zapata and Broad 2001). Despite these problems, the proactive response, facilitated by advances in forecasting, averted suffering of the magnitude of the 1982–83 event, when many hundreds of people died and thousands were permanently displaced.

Prologue

It is worth emphasizing that the El Niño event unfolded in the context of recent Peruvian political instability, characterized by radical shifts from the military government that took power in 1968 to the return of elections dominated by older political parties and the emergence of new ones in the 1980s. These shifts resulted in increases in local political autonomy, particularly evident in the election of municipal officials who formerly were appointed. The 1980s are commonly referred to throughout Latin America as "the lost decade." In Peru they saw deep economic recession and hyperinflation; they also saw great violence associated with the extremism of the Shining Path and the military counterterrorism programs. Such instability set the stage for the rise of a political outsider, Fujimori, an agronomist of Peruvian–Japanese descent and rector of a major public university. Given the nickname "El Chino" (a familiar and somewhat pejorative term perhaps best translated as "the Chinaman") because of his Asian origins, Fujimori presented himself as a technocrat who could bring global prosperity to Peru. He promoted the 1993 constitution that strengthened the executive, weakened the legislature, and allowed him to run for a second term. He won by a large margin in 1995 because of the partial economic recovery, his popular public works and relief programs, and his success in defeating the Shining Path. In 1996, he set up the Ministry of the Presidency (MIPRE), a kind of executive branch within the cabinet, which came to control over one-fifth of the national budget, with little oversight from congress (Zapata and Suiero 1999: 61). With congress weakened, much of the opposition to Fujimori was concentrated in municipal governments and in NGOs, which often worked in concert (Markowitz 2001).

March–May 1997: Early rumors

The first Peruvians to become aware that an El Niño event might be underway were the fishers on the north coast, who noticed that the ocean waters were warmer than usual and that tropical species, ordinarily found only in the warmer waters off Ecuador, were appearing in large numbers. Stories of these unusual phenomena spread quickly through the coastal fishing towns.

The first images of the El Niño event to reach Peru arrived in April 1997, when staff in Peruvian scientific agencies accessed the Internet website of the U.S. National Oceanographic and Atmospheric Agency (NOAA). This site presented maps that showed sea surface temperatures characteristic of such events. We note that the Peruvian scientists actively sought out these images on their own, rather than receiving them passively as e-mail attachments, as bulletins or alerts, or in other forms. By May, newspapers carried stories of the unusually warm weather and sea temperatures and suggested that an El Niño event might soon begin.

June–August 1997: The search for a plan

The public awareness of the impending event increased sharply in early June, when *El Comercio*, the most cosmopolitan of Peruvian newspapers (and sometimes identified with the opposition), published an image of the pool of warm water that was moving eastward across the Pacific (see Figure 22.1). This newspaper has a narrow, although influential, readership among wealthy and educated urban residents. The striking image was seen by many people who did not purchase the paper, because it was prominently displayed in newspaper stands throughout the country. This image, downloaded from the NOAA website, was edited to make the threat of the event more palpable, as indicated by its caption, which can be translated as "Strong Blow"

Golpe contundente

Este es el panorama definitivo hecho ayer por el Servicio de Meteorología e Hidrología para mostrar la llegada de 'El Niño'. Las temperaturas del mar ya están 4 grados por encima de lo normal en la costa norte y desde el sur se avecinan corrientes elevadas.

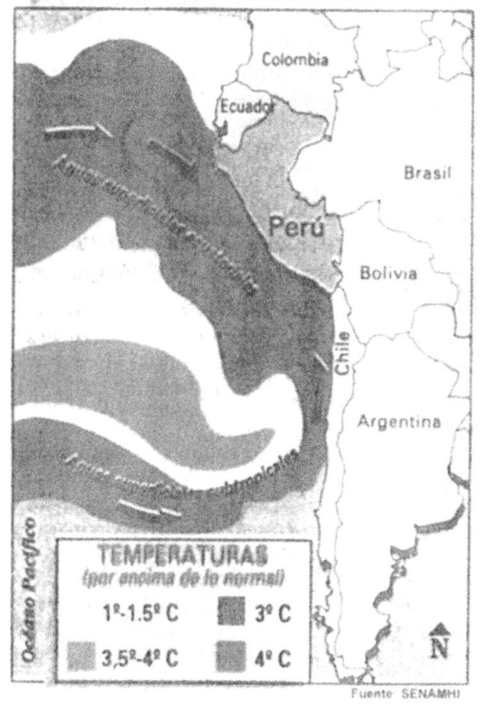

Figure 22.1 *Map on the front page of a major Lima newspaper, June 1997.*

or "Devastating Impact." A broad range of tones—dark blue, light blue, orange, and red—was used to indicate the different temperatures of water. Although this image presented some details of the ocean conditions differently from the original NOAA image—the temperature ranges for the bands of warm water that were indicated in the legend included discontinuities that are physically impossible—the use of color succeeded in underscoring a key point, the unusual heat of the warmest water. Arrows, absent in the website image, were added to give a narrative dimension to the newspaper image. Above all, the visual power of the image derived from its vivid depiction of a red blob about to hit Peru, whose national territory was shaded in a different color to make it more noticeable within South America. Building on the attraction of this and other related images, *El Comercio* developed a section later in the year that was dedicated to El Niño and included scientific and personal-interest coverage.

The wide appeal of this image constitutes the first example of channeling globality that we discuss. It is worth noting that the image did not simply flow from some cosmopolitan center to Peru but, rather, that specific actors—members of the staff of a newspaper—located the image, brought it to Peru, and modified it before circulating it. (This transformation of a global image into a more comprehensible form is broadly similar to the "translation" of universalistic discourses into local terms that Timothy K. Choy [2005] discusses in his account of environmental politics in Hong Kong.) The national scale of the public for which they reworked it was shown by the circulation of the newspaper in major towns throughout the country, by crediting the map to a government agency (SENAMHI, the national weather and hydrology service), and by the emphasis on Peru on the map.

Of equal importance to the production of the image was its reception. The urgency of the situation and novelty of this sort of map accounted for some of its appeal, as did the authority it derived from its links to international scientific agencies, mentioned in the article that accompanied it. We note that the image did not simply present itself to passive readers but, rather, quickly elicited the great attentiveness of Peruvians to news and to other sources of information and evaluation of public life. As Michael Warner points out in his discussion of publics, individuals can belong to a social class or a nation because of their position in a social structure or system of classification "whether they are awake or asleep, sober or drunk, sane or deranged, alert or comatose" (2002: 61), but they are part of a public only when they pay at least some attention to its discourse. Warner talks broadly about the reflexive circulation of discourse, the temporality of this circulation, and the possibility of alternative, oppositional, or counterpublics (cf. Hayden 2003). Our interest here is to note, in the context of discussions of globality, that this attentiveness is an important element to the popularity of the image. These points—that attentiveness is variable and that some images are widely taken up whereas others are not—are congruent with Tsing's discussion of friction and show its difference from earlier conceptions of flow. We return to some specific features of this attentiveness.

Also in the month of June, three government scientific agencies, one connected with meteorology and hydrology, a second with oceanography and fisheries management, and a third with geosciences and seismology, gave contradictory forecasts about the possibility of an El Niño (Broad et al. 2002). These agencies faced two opposed influences: The information from NOAA and other sources suggested that an El Niño event was impending, but political forces pressed the government agencies to avoid announcing environmental problems that could lead national and international banks to cut credit. Moreover, the agencies used different sorts of analyses, reflecting their specific areas of expertise. The leaders in each of the agencies also felt a concern to distinguish themselves from the other two. This complex situation led their forecasts to vary, adding to the public concern to anticipate the coming event.

As warming continued in June and July, news of the event rapidly spread. A second example of channeling globality occurred around this time, in the rumors that circulated widely in many sectors of society. We heard them in small fishing villages on the desert coast and in cafes in exclusive neighborhoods in Lima, in provincial government offices and in urban fish markets, and in other settings as well. Some stories explained that the anomalously warm waters nearing the Peruvian coast were a result of French or Chinese nuclear bomb testing in the Pacific and that powerful groups had succeeded in suppressing this information, and others attributed the warm waters to an unspecified intervention by the United States and Europe to keep Peru underdeveloped. Other stories mentioned that Fujimori had access to private sources of information in Japan that he used to track the event, or that he had a unique access to Internet information through his special computer experts, popularly called "cibernautas": The word combines the roots *ciber-* (the Spanish equivalent of the English *cyber-*) and *-nauta*, corresponding to the *-naut* in *astronaut* and *aeronautics*. These rumors illustrate one kind of friction, in this case an ambivalence on the part of many Peruvians toward developed countries that includes admiration, or envy, of their greater power and fear that they will exploit poor countries.

Rumors are a near-constant feature of conversation in Peru as in so many other countries. At this time, some rumors centered on a variety of topics set in Peru or neighboring countries, such as corruption in congress, secret arms deals between the Peruvian military and left-wing Revolutionary Armed Forces of Colombia (FARC) guerrillas, and the whereabouts of Abimael Guzmán, the then still-uncaptured leader of the defeated Shining Path movement. In contrast, the rumors of the El Niño event all located Peru in a global context and presented the country as the passive victim of more powerful foreign countries. It was the Peruvian public itself that channeled globality in creating and circulating these stories that made sense of the new forecasts.

A sign of how rapidly the event became part of everyday reality can be observed in an advertisement produced in July by an upscale department store in Lima, Saga Falabella. Reproduced in Figure 22.2 are the two sides of the advertisement, printed in color on heavy stock and delivered by hand to residences in prosperous neighborhoods in Lima. On one side, the reader could complete the phrase "El niño nos ca ..." with the syllable *gó*, "El niño nos cagó" [The child shat on us]. This phrase is inescapable given the image of the chamberpot in the advertisement. The reader who turns over the card, however, finds that the other side contains a different ending: "El niño nos cambió la moda" [The El Niño event changed fashions for us]. The humor lies not only in the near mention of the verb *to shit* but also in the link between the two sides of the advertisement; they evoke the theme of *engaño*, or trickery, because the advertisement tricks the viewer into completing the word incorrectly. The warmer temperatures associated with the El Niño event meant that clothing retailers in Lima were unable to sell their inventory of winter clothing,

Figure 22.2 *Advertisement distributed in Lima, July 1997.*

so they reduced prices greatly. The advertisement encouraged the reader to take advantage of this temporary situation and stock up on cheap winter clothes.

This humor represents an interesting counterpoint to the rumors. It shows that the attention that Warner discusses derives from more than an instrumental concern to follow events of political significance—the sort of interest that leads people to look closely at a map in a newspaper that depicts impending disaster. More broadly, this attention comes from a wish to keep up with the changing elements in public discourse. In Peru, as in many other countries, people like to tell rumors and jokes to others who have not heard them before, and people like to hear new rumors and jokes. Each telling has an intrinsic value (of information, titillation, and humor) and, an additional value of timeliness and alertness.

To gain insight into different forms of such public attention, an excellent starting point is the work of the art historian Michael Fried, in particular, his 1980 book about French painting in the middle of the 18th century, a period of interest not only because it led up to the French Revolution but also because painting itself was widely viewed and debated at that time, especially at the Parisian salons. These salons were annual events that lasted several weeks; men and women of divergent social backgrounds gathered in a large building to view and to discuss the hundreds of paintings that were displayed (Crow 1985). Fried describes two different modes of attention in this period, noting these types both in the people depicted in the paintings and in the beholders of the paintings. He terms the first "absorption," a sustained, engrossed, contemplative engagement, often directed at a self-contained object that evokes general or abstract themes. The second, "theatricality," is quite different: a more transient and self-conscious appreciation, often directed at an object that makes an effort to catch the viewer's eye. Fried shows that these two modes coexisted and were in competition. He traces many fascinating issues involving these modes, including the debates about their merits (some argued that the sincerity of absorption could tend toward sentimentality or banality, whereas others claimed that the sophistication of theatricality could lead toward artifice and frivolity) and the way that these two modes both supported the dominant view of the superiority of paintings of historical scenes over other genres such as portraiture and landscapes. Writing a few years later, Jonathan Crary (1990) considers more broadly the coexistence of multiple forms of observation; he traces other cultural theorists whose works can be understood as efforts to demarcate transitions in the history of attention, including Walter Benjamin's discussion of the flaneur and Michel Foucault's account of the spectacle.

For the purposes of our discussion, Fried's work is of use because it suggests the possibility of describing different sorts of public attention, a kind of parallel to Tim Ingold's (2000) more anthropological consideration of the many specific forms of attention. The examples of the map, rumors, and joke that we have just discussed demonstrate a third mode of public attention, one that we call "viveza." This word is commonly used by Peruvians to describe a specific sort of alertness that combines sharpness, wit, and a refusal to be tricked or fooled. This *viveza* can be considered a core cultural value, because Peruvians generally speak of it positively. The possibility of an El Niño event provided an excellent opportunity for many people to display this value by closely following the news and by circulating rumors and jokes. *Viveza* has a direct relation to *engaño*, discussed earlier, because the person who has more viveza can trick others and is less likely to be tricked. As we discuss below, this particular form of attentiveness was clearly evident in the public's active reception and circulation of images associated with the El Niño event. (We note the complexity of discussions of viveza within Peru, especially the apparently contradictory pride and embarrassment with which many Peruvians speak of this key element in a kind of national self-awareness [Herzfeld 1997].)

Our discussion can now return to the particular situation in August 1997. At that time, the Peruvian public recognized that the warm temperatures of the ocean and atmosphere meant a strong likelihood of heavy rains late in the year and early in 1998. This possibility evoked memories of the destructive floods during El Niño events in earlier decades. These concerns led to the question of what responses could be taken. Two major alternatives coalesced. In June, Fujimori set up a government commission, composed of representatives of four ministries, soon joined by members of the military's civil defense organization, to coordinate planning of mitigation programs. They focused attention on civil defense and planning to reinforce bridges and roads, to dig drainage canals to divert flooding rivers around major cities, and to strengthen sea walls in coastal towns against storm surges. These actions showcased Fujimori's technical skills and offered him political advantages by garnering him support in certain critical regions of the country and by allowing him to control a large budget. In addition, the government commission proposed public health campaigns to curb expected outbreaks of cholera and dengue fever and subsidized credit for homeowners to reinforce their dwellings.

Municipal governments, NGOs, and universities, all active in the political opposition, formed a loose network that developed an alternative program. Two of the most active NGOs were located in the northern coastal departments of Piura and Lambayeque, historically vulnerable to severe flooding in El Niño years. These NGOs and local universities had long been involved in studying historical records of El Niño events and formulating local response plans, although they did not have the extensive scientific background of the national agencies or the sophisticated familiarity with the Internet. The alternative program emphasized emergency food relief and credits to agricultural producers and homeowners. It included civil defense projects as well, although its plans suggested a larger number of somewhat smaller and more decentralized projects than the plan formulated by the government commission. It proposed channeling funds through municipal offices rather than national ministries, because this system would allow support to reach local actors more effectively, and because the opposition parties, although weak in congress, held over half the mayor's offices in provincial towns (Tuesta 1998).

A sense of uncertainty continued to pervade the country as the anticipated rains drew closer. The August 14, 1997, issue of *Caretas*, a national magazine associated with the opposition, carried a cover image of an Asian boy, clearly President Fujimori, standing under an umbrella; the caption reads, "Will he stay dry through the downpour?" suggesting that the readers of the magazine were aware of the forecasts of heavy rains to come. Implicit in the cover is the question of Fujimori's political stability. More broadly, the cover of this national magazine shows the naturalness of the assumption that the state was the key entity that would address the impending threat.

It was the president's plan, rather than the opposition plan, that triumphed in September 1997. In hindsight, it is not surprising that this plan won. Fujimori had tight control over his party, which held nearly 60 percent of the seats in congress. The remaining seats were split among a number of smaller parties. It is very difficult to imagine that a strong coalition of opposition parties and defectors from Fujimori's party could have defeated his plan. Nonetheless, there was less debate in congress over this proposal than over many of Fujimori's other programs.

Fujimori put considerable effort behind his proposal. It fit in well with his self-presentation as an engineer with great technical expertise, and it also met his concern to shore up support in critical regions, especially the north coast. He chose Daniel Hokama, the Japanese–Peruvian minister of the presidency, to present the program to congress on September 15. Hokama's speech contained many elements of channeling globality, which were reinforced by his own visible Asian ancestry.

The speech contained detailed figures and charts that emphasized Fujimori's access to scientific information; two congressional representatives had told us in August 1997 that they were sure that the president and his top ministers had personal sources of information on the event but that they kept that information secret so that they could develop a plan to respond to the event. The speech also emphasized the international basis of market economic rationality. In making the point that spending government funds on public works to prevent El Niño-related damage was an investment that would benefit the national economy by reducing future expenditures in reconstruction, Hokama gave figures in dollars rather than in the national currency of the *nuevo sol*. More broadly, he alluded to the economic model of neoliberalism that emphasized the integration of developing countries into global markets and that Fujimori strongly supported (Campodónico et al. 1993; Castillo 1997; Gonzales 1998). Ironically, other neoliberal policies of Fujimori's had increased the vulnerability of many Peruvians to the climate event. For instance, because of elimination of protections for labor (e.g., minimum-wage laws and minimum periods for work contracts), industrial fisheries workers were fired before catches declined; during earlier El Niño events, catches also declined, but workers retained some wages and insurance. In the agricultural sector, the heavy rains favored rice producers who produced a bumper crop, but imports of inexpensive Asian rice dramatically deflated prices, leading to a massive supply glut. In most sectors, subsidized credit to take preventative measures was unavailable so that, even with advance warning of impending floods, those most vulnerable were unable to take significant action.

Fujimori's plan could be described as an instance of what Tsing has termed "globalist projects." In Tsing's terms, the global scale of Fujimori's reach was not merely preexisting but, rather, was "brought into being" (2005: 58) through specific acts that drew some elements at a global scale together with other elements at different scales. Like the entrepreneurs Tsing describes, Fujimori's combinations seem a kind of conjuring, carried off in an audacious and imaginative fashion. Tsing's accounts emphasize the varieties of scales (she writes, partly tongue in cheek, about "articulations of partially hegemonic imagined different scales" [2005: 76]). In the case of Fujimori, though, it is the importance of the national scale, rather than the multiplicities of scales, that is most striking, not merely because Fujimori drew on his control of national institutions but also because of the importance of national scales of attentiveness. Fujimori showed his viveza in accessing distant, novel information sources through advanced technology and commanded the attention of the public, which also displayed its viveza in creating images, jokes, and rumors about him. This public engagement with Fujimori is an example of a broader aspect of state power that Begoña Aretxaga describes as the way that the state evokes an insatiable fascination because of its nature as "a screen for political desire as well as fear" (2003: 394).

We note a more specific aspect of state power as well. The nature of the El Niño event places state organizations in a position from which they are well poised to channel globality. Indeed, there is a kind of affinity between El Niño events and national governments. The large spatial scale of the events means that the responses have to be coordinated over a large area; many NGOs have proven themselves most effective at the local level, rather than at a provincial or national level. Moreover, the global division of labor has assigned climate-related concerns such as aviation and national security to national institutions. Because of their long temporal continuity, wide spatial coverage, and access to public funding, these institutions are better suited to these tasks than smaller public agencies or private institutions. By working together, national weather services in different countries can develop common standards of measurement and share data, allowing them to perform their tasks more effectively. The major climate organizations in Peru, as in most countries, are national agencies and provide key outputs—data and forecasts—to the central government rather than to NGOs. El Niño events are not the only atmospheric issues that have been addressed largely through national governments; others include air pollution and acid rain (Morag-Levine 2003), ozone thinning (Benedick 1991; Litfin 1994), and, more recently, global warming. National governments are the central actors in the UN Framework Convention on Climate Change and its key agreement, the Kyoto Protocol, and in the somewhat more activist-oriented mobilization around climate change in the Arctic. The Arctic Council, composed of the United States, Canada, Iceland, Norway, Denmark, Sweden, Finland, and Russia, has sponsored major studies of impacts of global warming on natural and social systems at high latitudes and has pressured other international agencies to take steps to reduce greenhouse gas emissions. In sum, this general association of national governments and weather services gave Fujimori an additional advantage, one that he seized very effectively.

September 1997–November 1997: The president takes charge

With a clearly formulated and well-funded program in place, the Peruvian public might have settled down after September, trusting that their president had used forecasts, which had been unavailable in previous El Niño events, to protect them from the rains that were projected to begin in December or January. Instead, the event continued to receive media coverage and remained a major topic of conversation. In part, this attention came from structural causes. Media institutions and government organizations offered different forecasts because the scientific models were not in complete agreement and because a novel forecast attracted more attention than a familiar one. This attention also came from the public's insatiable appetite for information, a quality that we discuss in subsequent sections.

In particular, rumors continued to circulate in many different social sectors. Members of the public were eager to obtain new information that would allow them to develop plans to face the coming event. One woman who worked part-time in a fish processing plant thought that the plant managers had been told that the El Niño event would reduce fish stocks but that they would not pass that information on to workers. Instead, the plant managers wanted the workers to keep hoping that they would remain employed. A midlevel bank employee was denied information by some scientists; he explained that certain government scientists withheld the truth because of industry pressure or their wish to act as consultants to private firms. Distrust associated with the flow of climate information was also evident at the household level. In interviews, spouses of both artisanal fishermen and industry executives expressed doubt about whether their husbands were telling the truth about the imminent arrival of a damaging El Niño or were merely using the event as an excuse to reduce the wives' allowances. These very different people had all discussed the El Niño event with coworkers and others and had all arrived at the same belief: Their scope of action was limited by the hidden actions of powerful figures. Aretxaga points out that such views of states are common; she writes of the "shroud of secrecy surrounding the being of the state" (2003: 400). This general perception is particularly strong in Peru, a setting in which viveza is emphasized; conversations often turn to the possibility that someone is concealing information, or that someone is attempting trickery, or engaño.

The nature of public attention during this time of uncertainty can be seen by the reception of a major Climate Outlook Forum that took place in Lima in October. We discuss this conference in some detail because it represents an interesting example of a failure to channel globality. Sponsored by the governments of Peru, Ecuador, Colombia, Chile, and Bolivia, with support from NOAA and Peruvian private industry, it was widely covered by the Peruvian media. The topic of the day was whether the impending El Niño event would be as severe as the 1982–83 event, the most severe in recent history. The Climate Outlook Forum was publicized as a meeting of experts intent on answering the question "Is this the El Niño of the century?" This conference, organized like other Climate Outlook Forums held in Africa, Asia, the Pacific, and elsewhere in Latin America, illustrates the transformation of globalized environmental knowledge as it enters a national setting. Following the emphasis in current development culture on participatory approaches, international agencies were balanced by local expertise from western South America; there were efforts to involve "stakeholders" at the meeting and to disseminate information broadly. These seemingly altruistic, although perhaps paternalistic, goals were largely thwarted by political rivalries and individual greed. No NGOs were represented, and several key governmental groups were excluded because of competition with other agencies. Participation from provincial areas was limited, and high entrance fees further reduced attendance. A powerful subset of the fishing industry dominated the meeting and pressed for interpretations of evidence in accord with its interests.

Rumors circulated to discount the conference. Many observers, aware of pressure from fishing concerns and of the eagerness of individual scientists to obtain consultancy positions with private industry, discounted the conference's conclusions about El Niño's likely impacts; some of these observers told crude jokes that suggested the intimate nature of the relations between the state and private industry and that evoked the engaño, or deceit, with which the state and industry tried to trick the public. The international climate scientists who participated in the meeting assumed that the globalized types of information brought to the forum—climate information derived from a worldwide environmental observational network and multinational computer modeling collaboration—was nonpolitical; these scientists were surprised by the way that these data were reinterpreted in the context of the national meeting to serve private interests.

The clearest failure to channel globality can be seen in the final product of this conference, a map that presented the consensus forecast developed by the scientists in attendance. This map divided western South America into subregions (see Figure 22.3). For each of these subregions, the map provided three numbers, the probabilities that rains in coming months would be above normal, near normal, or below normal, with these probabilities always totaling 100 percent. It is here that the friction that shapes global projects can most clearly be seen, as this map closely resembled those prepared with NOAA direction at Climate Outlook Forums elsewhere in the world. The media, however, did not publish this map. This was perhaps the most globalized of images in terms of the range of inputs into its creation. Its contents—the outcome of two days of intense debates among international experts armed with the latest computer-generated predictions—represented the

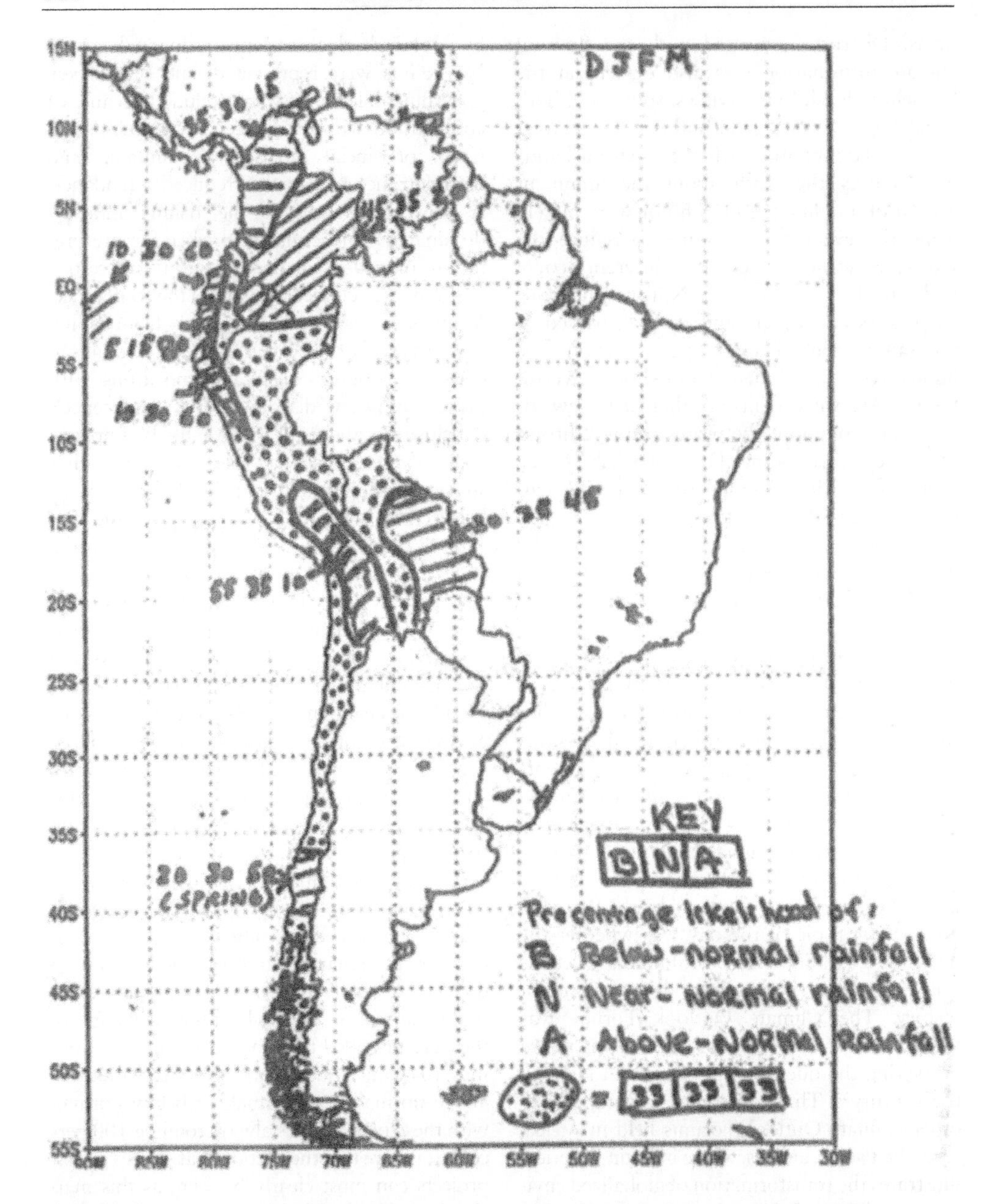

Figure 22.3 *Map produced by atmospheric scientists at Climate Outlook Forum, October 1997.*

newest developments in climate science, although it was hand-drawn with felt-tip pens. Newspaper and television reporters covered this conference extensively, expecting it to resolve debate over the event's magnitude, but they never reproduced or referred to this visually confusing map. The lack of attention to this map, which contrasts so sharply with the interest in other images, suggests that the Peruvian media and public did not passively receive all images from global sources but, rather, selected the ones that they found most attractive. Some media representatives expressed surprise (and even laughter) at the technological crudeness of the hand-drawn map, implying that

Peruvian audiences expect graphic sophistication and forceful presentation on a par with the developed world. The lack of concrete, vivid description of the consequences of the projected rainfall may also explain the lack of popularity of this image, given that much of the reporting accompanying this El Niño was sensationalistic (Zelada 1998). We note that the conference map was provided by the same agency that produced the map that was redrawn by Peruvian graphic specialists and printed in *El Comercio*, reproduced here as Figure 22.1. It is not the source of the image that grants it authority but, rather, its usefulness and legibility to particular people at a particular time that give it potential interest.

The Climate Outlook Forum only increased public confusion and concern. It underscored Fujimori's lack of confidence in his own scientific agencies. In response, he intensified personal involvement in the prevention efforts and turned to a small group of close advisers, reinforcing the public perception that he had a monopoly on the latest technology, information, and expertise.

This failed conference resembles the global assemblage discussed by Ong and Collier, in that it drew together global forms and other entities to occupy "a common field in contingent, uneasy, unstable interrelationships" (2005: 12). A multiplicity of actors and images came together in networklike fashion, very different from the flow or diffusion of information outward from centers envisaged by many globalization theorists. The universality of the global that Ong and Collier describe was present as well, at least from the perspective of the international agencies, who saw the Climate Outlook Forum in Peru as one instance of a generic structure, with Latin American "stakeholders" easily substituting for African or Asian counterparts in other instances.

Peru's oldest and largest television network, America-TV, was more successful in satisfying the public thirst for high-tech images of the El Niño event. This network, founded in Peru in 1958, established a new television feature in November 1997. It aired twice a day in Peru throughout the event. The name of the program, *SATEL*, was an acronym for "Sistema de Alerta Temprano para El Niño" (Early Warning System for El Niño) and referred to the satellites that provided real-time images of El Niño-related storm systems. Its daily updates on the event—accompanied by colorful satellite maps of the country—attracted many viewers. Some of its success came from the trust many Peruvians place in television show hosts, who are often seen as being unbiased, and from the broad perception that the weather announcer was not aligned with government agencies or political parties. This success may further be attributed to the belief that the images, beamed down to Peru from international satellites, somehow escaped the interference of powerful forces in Peru. The references in the show to current temperatures in cities in Peru also confirmed the belief that the show was broadcast live and so was not open to manipulation. (Indeed, talk shows, broadcast live, are generally popular in Peru [Fowks 2000].)

This emphasis on global technologies was a key element in the program's popularity, even though the program continued to rely on established land-based weather stations, run by the national meteorological agency SENAMHI, for much of its information, such as daily high temperatures in cities throughout Peru. It is interesting to note that a national media organization channeled globality so effectively. Even the richest of the provincial television stations did not have the resources to produce and distribute a program on this scale, but, more importantly, Peruvians perceive weather as a phenomenon that is national (it makes sense to listen to the daily high temperatures and precipitation levels in different cities, much as one might follow soccer scores or election results in other parts of the country) and global (it can be observed on a planetary scale from space).

In her thoughtful account of Egyptian television, and of the ways that it reflects and shapes understandings of the nation, Lila Abu-Lughod (2005) examines a particular form, the serial drama that focuses on a particular theme or situation and that runs for a fixed number of episodes rather than for one or more yearly seasons, as most U.S. television programs do. This focus, short time span, and novelty concentrate attention and allow viewers to engage deeply with the complex issues of addressing the profound differences within Egypt and the location of Egypt within wider regional and global contexts and with universal fields such as development and education. Despite its links to the genre of news rather than drama, *SATEL* had many of the attributes of such serials in its focus, temporal limits, and novelty; the distinctive personality of the announcer, the videos of moving storm systems, and the repeated references to particular cities and regions added significant dramatic elements as well. Large audiences tuned in to *SATEL* for emotional engagement as well as receipt of information. The program allowed them to locate their nation in the universal fields of progress and technoscience.

The Main Event

December 1997–April 1998: The rains arrive

In December 1997, the heavy rains began. Flooding occurred on the north coast and, surprisingly, in the south–central department of Ica as well. Tens of thousands of houses were damaged or destroyed, and hundreds of people were killed. Despite the extensive prevention works, these rains rendered sections of the major coastal highway impassable and disrupted transportation nationwide. The major commercial fisheries were heavily impacted. The highland areas received a mix of drought and flood, leading to mudslides, damage to crops, and the inability to transport goods to market (Colegio de Ingenieros del Perú [CIP] 1999).

Fujimori continued to lead the government response. Although he had initially stated that Peru could handle El Niño alone, by the end of 1997 he had actively sought aid from the World Bank and the Inter-American Development Bank, receiving $450 million in foreign loans while spending approximately $162 million out of the Peruvian treasury (Zapata and Suiero 1999). This international flow of capital occurred after global images had circulated widely in national media and discourse and political alternatives had been debated. Although international financial institutions are widely understood to have a strong hand in shaping policy in developing countries, this instance, at least, shows the national dimensions of politics and supports a view of globalist projects, or global assemblages, rather than of a rapid flow of capital in an era of globalization. To gain support for his programs in September, Fujimori channeled globality by drawing on global imagery, rather than by receiving promises of support from international institutions. Despite the common perception that international institutions had shifted their focus to community-level projects, often run by NGOs, money for El Niño-related activities flowed directly to the central government.

Fujimori and his advisers in the Ministry of the Presidency organized another international meeting, held in January 1998. Recognizing the need to assure the continuity of external funding for reconstruction projects, they wished to show off a level of technical expertise and national consensus that may not have been entirely genuine. They invited the national oceanographic agency, IMARPE, and the UN Development Programme to review work completed in the different provinces. International experts were brought in to lecture and to provide legitimacy to the event. Video teleconferencing linked conference attendees to the provinces, where mayors and civil defense leaders reported on the status of prevention efforts and current climate conditions. The immediacy of this novel communication form, although awkward for the participants, was a first attempt at involving the provinces in the centralized show of competence. The modes of information flow point out, ironically, the relative ease in obtaining information from international sources and the difficulty of utilizing information within national borders. In this venue, some participants in the provinces fleetingly sensed shifts toward cosmopolitanism, along the lines envisioned by some globalization scholars; for once they were contributing to a discussion rather than being the passive receptors of information generated in Lima, joining, at least momentarily, the imagined community of El Niño researchers. Although its linkages were somewhat shaky, this global assemblage held together, for the span of a few days at

Figure 22.4 *Cartoon from popular Lima newspaper, January 1998.*

least. The friction from this meeting helped Fujimori receive approval for funding a few months later.

The El Niño event continued to receive attention in television and other media once the heavy rains began in December 1997. The competition between media outlets to appear as having privileged access to information is exemplified by the following headlines that appeared within a day of each other in two major newspapers: "El Niño may be arriving at its end" (*El Comercio* 1998: 1) and "The worst of El Niño has not happened, it will occur in next month" (Castillo 1998: 3). *Caretas* devoted five of its covers in January and February of 1998 to the event—an unusual number for a magazine that usually features politicians, movie stars, and soccer players on its cover. Radio discussions of the event expanded as well.

A more overtly comical image appeared on the front page of the January 9, 1998, issue of *Chesu!*, a newspaper published in Lima and sold throughout Peru (see Figure 22.4). This newspaper was displayed and sold widely in newsstands, so many pedestrians who did not buy it still saw the cover. It plays on the crude humor typical of the tabloid press, known in Peru as *la prensa chicha*, the press that appeals to the lower classes, especially of mixed-race background, and that Fujimori often used to run smear campaigns against opposition leaders. The image, in full color, shows a woman and a man standing in the ocean. The woman says, "Oh, the ocean is really warm," and the man replies, "Yes, that's because of the 'El Niño current.'" The alternate meaning of the "El Niño current" is evident in a boy urinating in the water as he whistles and averts his eyes in an unsuccessful effort to avoid being noticed. This cartoon plays on the favorite theme of engaño, deception; here, the man has fooled the woman about the nature of the water in which they are standing.

It is interesting to compare this image with that of the advertisement in Figure 22.2. They appeal to different classes, and they also make reference to different races. The child in the advertisement is clearly blond, whereas the relative Indianness of the people on the cover of the tabloid is marked by the woman's dark skin, the man's short stature, and the boy's cheap sandals. This range suggests the wide appeal of humor across social divisions. The two portrayals show the dramatic shifts from the original religious association between the El Niño current with the Christ child to the newer incorporation of El Niño into Peruvian popular culture.

Fujimori also used visual media to display cleaned waterways, new dams, and reinforced bridges. Other elements in the popular press, however, incorporated the El Niño event into critiques of Fujimori's other problems, such as high unemployment rates, human-rights violations, and widespread corruption. The tongue-in-cheek cover of a leading national news magazine shows these connections. The headline reads, "Watch out! The attack of the Super-Niño, or the kryptonite of being the center of attention." It suggests that if Fujimori wishes to present himself as a kind of hero with superpowers, he is also vulnerable to a kind of kryptonite. The muddy, bare foot of a brown-skinned person suggests an ordinary Peruvian, the victim of flooding associated with the El Niño event (see Figure 22.5).

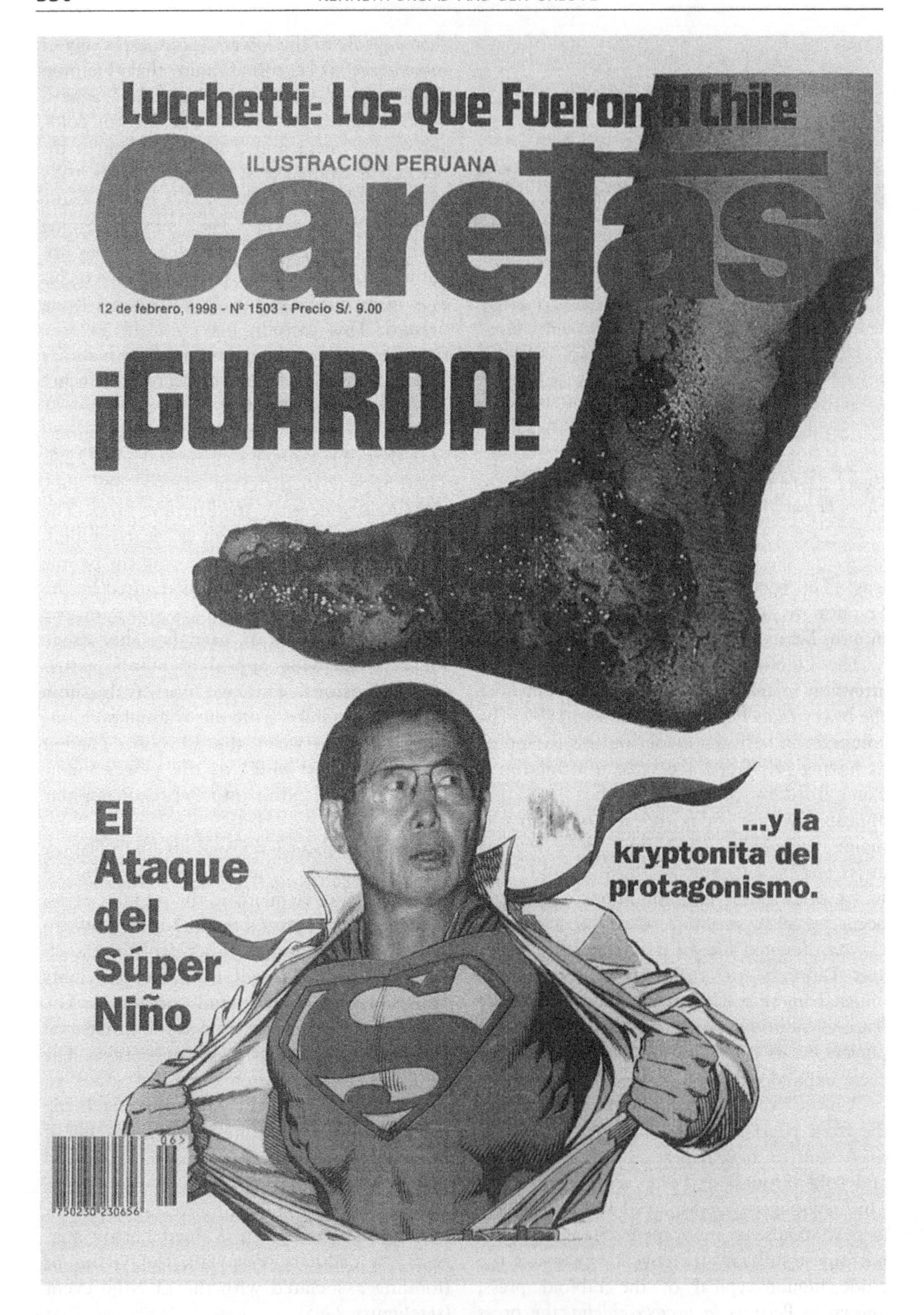

Figure 22.5 *Cover of major Peruvian news magazine, February 1998.*

May 1998–December 1998: The nation cleans up

The connection of distant locales within Peru was by no means wholly illusionary. The country witnessed countless stories of individual and community heroism during and after the floods, in which blocked roads were cleared, survivors were dug out from landslides, and neighbors assisted one another. The government had averted a disaster of the magnitude of the 1982–83 event through its short-term infrastructure improvements, allowing Fujimori to gain some popular approval. The reconstruction period, however, further highlights the place-based political underpinnings of societal vulnerability to recurring environmental shocks.

In June 1998, Fujimori formed the Executive Reconstruction Committee for El Niño (CEREN) to oversee reconstruction of damaged infrastructure, using loans from international organizations and some matching funds from the government. The reconstruction process was marked by periods of sporadic intensity that seemed to correlate with the lobbying and campaigning leading up to the municipal elections in October 1998 (Zapata and Suiero 1999). Unsurprisingly, the reconstruction process followed many of the patterns previously described: petty corruption, political favoritism, interagency squabbling, and bureaucratic inefficiency. Rebuilding began in the shadow of the torrential rains and floods, and the government, claiming pressure from the IMF, partially held back on its promise of matching funds, thus reducing the amount dispensed by the international agencies.

The sense of distrust reflected in both jokes and rumors, particularly the sensitivity to engaño, or deception, proved to be reasonable, given that the public statements of major officials differed from their actions. Although they promised to address the needs of all the Peruvian people, the funding was concentrated in infrastructure projects in large cities (reinforcing the Pan-American Highway and other major roads and clearing riverbeds in coastal capitals) and in credit to large agricultural firms. These actions were certainly a necessary step to avoid the major disruptions akin to those in the 1982–83 El Niño event. Nevertheless, residents of smaller towns and poor neighborhoods expressed a sense of their neglect by the central government, which did not address the impacts that most directly affected them.

The period of reconstruction after the rains ended in April provided Fujimori and his associates with opportunities to be photographed as they led the nation out of the disaster—and gave the opposition visual moments as well, for example, when poorly constructed bridges and retaining walls collapsed. Attention tapered off late in 1998, especially after the municipal elections in October; the *SATEL* program was terminated at the end of the year.

Epilogue: The Limits of Perception Management

Ideologically, no dramatic shifts occurred in the established political forms as a result of Peru's location at the epicenter of this global environmental phenomenon. The strength of the state was not undermined by the new imagery and knowledge, and no shifts in identity took place. And although El Chino may have beaten El Niño, his glory was short-lived, and the same visual media that he had used to manage his image as the efficient technocrat triggered the end of his rule. The event that precipitated Fujimori's fall from power was the release of a video in September 2001 showing his secretive adviser, Vladimiro Montesinos, bribing an opposition leader. Further investigation revealed that thousands of similar tapes existed, tools of extortion used by Fujimori to retain political advantage. (Later that year, the owners of America-TV, the channel that developed the successful *SATEL* program, fled to Argentina after a video showed them receiving bribes from Montesinos to air pro-Fujimori programs.) Fujimori's flight to Japan rejuvenated claims about his Asian, non-Peruvian identity and opened space for an interim government that led to the 2002 election of Alejandro Toledo, who played on his globalized identity with Indian peasant roots and Stanford education, at ease in both the remote countryside and in the boardroom.

Conclusions

The 1997–98 El Niño climate event in Peru offers a useful case for considering globality. Descriptions of this event, written closer in time to its occurrence, might have drawn on earlier models of globalization and emphasized the importance of flow: the flow of satellite images from major international centers to a peripheral country such as Peru that would have built awareness of the event, the flow of experts who would have traveled to conferences in Lima and who would have shaped the perception of the event as very serious, the flow of funding for relief programs that would have favored political organizations in Peru best suited to receive such support. In contrast, our account emphasizes the agency of actors in Peru in selecting and, moreover, in seeking out certain global connections (and in recasting these connections in their own terms) while refusing other such connections. The Peruvian press and public adopted, and reworked, one NOAA image (see Figure 22.1) but rejected another (see Figure 22.3). Two sets of political forces in the country developed plans to protect against anticipated flooding and other impacts; only after one of these forces triumphed were international funds solicited. This emphasis on a contingent interweaving of elements at a global scale with elements at other scales is much closer to Tsing's discussion of globalist projects or Ong and Collier's presentation of global assemblages than to an earlier view of globalization as a process of unimpeded flow of images, capital, people, and objects. We use the term *channeling globality* to refer to efforts to establish linkages with entities on a global scale. This term indicates the active nature of efforts that are made to establish connections at a global scale. It also draws on another sense of the word *channeling* to suggest that some actors gain power by claiming special personal powers in linking with distant sources of value. Finally, the term *channeling globality* also serves to suggest that actors make connections on established routes, drawing on specific modes of attending to images.

These reflections raise the question of the autonomy of actors in selecting the terms by which they link with global entities. Some recent works have proposed two major constraints on actors: the importance of universal discourses and the role of immediate conjunctures. Ong and Collier suggest that discourses that can be seen as universal, such as technoscience and neoliberalism, often play a crucial role in supporting global assemblages; by their great abstractability, they facilitate the connection of diverse elements across spatial scales. Capable of being decontextualized, they can be recontextualized in new settings. In a related vein, Tsing discusses universals such as prosperity and freedom. These abstractions rest on basic elements of human existence and, because of this, can link elements together in global projects. But, Tsing reminds us, "Universals are effective within particular historical conjunctures that give them content and force" (2005: 8). She suggests that this conjunctural feature of universals can be called "engagement," and discusses what she terms "engaged universals." Indeed, the world is replete with examples of such engaged universals. To take only one, Nils Bubandt (2005) discusses the growing global discussion of security, especially since the end of the Cold War and after September 11, 2001, and traces the development of specifically Indonesian forms of this concern. He considers the presence of the Indonesian military in villages in the remote eastern island of Helmahera and shows the ways that local and regional actors draw on this universal focus on security and create their own Indonesian variety of it, which he terms "vernacular security."

We note that the case of the El Niño event could be read in terms of such specific conjunctures. They could help explain why the universals in this case—human security against disasters and technoscience, in particular—have selective appeal. The *SATEL* television program drew large audiences for months, whereas the equally sophisticated NOAA map of the Climate Outlook Forum was rejected by the Peruvian media and public, who found it visually crude and quite literally illegible. Fujimori's program to address security needs received broad support, whereas the program articulated by the opposition failed. Other elements of universal security and technoscience, such as climate forecasts, were recombined with established cultural forms, such as rumors and salacious cartoons in the popular press, very much the kind of recontextualization that Ong and Collier discuss. In other words, the effectiveness of the universals could be seen as quite specific and conjunctural, as was made so explicit in the ways that concrete social and cultural formations shaped the use of discourses of citizenship and human entitlements in the struggles on the Gulf Coast following Hurricane Katrina (Lipsitz 2006). We would note two particular elements of the historical conjuncture in the Peruvian case, the salience of the national scale and the concrete form of public attention. The salience of the national scale is striking in the basic framing of the Peruvian nation as the object of the event and the Peruvian state as the key actor in the response to it. We have discussed as well the state's use of forms of public attention. Tsing notes that engaged universals "convince us to pay attention to them" (2005: 8). Indeed, an aspect of the historical conjuncture would be the pattern of public attention itself, the viveza (a kind of clever alertness that guards against trickery) that underlay the rapid spread of news of the El Niño event beyond major media and official pronouncements into the everyday world of conversation, rumors, and jokes, that created a concern to stay up-to-the-minute and to be among the first to hear news, and that Fujimori and the Peruvian media drew on so successfully.

We suggest, however, that this analysis would press the term *conjuncture*, with its associations of temporal brevity, to serve more broadly than it readily can. To be sure, some specific situational elements can appropriately be described as conjunctural: the strength of Fujimori's party at the time and his effective use of the well-funded Ministry of the Presidency, the concentration of media organizations in the national capital, and the like. Nevertheless, powerful state institutions and deeply rooted patterns of attentiveness have long histories; they are woven through the everyday worlds of experience or, if one prefers, of habitus. They exist in some middle ground between the universal and the conjunctural, a ground once unselfconsciously described by such terms as *structure* and *culture* and that might now more neutrally be called "context."

Whether one wishes to use the word *conjuncture* broadly or, as we suggest, to distinguish between a more temporary frame of conjuncture and a more enduring frame of context, we suggest that these elements have their counterparts in many settings around the world. Climate events create various forms of engaged universals, through their connection with conjunctural features in specific projects, with contextual elements of states and forms of attention, and with the abstract systems of technoscience and human security. Fujimori and his administration played one climate event brilliantly. George W. Bush and his counterparts proved far less able to turn public attentiveness and state institutions in the United States to similar advantage when another event, Hurricane Katrina, approached and struck the Gulf Coast far more rapidly than the warming ocean and atmosphere impacted Peru. It remains to be seen how public attention and state institutions will respond, in the midst of other pressing conjunctures, to the much slower, but much more powerful, planetary warming that is bearing down on us all.

REFERENCES

Abu-Lughod, Lila. 2005. Dramas of Nationhood: The Politics of Television in Egypt. Chicago: University of Chicago Press.

Appadurai, Arjun. 1996. Modernity at Large: Cultural Dimensions of Globalization. Minneapolis: University of Minnesota Press.

Aretxaga, Begoña. 2003. Maddening States. Annual Review of Anthropology 32: 393–410.

Benedick, Richard E. 1991. Ozone Diplomacy: New Direction in Safeguarding the Planet. Cambridge, MA: Harvard University Press.

Brenner, Neil. 1999. Beyond State-Centrism? Space, Territoriality, and Geographical Scale in Globalization Studies. Theory and Society 28(1): 39–78.

Broad, Kenneth, Alex P. Pfaff, and Michael H. Glantz. 2002. Effective and Equitable Dissemination of Seasonal-to-Interannual Climate Forecasts: Policy Implications from the Peruvian Fishery during El Niño 1997–98. Climatic Change 54(4): 415–438.

Bubandt, Nils. 2005. Vernacular Security: The Politics of Feeling Safe in Global, National, and Local Worlds. Security Dialogue 36: 275–296.

Campodónico, Humberto, Manuel Castillo, and Andrés Quispe. 1993. De poder a poder: Grupos de poder, gremios empresariales, y política macroeconómica. Lima: DESCO.
Caretas. 1997. Cover image. Caretas, August 14.
Castillo, J. 1998. El peligro de El Niño no ha pasado, lo peor ocurrirá en el próximo mes. La República, January 9: 3.
Castillo, Manuel. 1997. El estado post ajuste, reforma estatal, institucionalidad y sector privado. Lima: DESCO.
Choy, Timothy K. 2005. Articulated Knowledges: Environmental Forms after Universality's Demise. American Anthropologist 107(1): 5–18.
Colegio de Ingenieros del Perú (CIP). 1999. Informe del fenómeno del Niño 1997–98. Lima: CIP.
El Comercio. 1998. El Niño puede estar llegando a su fin. El Comercio, January 8: 1.
Crary, Jonathan. 1990. Techniques of the Observer: On Vision and Modernity in the Nineteenth Century. Cambridge, MA: MIT Press.
Crow, Thomas. 1985. Painters and Public Life in 18th Century Paris. New Haven, CT: Yale University Press.
Fowks, Jaqueline. 2000. Suma y resta de la realidad. Lima: Friedrich Ebert Stiftung.
Fried, Michael. 1980. Absorption and Theatricality: Painting and Beholder in the Age of Diderot. Berkeley: University of California Press.
Glantz, Michael H. 2001. Currents of Change: Impacts of El Niño and La Niña on Climate and Society. Cambridge: Cambridge University Press.
Gonzales, Efraín. 1998. Neoliberalismo a la peruana: Economía política del ajuste estructural. Lima: Instituto de los Estudios Peruanos.
Hannerz, Ulf. 2003. Macro-Scenarios. Anthropology and the Debate over Contemporary and Future Worlds. Social Anthropology 11: 169–187.
Hayden, Cori. 2003. When Nature Goes Public: The Making and Unmaking of Bioprospecting in Mexico. Princeton: Princeton University Press.
Herzfeld, Michael. 1997. Cultural Intimacy: Social Poetics in the Nation-State. London: Routledge.
Ingold, Tim. 2000. The Perception of the Environment: Essays on Livelihood, Dwelling and Skill. London: Routledge.
Kempton, Willett, James S. Boster, and Jennifer A. Hartley. 1996. Environmental Values in American Culture. Cambridge, MA: MIT Press.
Lipsitz, George. 2006. Learning from New Orleans: The Social Warrant of Hostile Privatism and Competitive Consumer Citizenship. Cultural Anthropology 21(3): 451–468.
Litfin, Karen. 1994. Ozone Discourses: Science and Politics in Global Environmental Cooperation. New York: Columbia University Press.
Markowitz, Lisa. 2001. Finding the Field: Notes on the Ethnography of NGOs. Human Organization 60(1): 40–46.
Morag-Levine, Noga. 2003. Chasing the Wind: Regulating Air Pollution in the Common Law State. Princeton: Princeton University Press.
Ong, Aihwa, and Stephen J. Collier, eds. 2005. Global Assemblages: Technology, Politics, and Ethics as Anthropological Problems. Malden, MA: Blackwell.
Tsing, Anna L. 2005. Friction: An Ethnography of Global Connection. Princeton: Princeton University Press.
Tuesta, Fernando. 1998. Instituciones, reformas y represntación política en el Perú. Paper presented at the 21st International Congress of the Latin American Studies Association. Chicago, September 24–26.
Warner, Michael. 2002. Publics and Counterpublics. Public Culture 14(1): 49–90.
Zapata, Antonio, and Kenneth Broad. 2001. Peru Country Case Study: Impacts and Responses to the 1997 El Niño Event. *In* Once Burned, Twice Shy? Lessons Learned from the 1997–98 El Niño. Michael H. Glantz, ed. Pp. 186–199. Tokyo: United Nations University Press.
Zapata, Antonio, and Juan Carlos Suiero. 1999. Naturaleza y política: El gobierno y el fenómeno el Niño en el Perú, 1997–1998. Lima: Instituto de Estudios Peruanos, CooperAccion.
Zelada, Angela. 1998. Medios de comunicación, sensacionalismo o doctrina. Prevención 5(11): 70–72.

Index

The Anthropology of Climate Change: An Historical Reader, First Edition. Edited by Michael R. Dove.

Printed and bound by CPI Group (UK) Ltd, Croydon, CR0 4YY
09/06/2025
14685921-0002